Compound Histories

Cultural Dynamics of Science

VOLUME 2

The titles published in this series are listed at *brill.com/cds*

Compound Histories

Materials, Governance and Production, 1760-1840

Edited by

Lissa L. Roberts
Simon Werrett

BRILL

LEIDEN | BOSTON

Cover illustration: "The Dissolution, or The Alchymist producing an Aetherial Representation." An alchemist using a crown-shaped bellows to blow the flames of a furnace and heat a glass vessel in which the House of Commons is distilled; satirizing the dissolution of parliament by Pitt. Coloured etching by J. Gillray, 1796. COURTESY OF THE WELLCOME LIBRARY, LONDON. Library reference: ICV No 11565. CC BY 4.0 (The title of the work situated below the etching is removed).

Library of Congress Cataloging-in-Publication Data

Names: Roberts, Lissa, 1954- editor. | Werrett, Simon, 1971-editor.
Title: Compound histories : materials, governance, and production, 1760-1840 / edited by Lissa Roberts, Simon Werrett.
Description: Leiden ; Boston : Brill, [2018] | Series: Cultural dynamics of science, ISSN 2351-9932 ; 2 | Includes bibliographical references.
Identifiers: LCCN 2017046831 (print) | LCCN 2017049252 (ebook) | ISBN 9789004325562 (E-book) | ISBN 9789004325494 (hardback : alk. paper)
Subjects: LCSH: Chemical engineering--France--History--18th century. | Chemical engineering--France--History--19th century. | Chemical engineering--Great Britain--History--18th century. | Chemical engineering--Great Britain--History--19th century. | Chemical engineering--Europe--History--18th century. | Chemical engineering--Europe--History--19th century. | Chemistry, Technical--France--History--18th century. | Chemistry, Technical--France--History--19th century. | Chemistry, Technical--Great Britain--History--18th century. | Chemistry, Technical--Great Britain--History--19th century. | Chemistry, Technical--Europe--History--18th century. | Chemistry, Technical--Europe--History--19th century. | Chemistry--France--History--18th century. | Chemistry--France--History--19th century. | Chemistry--Great Britain--History--18th century. | Chemistry--Great Britain--History--19th century. | Chemistry--Europe--History--18th century. | Chemistry--Europe--History--19th century.
Classification: LCC TP155 (ebook) | LCC TP155 .C6175 2018 (print) | DDC 660.0944/09033--dc23
LC record available at https://lccn.loc.gov/2017046831

Typeface for the Latin, Greek, and Cyrillic scripts: "Brill". See and download: brill.com/brill-typeface.

ISSN 2351-9932
ISBN 978-90-04-32549-4 (hardback)
ISBN 978-90-04-32556-2 (e-book)

This book is printed on acid-free paper and produced in a sustainable manner.

Contents

Acknowledgments

This volume is the result of a collaborative project, built on the support and participation of a number of institutions and valued colleagues. It is a pleasure to acknowledge them here. Without a grant from the Netherlands Organization for Scientific Research (NWO) to fund the research network that gave rise to this book, our collaboration would have remained a dream. Bringing together colleagues from across Europe and the United States was an expensive proposition; a number of institutions and organizations supplemented our NWO grant, providing both financial and material support. A word of thanks to the Wellcome Trust, Society for History of Alchemy and Chemistry (SHAC), Maison Française – Oxford University, Catholic University of Leuven, Museo Galileo (Florence), Max Planck Institute for the History of Science (Berlin), Thyssen Foundation, European University Institute (Florence), Regione Toscana, Museum Boerhaave (Leiden), Teylers Museum (Haarlem) and the Chemical Heritage Foundation (Philadelphia).

While all the members and fellow travelers of our research network also deserve thanks for their support and contributions, I would like to single out a few colleagues for special thanks. Following our inaugural meeting in Oxford in 2012, Geert Vanpaemel offered to host our first workshop in Leuven in 2013. His hospitality was essential to the meeting's overwhelming success. Simon Werrett followed suit by hosting a one-day workshop in London later that year. In early 2014, Elena Serrano spearheaded the funding and organization of our next workshop at MPI in Berlin. Following a small gathering in Oxford three months later that focused on how to incorporate digital methods into our project, a larger group met that autumn in Florence, thanks to the amazing hospitality of Stéphane Van Damme and Marco Beretta. It was then my pleasure to host our network's final workshop in Haarlem and Leiden in June 2015. Present there was Ernst Homburg, whose advice throughout our project was invaluable. Whatever success we have achieved would not have been possible without the generous efforts of these colleagues, for which great thanks are in order.

The project is especially indebted to John Perkins. He was my initial partner and network co-organizer. Despite increasing ill health, he tirelessly shared his dedication and enthusiasm, which inspired us all. Though he finally felt unable to contribute an essay to this volume, his encouragement and sage advice have been unmissable. For that reason, it is an honor and privilege to dedicate this volume to him.

Finally, I want to acknowledge what an enormous pleasure it has been to work with Simon Werrett as my co-editor. Simon is the perfect partner: kind, generous, patient and brilliant. I think we made a good team.

Lissa L. Roberts
July 2017

List of Illustrations

Figures

Tables

Graphs

Notes on Contributors

Robert Anderson
is president and CEO of the Chemical Heritage Foundation, Philadelphia. He previously served as director of the National Museums of Scotland (1984-92) and the British Museum (1992-2002). He is currently a fellow of Clare Hall, Cambridge and chairperson of the Society for the History of Alchemy and Chemistry (SHAC). His publications include *The Correspondence of Joseph Black* (London, 2012) and *The Cradle of Chemistry: The early years of chemistry at the University of Edinburgh* (Edinburgh, 2015).

Bernadette Bensaude Vincent
is a philosopher, historian of science and technology, and emeritus professor at Université Paris 1 Panthéon-Sorbonne. Her research topics span from the history and philosophy of chemistry to materials science and nanotechnology. Among her publications are *History of Chemistry* (with I. Stengers, 1997); *Lavoisier. Mémoires d'une révolution* (1994); *Chemistry. The impure science,* (with J. Simon, 2008); and *Fabriquer la vie. Où va la biologie de synthèse* (2011).

José Ramón Bertomeu-Sánchez
is director of the Institute for the History of Medicine and Science (University of Valencia). His research has been focused on the history of science education, material culture of science and other topics related to nineteenth-century chemistry in France and Spain. He has also published on the history of forensic medicine. His current project is about the emergence of scientific policing in twentieth-century Spain. His last book is *La verdad sobre el caso Lafarge* (Barcelona, 2015).

John R.R. Christie
is a historian of science and medicine who spent much of his career in the History and Philosophy of Science Department at the University of Leeds. He has worked latterly at the University of Oxford, and at the Max Planck Institut für Wissenschaftsgeschichte, Berlin. His main area of research concerns Scottish and English cultures of science and medicine, with a particular focus on chemistry, during the period of the Enlightenment.

Joppe van Driel
received his PhD from the Department of Science, Technology and Policy Studies (STePS), University of Twente, the Netherlands. He is currently revising

his *cum laude* dissertation, *The Filthy and the Fat: Oeconomy, chemistry and resource management in the age of revolution*, for publication as a monograph. His dissertation and a number of related publications focus on the history of chemistry and resource management during the period of the Industrial Revolution.

Frank A.J.L. James
is Professor of the History of Science at the Royal Institution and University College London. His has written widely on nineteenth-century science and technology and their social and cultural contexts. He edited *The Correspondence of Michael Faraday* (London, 2011), wrote *Michael Faraday: A very short introduction* (Oxford, 2010) and is currently studying Humphry Davy's practical work. He has been President of both the Newcomen Society and the British Society for the History of Science.

Thomas Le Roux
works at the National Centre for Scientific Research (CNRS – France) in Paris. His work deals with the impact of early industrialization on the environment from 1700 to 1850. He is currently working on a comparative history of industrial nuisances between Paris and London, as well as on the history of occupational health, industrial accidents, and mining.

Christine Lehman
formerly taught physics and chemistry. She is the author of a Ph.D. dissertation entitled *Gabriel-François Venel (1723–1775): Sa place dans la chimie française du XVIIIe siècle*, which she defended at Paris X-Nanterre University in 2006. She published the transcription of Venel's chemistry course taught in the year 1761 (Dijon 2010). Her main research interest is eighteenth-century chemistry.

Lissa L. Roberts
is Professor of Long-Term Development of Science and Technology at the University of Twente, the Netherlands, and has published widely on subjects including the history of chemistry, scientific instruments, steam engines, the history of the senses and global history. Her recent publications include *Accumulation and Management in Global Historical Perspective* (special issue of *History of Science*, 2014) and *Practicing Oeconomy During the Late Eighteenth Century* (special issue of *History and Technology*, 2014).

Elena Serrano
received a PhD from Universitat Autònoma de Barcelona for her dissertation *Science for Women in the Spanish Enlightenment* (2012), which she is currently revising for publication as a book. In 2012-2015 she was postdoctoral fellow at the Max Planck Institute for the History of Science, Berlin where she currently coordinates the project *Convivencia: From the Iberian Península to Global Dynamics (500-1750)*. Her research focuses on the interplay between knowledge, politics, societies, material culture and gender.

Anna Simmons
is an Honorary Research Associate in the Department of Science and Technology Studies at University College London. Her research and numerous publications explore various aspects of the history of British chemistry and pharmacy from c.1700 onwards, including the wholesale production of pharmaceuticals and the origins of the pharmaceutical industry, with a particular focus on the laboratories and pharmaceutical trade at the Society of Apothecaries.

Marie Thébaud-Sorger
is permanent Researcher at the C.N.R.S in History of Science and Technology, Centre Alexandre Koyré, Paris. She published several books on early ballooning, including *L'aérostation au temps des Lumières* (Paris, 2010, awarded by the *Académie française*). Through the study of artifacts and innovative objects that induce a new relationship with nature in urban contexts, her research explores inventive practices in eighteenth-century Europe, at the crossroad of knowledge of materials, market activities, social reform and technological hazards management.

Sacha Tomic
is Researcher at the Université Paris 1 Panthéon-Sorbonne. His work focuses on the history of chemistry in France during the long nineteenth century, with special attention to teaching and practices of academic chemistry, the relationships between chemistry and other disciplines, and the professionalization of chemistry in industrial society (expertise, chemical hazards, legal chemistry). He is the author of numerous studies, including *Comment la chimie a transformé le monde. Une histoire en sept tableaux* (Paris, 2013).

Andreas Weber
is an assistant professor in the department of Science, Technology and Policy Studies (STePS) at the University of Twente in the Netherlands. He is currently

finishing a monograph that examines the historical relationship between science and colonial governance through the lens of everyday materials. He is also involved in a digital humanities project, "Making Sense of Illustrated Handwritten Archives," directed toward developing search capability for illustrated archival material.

Simon Werrett
is the author of *Fireworks: Pyrotechnic arts and sciences in European history* (University of Chicago Press, 2010). Since 2012 he has been a member of the Department of Science and Technology Studies at University College London. He is currently working on a book entitled *Thrifty Science*, an exploration of ingeniously improvised experiments performed in domestic settings in the seventeenth and eighteenth centuries, and their value for thinking about the sustainability of science today.

Introduction: "A More Intimate Acquaintance"

Lissa L. Roberts and Simon Werrett

Power, transformation, promise, subjugation: terms that might easily be invoked to describe the decades between 1760 and 1840. Together they point toward the multi-faceted developments through which Europe took on its modern character and dominant position in the world – what this volume refers to as 'compound histories'. Simultaneously linked to the Baconian dictum that 'knowledge is power' and the brute facts of power-driven conquest and exploitation, this period is characterized by the historical tensions through which the promise of progress and subjugation of regions and resources around the world fed off and gave rise to social, political, economic, cultural, scientific, technological and environmental transformations. It was a time marked by the interactive appearance of new, janus-faced forms of political organization, scientific and technological capabilities, social and economic configurations: the growth of democracy coupled with empire; increasing abilities to harness the material world and its forces for productive ends coupled with destructive wars and environmental degradation; opportunities for great wealth creation coupled with new strains of poverty and deprivation.

It is this complex weave and the question of what binds its threads together that continue to make the 'age of revolution' so intriguing to historians.[1] While there is certainly no single answer to this question, which requires insights drawn from multiple subdisciplines of history, the contention undergirding this volume is that one key element has been insufficiently explored and integrated into the larger picture of historical development. Rather than baldly state what that is, let us turn to a voice from the period itself. In 1805 John Playfair, Edinburgh professor of natural philosophy, wrote:

> Nature, while she keeps the astronomer and the mechanician at a great distance, seems to admit [the chemist] to a more intimate acquaintance with her secrets. The vast powers which he has acquired over matter, the astonishing transformations which he effects, his success in analysing almost all bodies, and in reproducing so many, seems to promise that he

1 E.J. Hobsbawm, *The Age of Revolution: Europe 1789-1848* (London: Weidenfeld and Nicholson, 1962).

> shall one day discover the essence of a substance which he has so thoroughly subdued.[2]

Playfair viewed chemistry as the foremost scientific agent of the terms we have identified as defining this period of history. The growing powers chemists exercised over the material world, he declared, were leading to its subjugation, yielding "astonishing transformations" and the promise of understanding and absolute control. Though Playfair limited his remarks to the relations between humans and the material world, he and countless others recognized and engaged with chemistry in ways that brought the material and social realms together. Through their manipulative interactions with an increasing range of materials, chemists and chemistry left their mark virtually everywhere: increasing agricultural yields, expanding the range and scale of industrial production, extending the reach and precision of governance programs and practices, spearheading social improvement and public health. But so too did they contribute to environmental degradation through the unbridled exploitation of resources and aggravated industrial pollution, as well as to unsafe labor conditions and misery, the ferocity of warfare and the rapacious practices of empire.

The purpose of this volume is to raise broader attention to the position that chemistry was once recognized to hold as an active component of the great economic, social, and political developments of the period 1760-1840. It aims to do two things. First, by exploring the historically intertwined realms of production, governance and materials, it places chemistry at the center of processes most closely identified with the construction of the modern world. This includes chemistry's role in the interactive intensification of material and knowledge production; the growth, direction and management of consumption; environmental changes, regulation of materials, markets, landscapes and societies; and practices embodied in political economy. Second, the volume moves away from a narrative structured by a revolutionary break at the end of the eighteenth century and the primacy of innovation-driven change. Instead it aims to highlight the continuities and accumulation of less momentous changes that framed historical development over time and across the various spheres (the academic world, manufactures, public health and medicine, governmental administration, civil society and agriculture) in which chemists and chemistry operated.

Standard historical surveys tend to ignore eighteenth and early nineteenth-century chemistry – at best mentioning Lavoisier and the Chemical Revolution – or to subordinate it to physics and the mathematical sciences. Mechanization and quantification are often privileged as prime movers of historical change,

2 John Playfair, "Biographical Account of Hutton," *Transactions of the Royal Society of Edinburgh* 5 (1805): 39-99, on 74.

joined (finally) by chemistry in a 'second industrial revolution' during the final decades of the nineteenth century.[3] Switching from generalities to the more detailed practices of governance and production in the period 1760 to 1840, however, reveals a different story. This volume recognizes chemistry as broadly integrated in daily life, as essential to industrial development and agricultural improvement, and as fundamental to the governance of both society and the environment.

Crucial to such discussions is the question of who was a chemist. Should this label be applied only to those attached to universities and scientific academies or to a broader range of actors who engaged in chemical practices? If the latter, apothecaries, mining officials, manufacturers, inventors and others should and have been investigated as part of the history of chemistry.[4] Historians have recently asserted the existence of 'chemical experts', who served as consultants or held administrative and management positions.[5] This has brought the history of chemistry into closer contact with the history of governance (the stimulation and management of both public and private enterprises), a central theme of this volume. Attention to the ambiguous legacy of chemist-consultants heightens our awareness of chemistry's equivocal hold on public authority. Sometimes identified as arbiters of product purity and with improving public health and welfare, chemists were also viewed with distrust for representing the interests of industry and furthering environmental degradation.[6]

The expanding franchise of chemical practitioners reflects another critical change in how the history of chemistry is and should be examined. Historians

---

3 Tore Frangsmyr, J.L. Heilbron, and Robin E. Rider, eds., *The Quantifying Spirit in the Eighteenth Century* (Berkeley: University of California Press, 1990); Pat Hudson, *The Industrial Revolution* (London: Bloomsbury, 2014); Daniel Roche, "Encyclopedias and the Diffusion of Knowledge," Mark Goldie, Robert Wokler, eds., *The Cambridge History of Eighteenth-Century Political Thought* (Cambridge: Cambridge University Press, 2006), 172-194 on 175; But see Archibald and Nan Clow, *The Chemical Revolution: A contribution to social technology* (London: Batchwork Press, 1952; John Graham Smith, *The Origins and Early Development of Heavy Chemical Industry in France* (Oxford: Clarendon Press, 1979).

4 Hjalmar Fors, *The Limits of Matter: Chemistry, mining and enlightenment* (Chicago: University of Chicago Press, 2015); David Philip Miller, *James Watt, Chemist: Understanding the origins of the steam age* (London: Pickering and Chatto, 2009); Jonathan Simon, *Chemistry, Pharmacy and Revolution in France, 1777-1809* (Aldershot: Ashgate, 2013).

5 Ursula Klein, "Chemical Experts at the Royal Prussian Porcelain Manufactory," *Ambix* 60 (2013): 99-121.

6 See Thomas Le Roux' contribution to this volume; idem., "Chemistry and Industrial and Environmental Governance in France, 1770-1830," *History of Science* 54 (2016): 195-222; Christopher Hamlin, "The City as a Chemical System? The chemist as urban environmental professional in France and Britain, 1780-1880," *Journal of Urban History* 33 (2007): 702-728.

of chemistry are increasingly interested in the materials and material objects with which chemists interacted. Beyond a focus on the instruments, vessels and tools that furnished chemistry laboratories and workplaces, this includes chemistry's involvement in the histories of diverse materials through experimentation, consultation, regulation, and production improvement.[7] Materials and material objects were critical sites where chemistry met with governance and production between 1760 and 1840. Insofar as chemists could claim to manage the powers of airs, acids, minerals, metals, dyes and waters, so could they determine the habits, practices, and positions of those who engaged with these materials as manufacturers, regulators or consumers. When publicly recognized, chemists' knowledge and skill provided a platform for their claims of authority and expertise, which warranted their interventions in matters of governance. The essays in this volume work to identify such stances, positioning chemistry at the heart of the organization of social order.

To achieve these aims, this volume is divided into three sections.The first accentuates materials and material objects, along with the resources they provided chemical practitioners for building and exercising knowledge and expertise. The entangled nature of the social and material is also evident in the second section, on chemical governance, whereby chemists became involved in both the governmentally sanctioned and privately organized management of resources, people and environments. Finally, drawing on lessons from these two sections, the third revisits the classic theme of production, understood to include material and knowledge production, as well as their relation.

Materials and Material Objects

Given the centrality of materials and material objects to chemistry, attending to materiality as a key element of its history is bound to be a frutiful approach.[8] The traditional historiography of the "Chemical Revolution" certainly spoke of substances (phlogiston, calxes, oxygen, caloric) and instruments (the ice calorimeter, the balance), but too often in relation to the development of revolutionary ideas and concepts. Explorations that focus on their materiality have much to tell us about the details and consequences of chemical practice, which linked chemistry to broader historical developments.[9] Happily, historians are

7 Ursula Klein and Emma Spary, eds., *Materials and Expertise in Early Modern Europe: Between market and laboratory* (Chicago: University of Chicago Press, 2010).

8 Tom Ingold, "Materials Against Materiality," *Archeological Dialogues* 14 (2007): 1-16.

9 John G. McEvoy, *The Historiography of the Chemical Revolution: Patterns of interpretation in the history of science* (London: Routledge, 2010), 23-52; But see Lissa Roberts, "The Death

increasingly engaging with materials qua materials. Historians of early modern alchemy have used experiments and restagings to assess the particulars of alchemical recipes and procedures.[10] Scholars have long recognized the importance of instruments, and archaeology is now shedding light on historical chemical and alchemical instrumentation.[11] Catherine M. Jackson has analyzed how glassware altered nineteenth-century chemistry's laboratory practices and possibilities.[12] Historians have also expanded the repertoire of what counts as a chemically relevant material object by considering practitioners' interactions with a variety of commodities and substances.[13]

There is still, however, much to do. While there are now many histories of particular instruments and some distinct substances and spaces, we need to reflect more deeply on how to frame our inquiries. How should we approach the materiality of the substances and objects that populated and moved between the spaces in which historians are interested? Can we identify the broader practical and conceptual regimes of which these socio-material transformations were a part? How can we approach this subject without an *a priori* assumption that chemistry's development during this period depended on the introduction of innovative instruments and devices?[14]

---

of the Sensuous Chemist: The 'new' chemistry and the transformation of sensuous technology," *Studies in History and Philosophy of Science Part A* 4 (1995): 503-529.

10 Pamela H. Smith at al, "The Making and Knowing Project" <http://www.makingandknowing.org/> accessed March 16, 2016; Lawrence M. Principe, "Apparatus and Reproducibility in Alchemy," Frederic L. Holmes and Trevor H. Levere, eds., *Instruments and Experimentation in the History of Chemistry* (Cambridge, MA: MIT Press, 2000), 55-74.

11 Trevor H. Levere, "The Role of Instruments in the Dissemination of the Chemical Revolution," *Endoxa* 19 (2005): 227-242; Frederic L. Holmes, *Eighteenth-century Chemistry as an Investigative Enterprise* (Berkeley, CA: Office for History of Science and Technology, University of California at Berkeley, 1989); Jan Golinski, "Precision Instruments and the Demonstrative Order of Proof in Lavoisier's Chemistry," *Osiris* 9 (1994): 30-47; Lissa Roberts, "A Word and the World: The significance of naming the calorimeter," *Isis* 82 (1991): 199-222; Marcos Martinón-Torres, "Inside Solomon's House: An archaeological study of the old Ashmolean chymical laboratory in Oxford," *Ambix* 59 (2012): 22-48; Simon Werrett, "Matter and Facts: Material culture in the history of science," Robert Chapman and Alison Wylie, eds., *Material Evidence: Learning from archaeological practice* (New York: Routledge, 2014), 339-352.

12 Catherine M. Jackson, "The "Wonderful Properties of Glass: Liebig's Kaliapparat and the Practice of Chemistry in Glass," *Isis* 106 (2015): 43-69.

13 Klein and Spary, eds., *Materials and Expertise* (see note 7).

14 See Werrett's forthcoming *Thrifty Science: Making the most of materials in the history of experiment.*

To take the last first, chemical practitioners between 1760-1840 were just as likely to use ready-to-hand objects in adapted spaces as to introduce new and specialized instruments and dedicated spaces for experiment. Adaptibility, bricolage and repair were hallmarks of chemical practice. As Simon Werrett discusses in this volume, many chemical practitioners set up laboratories in their homes where they adapted tea cups, saucers, clay pipes, gun barrels and household furniture to chemical ends.[15] Even Lavoisier, famous for using new and prohibitively expensive instrumentation, sometimes cobbled together experimental set-ups from objects originally intended for other purposes; his epoch-making demonstration of the decomposition of water, for example, featured an adapted gun barrel. In practice, chemistry relied at least as much on adaptation, knowedge of lutes and luting, awareness of the most appropriate amalgam, and artisanal proficiency, as it did on theory.

Considering chemists as innovators, bricoleurs and reparateurs is not only apt because they were sometimes one and sometimes the other. Innovation, bricolage and repair often went hand in hand. As Elena Serrano illustrates in this volume, novel instruments and devices were often hybrid compositions of new and recycled or innovative and mundane components. This was especially the case when novel apparatus were commodified for wider distribution; simplified use and repair were important considerations when designing for a broader public.

This sort of adaptive design and use was often discussed in terms of 'oeconomy'. Manuals on household management or 'domestic oeconomy' circulated since the sixteenth century, promoting a balance between excess and conservation, saving and expense, using the old and investing in the new.[16] This was not only a call for thrifty management for its own sake. Oeconomy was widely taken to cover a broader set of meanings and practices by the mid-eighteenth century. Alongside material and financial considerations, oeconomy spoke to the virtues of order, prudence and moral responsibility.[17] Exploring the meanings and practices associated with the word's contemporary uses reveals how actors at the time framed their understanding of and engagement with the world around them.

---

15 Simon Werrett, "Recycling in Early Modern Science," *British Journal for the History of Science*, 46 (2013): 627-646.

16 Karen Harvey, *The Little Republic: Masculinity and domestic authority in eighteenth-century Britain* (Oxford: Oxford University Press, 2012).

17 Lissa Roberts, "Practicing Oeconomy During the Second Half of the Long Eighteenth Century: An introduction," *History and Technology* 30 (2014): 133-148.

Whether linked to the human or animal body, private households, the state, nature, or chemistry, oeconomy spoke to the maintenance of a well-balanced order. Often associated with 'improvement', oeconomy pointed to productivity, but never in exclusive terms of maximizing material production and profit. Invariably, it also carried a moral connotation, placing the improvement of agricultural and manufacturing yields in the context of stimulating stewardship of material *and* social resources – whether within the individual, regional or national household.[18] This variously entailed tying educational programs to the goals of cameralist administration; integrating programs of experiment, communication and engagement aimed at public education with the improved production of domestic goods; bringing education and practice together to stimulate the circulation and use of rural waste products and industrial leftovers to further production in both agriculture and manufacturing; and tying educational programs for chemical practitioners to the ideals of good citizenship.[19]

The ideals and practices of oeconomy receded from their once prominent position in European cultural, institutional and political realms by the mid-nineteenth century. It is beyond the bounds of this study to explain why or fully how this occurred. But surely the mismatch between oeconomy's idyllic projections of balance and order and the often disruptive circumstances that marked the years 1760-1840 were involved. War, political upheaval, the growth of manufactures and social displacement constantly challenged the idealized harmonies of enlightened society. Scales of operation were transformed in the armed forces, the civil service, and in industry.[20] While domestic, artisanal modes of production continued, and while agriculture remained the largest employment sector until at least 1850, industrial manufactures grew, urban populations burgeoned and peoples traversed regions and continents *en*

18 Joppe van Driel, "The Filthy and the Fat: Oeconomy, Chemistry and Resource Management in the Age of Revolution," PhD Thesis, University of Twente, 2016.

19 Christophe Meinel, "Reine und angewandte Chemie," *Berichte zur Wissenschaftsgeschichte* 8 (1985): 25-45; Andre Wakefield, "Police Chemistry," *Science in Context* 13 (2000): 231-267; Elena Serrano, "Making Oeconomic People: The Spanish Magazine of Agriculture and Arts for Paris Rectors (1797-1808)," *History and Technology* 30 (2014): 149-176; Joppe van Driel, "Ashes to Ashes: The stewardship of waste and oeconomic cycles of agricultural and industrial improvement, 1750-1800, *History and Technology* 30 (2014): 177-206; Le Roux, "Chemistry and Governance" (see note 6); Lissa Roberts, "P.J. Kasteleyn and the "Oeconomics" of Dutch Chemistry," *Ambix* 53 (2006): 255-272.

20 John Brewer, *The Sinews of Power: War, money and the English state, 1688-1783* (London, 1989); Anna Simmons (this volume) discusses how changes of scale affected chemical production.

masse.[21] The ranks of chemical practitioners, often trained in newly autonomous laboratories, swelled accordingly. Changing scales and approaches to materials, production and governance, were undoubtedly tied up with changing attitudes and practices. The challenge is to explore and understand these connections and their consequences without the taint of teleology.

This requires rethinking the historical relationship between oeconomy and oeconomic practices, on one hand, and interpretive categories drawn from economics and economic history, on the other.[22] Given the performative impact and results of oeconomic formulations and activities, it is misleading to dismiss oeconomy as a cultural conceit of elite amateurs, a doomed project or rhetorical side-show to economic development. Neither was oeconomy a proto-concept that gave way to economic analysis as the latter's concepts matured during the early nineteenth century.[23] Such views move between [1] distinguishing between oeconomy as a cultural expression and economics as expressive of 'real world' activities and [2] isolating oeconomy as a concept and placing it under a larger rubric of economic concepts, which is teleologically structured by a movement toward modern economic organization and understanding. The first privileges economics as reflective of material reality; the second grants it conceptual priority, implying that economic (re-) conceptualization is a key motor of historical change.

This volume refuses both these options, arguing instead for a non-teleological perspective. By the mid-nineteenth century governing attitudes, policies (including colonial policies and taxation regimes) and practices in western Europe instantiated the market, social welfare and nature as distinct realms of conceptualization, activity and governance. But this development – which carved out a space for interacting with material and human resources in strictly calculative terms of economic value, cordoned off from issues of moral or environmental responsibility – was neither inevitable nor the result of a conceptual change. It arose in a historical landscape whose contours evolved over time as humans interacted with specific material substances and objects in various contexts of conceptually and administratively governed production, con-

21 Rondo Cameron, "A New View of European Industrialization," *The Economic History Review* 38 (1985): 1-23, 6.

22 Timothy Mitchell, "Rethinking Economy," *Geoforum* 29 (2008): 1116-1121.

23 Henry Lowood, *Patriotism, Profit, and the Promotion of Science in the German Enlightenment* (New York: Garland Press, 1991); Margaret Schabas and Neil Di Marchi, "Introduction to Oeconomies in the Age of Newton," Margaret Schabas and Neil Di Marchi, eds., *Oeconomies in the Age of Newton. Annual Supplement to History of Political Economy* 35 (2003): 1-13; Margaret Schabas, *The Natural Origins of Economics* (Chicago: University of Chicago Press, 2005), 1-21.

sumption and use – sociomaterial interactions often mediated by chemists and chemistry.[24]

But if historical change is a consequence of sociomaterial interaction, we need to understand what that entails, both in general historiographical and specific historical terms. Contributions to this volume offer a variety of approaches to this subject, as they explore histories of specific materials and material objects between 1760-1840. Lissa Roberts and Joppe van Driel tackle the case of coal in their essay, generally identified as the energy source that fueled the industrial revolution.[25] Demonstrating that coal's identity was actually far from settled at the time, they argue for understanding material identities – and the values associated with them – as historically open rather than ontologically fixed.

Scholars such as Hans-Jörg Rheinberger and Karin Knorr Cetina have made similar claims in their discussions of 'epistemic things' and 'epistemic objects', emphasizing the epistemic openness of objects – but only in the context of experimental investigation.[26] Rheinberger thereby distinguishes between 'epistemic objects' and "stable, technical objects that may define the boundary conditions of further epistemic objects." Knorr-Cetina contrasts the epistemic openness of objects that undergo scientific research with the stability of "commodities, instruments and everyday things."[27]

Simon Werrett's investigation of household chemistry in which ready-to-hand objects were pressed into experimental and productive service problematizes the distinction between research objects' epistemic openness and the stability of 'technical' and 'everyday' objects. Making do with impromptu equipment

24 William Ashworth, "'Between the Trader and the Public': British alcohol standards and the proof of good governance," *Technology and Culture* 42 (2001): 27-50; Joppe van Driel and Lissa Roberts, "Circulating Salts: Chemical governance and the bifurcation of "nature" and "society"," *Eighteenth-Century Studies* 49 (2016): 233-63; Joppe van Driel, "The Filthy and the Fat (see note 18).

25 E.A. Wrigley, *Energy and the English Industrial Revolution* (Cambridge: Cambridge University Press, 2010).

26 Hans Jörg Rheinberger, *Toward a History of Epistemic Things* (Palo Alto: Stanford University Press, 1997); Karin Knorr Cetina, "Objectual Practice," T.R. Schatzki, K. Knorr Cetina, & E. von Savigny, eds., *The Practice Turn in Contemporary Theory* (New York: Routledge, 2001), 184-197; Cyrus Mody and Michael Lynch, "Test Objects and other Epistemic Things: A History of a Nanoscale Object," *British Journal for the History of Science* 43 (2010): 423-458.

27 Hans-Jörg Rheinberger, "A Reply to David Bloor: 'Toward a Sociology of Epistemic Things'," *Perspectives on Science* 13 (2005): 406-10, 407; Knorr-Cetina, "Objectual Practice," p. 84 (see note 26).

involved a process of learning about what works. Which materials reacted with experimental substances and were therefore unusable in experimental setups? What was the best way to seal a make-shift container?[28] Roberts and Van Driel's discussion of the history of coal reveals further that the openness of "commodities, instruments and everyday things" is not only epistemic. The identities of materials and material objects are as much a matter of what they do as of what we know about them. But what they can do is neither simply a question of some essential capability or characteristic, nor only of human use. Philosopher Annemarie Mol writes, "[O]ntology is not given in the order of things [...] instead, ontologies are brought into being, sustained, or allowed to wither away in common, day-to-day, sociomaterial practices."[29] As represented by Elena Serrano's discussion of 'affordance' in her essay, our goal is to portray material identities and claims of agency in ways that recognize the historical interplay between the specificities of materials and material objects and the contexts in which they were investigated and put to work.[30]

The isolation and identification of qualitatively distinct 'airs' or 'gases' form a central focus of histories of the Chemical Revolution.[31] In her essay, Marie Thébaud-Sorger goes beyond considering them as epistemic objects to the question of how they became manipulable commodities that firmly attached chemistry to both the increasing commodification of society and the spectacles that celebrated this transformation. Substances such as coal and airs also linked chemistry to processes such as urbanization and (initially oeconomic) concerns over public health, which emerged as foci of governance in the period 1760-1840. Aiming to operate at a scale of whole populations, evolving regimes of cleanliness and hygiene, health and security depended on massive material investments and the disciplining of large populations. Key to this process was the construction of urban architectures enabling the circulation of clean air and water and the elimination of foetid smells and poisonous miasmas.[32]

28 Adele Clarke and Joan Fujimura, eds., *The Right Tools for the Job* (Princeton, NJ: Princeton University Press, 1992).

29 Annemarie Mol, *The Body Multiple: Ontology in medical practice* (Raleigh, NC: Duke University Press, 2002), 6; Ingold, "Materials," p. 1 (see note 8).

30 Affordance refers to "those functional and relational aspects of technology that frame but do not determine the possibilities for action in relation to an object." Brian Rappert, "Technologies, Texts, and Possibilities: A reply to Hutchby," *Sociology* 37 (2003): 565-80, 566.

31 Jan Golinski, "Chemistry," Roy Porter, ed., *The Cambridge History of Science: Volume 4, eighteenth-century science* (Cambridge: Cambridge University Press, 2003), 375-396.

32 Thomas Markus, *Buildings & Power: Freedom and control in the origin of modern building types* (London: Routledge, 1993), 146-158; Christopher Hamlin, "State Medicine in Britain,"

Chemists played a central role in such activities as experts on spa waters and urban water and gas supplies, as overseers of ventilation projects, and through fumigation practices using new chemical substances.[33] Elena Serrano explores one such case, focusing on how newly designed 'fumigating machines' were used to combat disease in France and Spain at the beginning of the nineteenth century. Importantly, these machines simultaneously transported a recently isolated air and new knowledge claims that explained it. They also embodied reformulated modes of governance that, mediated by chemical experts, transferred direct responsibility for public health from government agencies to the individual behavior of citizens who were charged with using such contraptions.[34] But material concerns were as crucial as the role played by chemists. For operating on a transnational scale, as this project did, necessitated adaptation; simplified designs and cheaper materials enabled the machine's mass manufacture.

José Ramon Bertomeu's exploration of arsenic in the 1830s and '40s offers one more example of how focusing on materials exposes the non-teleological co-construction of sociomaterial identities. His essay draws attention to the efforts waged in these decades to identify the notorious poison arsenic. If oeconomy tolerated – or even valued – material ambiguities and open-ended capacities for repurposing and re-use, the economic and social orders emerging by the late 1830s depended on various institutions – manufactories, government laboratories, law courts, public health bureaus – that required specific definitions and identities. Institutional attempts to know and thereby govern materials and their use nonetheless continued to be plagued by ambiguities of material definitions and application. The definition of arsenic, its presence and properties, thus emerged alongside the identities of chemical practitioners, productive sectors, uses and institutions that engaged with it, determining the agency of all the actors involved.

Dorothy Porter, ed., *The History of Public Health and the Modern State* (Amsterdam; Atlanta, GA: Rodolpi, 1994), 132-164.

33 Matthew D. Eddy, "The Sparkling Nectar of Spas; or, mineral water as a medically commodifiable material in the province, 1770-1805," Klein and Spary, eds., *Materials and Expertise*, 283-292 (see note 7); Christopher Hamlin, *A Science of Impurity: Water analysis in nineteenth-century Britain* (Berkeley, CA: University of California Press, 1990); Leslie Tomory, *Progressive Enlightenment: The origins of the gaslight industry, 1780-1820* (London; Cambridge, MA: MIT Press, 2012).

34 Simon Schaffer, "Measuring Virtue: Eudiometry, enlightenment and pneumatic medicine," Andrew Cunningham and Roger French, eds., *The Medical Enlightenment of the Eighteenth Century* (Cambridge: Cambridge University Press, 1990), 281-318.

What, then, was chemistry in the period 1760 to 1840? This section's focus on materials and material objects positions chemistry at the intersection of gradual yet impressive shifts in production, governance and their relationship. Chemistry flourished through its ability to use materials and multiply their varied affordances, but the manifestations, management and meaning of this ability gradually changed. An initially oeconomic orientation associated with household management and its prudent (re-) use of ready-to-hand objects and instruments posited the inseparably social and moral character of material order. By the late 1830s, the sociomaterial challenges of shifting scales and multiplying and increasingly various fruits of chemical production simultaneously fed and responded to efforts to govern them. Now against a view of materials as open-ended and capable of continuous revision, manufacturing – along with various governance practices (often mediated by chemists and chemical 'expertise') that regulated and taxed its materials, processes and products – divided phases of production and consumption, seeking to fix the identity of material objects as commodities. Chemical practitioners operated in a growing number of contexts, assessing the properties of materials and their suitability to manufactures, developing novel products and processes, and providing credit and controls for unfamiliar products. Managing this complex state of affairs increasingly relied on two mutually reinforcing loci of governance. One was situated in the specifying processes of governmental legislation and courtroom adjudication. The other resided in the organization and conceptualization of market oriented practices that translated social and material interplay into calculations and models, masking their multifaceted interactions as they transformed them.

The division of labor and specialization demanded by these processes and their requirement of strict definitions and identities proved a double-edged sword. On one hand they afforded chemistry's growing autonomy, professional identity and recognized expertise. On the other, they narrowed understandings of material and social identities to the point where their complex intersections and mutual constitution seemed to disappear. What remained was a sense not of interpenetrating oeconomies of materials, production and governance, but of separate spheres of agriculture, industry, chemistry, and government. Composing separate historical narratives of these spheres then served to reinforce their boundaries, raising chimerical puzzles over how one influenced the other.

Chemical Governance and the Governance of Chemistry

Our focus on 'chemical governance' might seem odd or anachronistic at first. In current parlance, it is understood either as a form of 'corporate governance', whereby chemical manufacturers assume responsibility for construing and policing their own ethical performance, or related more generally to the management of hazardous chemicals.[35] Almost invariably, chemical governance is currently invoked in relation to the environmental impact of chemicals used in specific industrial contexts. Behind this configuration is a specific – neo-liberal – rationality that calculates 'good' governance in terms of transaction costs, bracketing it off from other relations seen as involving 'externalities' whose consequences might call for governmental response or as extra-governmental concerns best left up to social and corporate organizations or 'the market'.

Michel Foucault and others have called on us to step back and recognize the historical character of this regime whose beginning, they argue, was in the period covered by this volume.[36] Such a move frees us from considering chemical governance – like governance more generally – as formed or constrained by the currently reigning rationality, warranting instead the historicization of its conceptualization and practices. This requires an umbrella definition of chemical governance that stands above the ways in which specific historical regimes framed it. Here we define it as entailing the privately initiated or government sanctioned employment of chemists and their practices to stimulate or inhibit productive activities and manage resources, people, activities, environments, and their relations, in accordance with specific norms and goals. The essays in this volume zero in on the historical specificities of chemical governance and how they evolved during the period 1760-1840.

It helps to recognize that 'governance' was an actors' category during this period. A survey of uses between 1760 and 1840 shows a cluster of related meanings. Governance referred to the duties of governing; that is, the management of a socio-political unit, institution or individual estate, often with a paternal character and directed toward 'improvement'.[37] It spoke to the influence one had over another's life and behavior, but could also involve exercising

35 Henrik Selin, *Global Governance of Hazardous Chemicals: Challenges of multilevel management* (Cambridge: MIT Press, 2010); Lissa Roberts, "Exploring Global History Through the Lens of History of Chemistry: Materials, identities and governance," *History of Science* 54 (2016): 335-361, on 350-356.

36 Michel Foucault, *Naissance de la biopolitique, Cours au Collège de France 1978-1979* (Paris: Gallimard Seuil Haute études, 2004).

37 William Bridle, *A Narrative of the Rise and Progress of the Improvements Effected in His Majesty's Gaol at Ilchester* (Bath: Wood, Cunningham and Smith, 1822).

control over materials; chemists, for example, were said to exercise governance over fire.[38] Finally, by analogy to 'divine governance', it entailed the maintenance of material and social order for the public good.[39]

Attention to chemical governance in this volume highlights the ways in which chemists and chemical practices were integral to a broad range of significant governance processes between 1760 and 1840. Though much more work needs to be done, the biographies of leading figures such as Lavoisier, Guyton de Morveau and Jean-Antoine Chaptal point to how the practices and institutions of chemical knowledge production in France were intertwined with industrial and administrative developments.[40] So too has recent work on 'artisanal-scientific experts' who served throughout Europe as administrative officials, consultants and inspectors for various state agencies involved with the stimulation and management of sectors such as mining, metal production, agriculture, porcelain manufacture and textiles – been helpful on a more international scale.[41]

The essays here identify chemical governance as a practice that goes beyond individual case studies. The essays by Christine Lehman and Thomas Le Roux explore the history of chemical governance in relation to the French state's regulation of chemical industry up to 1830. Through an examination of requests for state support in the production of *céruse* (white lead or compounds containing it), Lehman concentrates on how processes of chemically mediated governance helped steer industrial production, complicating claims about innovation along the way. Far from simply a matter of developing knowledge and practices in a drive to improve the quality, quantity and/or profitability of production, French chemists who served as consultants and administrators found their mediations situated within a complex web of interests. Producers seeking state support might be driven by the desire to protect a manufacturing process, capture a geographically based market or outflank a competitor. The demands of various ministries directed attention toward often-irreconcilable

38 Jean François Clément Morand, *L'Art d'exploiter les mines de charbon de terre* (Paris: Saillant et Nyon, 1768-1779), vol. 2, 1192, 1195, 1255; Basil Valentine, "The Stone of Fire," in Francis Barrett, *The Lives of Alchemystical Philosophers* (London: Macdonald and Son, 1815), 232-236, on 233.

39 *The Book of Common Prayer* (Oxford: T. Wright and S. Gill, 1771).

40 Charles Gillispie, *Science and Polity in France: The revolutionary and Napoleonic years* (Princeton, NJ: Princeton University Press, 2004)

41 Ursula Klein, ed., *Artisanal-Scientific Experts in Eighteenth-Century France and Germany*, special issue of *Annals of Science* 69 (2012): 303-433; Bruno Belhoste, *La Formation d'une technocratie. L'École polytechnique and ses élèves de la Révolution au Second Empire* (Paris: Belin, 2003), esp. 75.

questions of international competition, domestic commerce and social considerations. The state was far from monolithic. The formulation of chemically based advice and administrative decisions was thus always a matter of negotiating between various interests.[42]

Such negotiations, of course, were always situated in specific contexts. The social, political, commercial and financial dislocations associated with the French Revolution and Napoleonic era framed the pursuit of and changes in governance processes in France while highlighting the constitutive role of chemists and chemistry. As Le Roux shows for the formation of regulatory policies and practices concerning the environmental impact of chemical industry there, the abolition of both traditional corporations ranging from artisanal guilds to the *Académie des sciences* and the institutional apparatus responsible for governance during the *ancien régime* framed an intensification of longer-term historical developments that – as Tocqueville first pointed out – were transforming France from a corporate to a modern state.[43] Wartime exigencies and increasing international competition, coupled with domestic dislocation and change, simultaneously intensified demand for the products of chemical industry and a greater need to adjudicate between the operational requirements of industrial production and the public's experience of its environmental consequences. It was in this context that chemists were called upon to help encourage industrial development, as well as to determine and compare the relative values of productivity and public health and welfare.

The determination and comparative measurement of value in relation to the interactive triad of industrial development, public welfare and environmental sustainability were inevitably bound to competing norms and issues of trust that often remained untranslatable into 'objective' numbers.[44] One answer to the persistence of qualitatively heterogeneous issues was provided by the evolution of new analytical categories through which to define, organize and judge chemically construed phenomena.[45] Beyond the development of new nomenclatural and instrumentally mediated practices, Le Roux argues, this entailed reconfiguring the legally sanctioned definitions and boundaries

42 Timothy Mitchell, *Rule of Experts: Egypt, techno-politics, modernity* (Berkeley: University of California Press, 2002); Lissa Roberts, "Accumulation and Management in Global Historical Perspective: An introduction," *History of Science* 52 (2014): 227-246, 238.

43 Alexis de Tocqueville, *L'Ancien régime et la révolution* (Paris: Michel Lévy Frères, 1856).

44 Theodore Porter, *Trust in Numbers: The pursuit of objectivity in science and public life* (Princeton, NJ: Princeton University Press, 1995).

45 On classification as a form of governance, see Steve Woolgar and Daniel Neyland, *Mundane Governance: Ontology and accountability* (Oxford: Oxford University Press, 2015), 55-77.

between 'harmfulness' and 'harmlessness' – a process in which chemists played a key role. More fundamental still, chemists throughout the period covered by this volume were intimately involved in a process whereby 'the marketplace', 'society' and 'nature' became reified as essentially distinct categories through the very governance practices that were established to police their hybrid interactions.[46]

It needs to be stressed that the history of chemical governance was an open-ended one – neither simply the consequence of a battle of ideologies such as mercantilism or cameralism versus liberalism, nor directed toward any particular teleological end. Rather chemical governance and the effects with which it was associated are best understood by tracing how it evolved out of largely mundane processes. This approach is especially promising for places such as Sweden, Prussia and the Austrian Empire where the state's regulation of mining and industry relied on chemical expertise. A number of recent studies have emphasized the role of 'hybrid experts', who brought a marriage of chemical and bureaucratic training and experience to the performance of their duties.[47] We still need more fine-grained studies of their daily activities, however, to inform longer-term histories of industrialization in these lands. In place of studies that turn to the influence of 'Baconian empiricism', 'Newtonian physics' or 'rationalist inquiry' to explain the transformation of production techniques and sociopolitical institutions, we need accounts that build on the actual work carried out by those who used their chemical knowledge and know-how in their daily practices as mining officials, industry inspectors, excise officers and so forth.[48]

William Ashworth has charted the ways in which mundane instrumentally-mediated regulatory processes carried out by British excise agents worked, not

46 Joppe van Driel and Lissa Roberts, "Circulating Salts" (see note 24); David Wachsmuth, "Three Ecologies: Urban metabolism and the society-nature divide," *The Sociological Quarterly* 53 (2012): 506-523.

47 Ursula Klein, "Savant Officials in the Prussian Mining Administration," *Annals of Science*, Special Issue: *Artisanal-Scientific Experts in Eighteenth and Nineteenth-Century Germany and France*, 69 (2012): 349-374; Peter Konečný, "The Hybrid Expert in the 'Bergstaat': Anton von Ruprecht as a professor of chemistry and mining and as a mining official, 1779-1814," *Annals of Science* 69 (2012): 335-347; Ursula Klein, "The Prussian Mining Official Alexander von Humboldt," *Annals of Science* 69 (2012): 27-68; Hjalmar Fors, "The Knowledge and Skill of Foreigners: Projectors and experts at the early modern Swedish Board of Mines," Hartmut Schleiff and Peter Konečný, eds., *Staat, Bergbau und Bergakademie im 18. und frühen 19. Jahrhundert* (Stuttgart: VSWG, 2012), 53-62.

48 Eric Dorn Brose, *The Politics of Technological Change in Prussia: Out of the shadow of antiquity, 1809-1848* (Princeton, NJ: Princeton University Press, 1993), 13.

only to fix government revenues and establish standards for foodstuffs and alcohol, but also to suggest new products and production processes for distillers and others looking to minimize taxes and maximize profits. In turn, developments in the production and marketing of new products fed the further development of chemical instrumentation and testing for policing the composition and healthfulness of comestibles.[49]

The sort of chemical governance discussed by Ashworth was largely an urban matter. Cities increasingly became sites of chemical concern and governance, as their rising populations engaged in expanding networks of production, exchange and consumption. This brought urban and rural environments into closer contact through the interweaving of agricultural and industrial practices. As discussed by Joppe van Driel, urban elites joined with landowners and government officials in the Netherlands both to encourage and police the collection and circulation of urban wastes for use as agricultural fertilizers, and the return of industrial crops for urban-based manufacturing. This reminds us that chemical governance was not only a governmental affair. It also engaged private individuals who often joined together in oeconomic societies to encourage and monitor 'improvement'.[50]

As cities grew, observers became increasingly aware of the potential to study and need to govern them as chemical systems in their own right. While mediating between the encouragement of industry and the health of urban dwellers exposed to industrial toxins was part of the story, so too were problems such as sewage, water and food supplies, lighting, building supplies and the collection of vital materials such as saltpeter – all candidates for chemical governance.[51] Ernst Homburg has discussed proposals to establish urban chemical police in various German states from the 1820s.[52] Christopher Hamlin has examined the roles played by French and British chemists between 1780 and 1880, as they simultaneously aspired to the position of urban regulators and tied their increasing professional status to industrial consultation. Without a clear identity, he argues, chemists were never able to create "a matter-based science of urban management" as an authoritative tool of governance along the lines of

49 Ashworth, "'Between the Trader" (see note 24); See also the essays in this volume by Elena Serrano and Marie Thébaud-Sorger.

50 Driel, "Ashes to Ashes" (see note 19); Lissa Roberts, "Practicing Oeconomy" (see note 17).

51 André Guillerme, "Enclosing Nature in the City: Supplying light and water to Paris, 1770-1840," *Construction History* 26 (2011): 79-93; Sabine Barles, *L'invention des déchets urbains: France, 1790-1970* (Seyssel: Champ Vallon, 2005).

52 Ernst Homburg, "The Rise of Analytical Chemistry and its Consequences for the Development of the German Chemical Profession, 1780-1860," *Ambix* 46 (1999): 1-32, 19.

architecture and engineering.[53] Instead they remained a mixed community of, at one and the same time, educators, guardians of the public interest and consultants for maximizing industrial profit.

The period 1760-1840 was an age of networks and activity that encompassed the entire globe, often involving the movement of massive quantities of chemical substances and objects, and innumerable possibilities for the exercise of both chemical production and governance.[54] Chemical governance was dispersed across a multitude of practices, techniques and decisions traversing the planet. Such dispersal may be appreciated better by examining a case of the global reach of chemical governance. Andreas Weber examines an episode in the history of imperial Dutch monetary policy in this volume to demonstrate that even something as seemingly abstract as monetary policy was subject to chemical governance. Importantly, his analysis underscores the polycentric character of imperial governance by revealing its mundane dependence on local chemical practitioners and their practices situated both in metropolitan centers and in far-off colonial settings.[55] Beyond their initial pronouncement, policies actually took shape through their embodiment in coins and bank notes – the production of which required locally available chemical expertise. A colonially based assayer and mint master, therefore, could change the direction and consequence of imperial policy by minting coins with an alloy containing more silver than instructed. Rather than simply serving imperial masters, local practitioners harnessed their chemical know-how to colonize the policy-making process by which the empire was supposed to be ruled. Substituting locally determined value calculations for those that originated in The Hague recalibrated the policy they were supposed to pursue, shifting the fulcrum of practical power away from Dutch ministerial designs toward the geographically distant networks to which such values were beneficial.

The point of Weber's analysis is that both imperial policies and the Dutch treasury were affected by the chemically informed acts of individuals operating on the other side of the world. But if chemical practices could alter the course of financial management at the level of national and imperial govern-

53 Hamlin, "The City" (see note 6).

54 Pratik Chakrabarti, "Empire and Alternatives: Swietenia febrifuga and the cinchona substitutes," *Medical History* 54 (2010): 75-94; Gregory Cushman, *Guano and the Opening of the Pacific World: A global ecological history* (Cambridge: Cambridge University Press, 2013).

55 Simon Schaffer, "Golden Means: Assay instruments and the geography of precision in the Guinea Trade," Marie-Noëlle Bourget, Christian Licoppe and H. Otto Sibum, eds., *Instruments, Travel and Science. Itineraries of precision from the seventeenth to the twentieth century* (London: Routledge, 2002), 20-50.

ments, so was financial governance crucial to the development of chemistry at the institutional level. Without the allocation and monitoring of funds to support institutions of learning, for example, the course of chemistry's disciplinary and professional development would have been quite different during this period. Importantly, as Sacha Tomic reveals in his study of financial governance at the Paris École de pharmacie, financial records are more than just a ledger showing income and expenditures. They also bear the imprint of a guiding moral economy under whose regime rewards were calculated in accordance with loyalty and service. Hence could a laboratory assistant rise to the rank of professor, in direct contradiction to current generalizations about 'invisible technicians' and the role their perceived lack of scientific insight played in the achievement of 'mechanical objectivity'.[56]

In sum, if chemistry's practices, practitioners and educational institutions were subject to various acts of regulatory governance, so too were chemists and chemical practices integral to more general processes of governance between 1760 and 1840. Insofar as chemists managed both materials and socially-embedded processes, they could claim to be essential to good governance. Chemistry had long been connected to medicine and the management of individual health. In the eighteenth century, chemists promoted and participated in practices that posited chemistry as an important or essential element in the medical management of communities. As chemical products and contexts of material production, use and disposal proliferated, chemists assumed an increasing number of roles as expert consultants in efforts to govern them. They tested, measured, experimented with and informed legal adjudications of material goods, serving taxation regimes, for example, through the study of adulteration. Such roles were far from straightforward, however, as chemists found themselves inserted into complex negotiations between competing communities and interests. In this context, financial and legal demands overtook oeconomic assessments of social and moral order that had previously been linked to chemical interventions.

Out of these interactions emerged regularly renegotiated policies, regulations, legal judgments, restrictions and codifications. Simultaneously these processes transformed the identities of materials and chemists themselves. Such mundane procedures and routines were also, no doubt, consequential for changing iterations of chemistry, governance and production, which suggests the inadequacy of a narrative that unduly focuses on revolutionary change in this period.

56 Lorraine Daston, "The Moral Economy of Science," *Osiris* 10 (1995): 2-24, 20.

Revisiting the History of Production

Production looms large as a focus in history writing on the period covered by this volume. Often linked to revolution – whether political, scientific or industrial, it is hard to ignore the roles that material and knowledge production have been said to play in historical development. An obvious question for this volume, then, is what can be learned by considering the history of production in conjunction with materiality and governance. In some ways, this approach and the analysis to which it gives rise underscore current historiographical views. While it is still standard to speak of 'the Industrial Revolution', for example, the term is now generally recognized as covering a longer and more gradual process of change. So too have we let go of the belief that industrialization followed a single paradigm.[57] And while economic historians continue to focus on economic growth as the Industrial Revolution's defining characteristic, they have joined with other historians to consider the relations between material and knowledge production; between production, consumption and use; and between production and the environment.[58] This has created ground for raising questions that are only answerable by bringing various historical disciplines together. Especially given the braided connections amongst the environmental, political, social, cultural and economic issues revealed by the complex challenges we currently face, historians find themselves looking to the past in more collaborative, interdiscipinary ways.[59] A central argument of this volume is that the history of chemistry provides a particularly apt vehicle for this sort of collaborative inquiry because its subject matter so patently sits at the intersection of historical engagements through which humans participated in weaving the social, material, political, economic and environmental together.[60]

The essays in this section highlight the interpretive possibilities afforded by bringing historical questions about production together with attention to gov-

57 Jeff Horn, Leonard N. Rosenband and Merritt Roe Smith, *Reconceptualizing the Industrial Revolution* (Cambridge, MA: MIT Press, 2010); Emma Griffin, *A Short History of the Industrial Revolution* (London: Palgrave, 2010).

58 Joel Mokyr, *The Gifts of Athena: Historical origins of the knowledge economy* (Princeton, NJ: Princeton University Press, 2002); Jan de Vries, *Consumer Behavior and the Household Economy, 1650 to the Present* (Cambridge: Cambridge University Press, 2008); E.A. Wrigley, *Energy and the English Industrial Revolution* (Cambridge: Cambridge University Press, 2010).

59 Brett Walker speaks of 'hybrid causation'. *Toxic Archipelago: A history of industrial disease in Japan* (Seattle: University of Washington Press, 2010).

60 Lissa Roberts, "Producing (in) Europe and Asia, 1750-1850," *Isis* 106 (2015): 857-865.

ernance and materials. Among other things, this approach underscores that 'production' is not simply a synonym for industry, translatable into measurable economic indicators. Along with material goods, producers make, use and consume knowledge, culture and political goods. The relations amongst all these elements require investigation.[61] Accordingly, Bernadette Bensaude-Vincent and Frank James examine the various ways in which governance mediated between the supposedly distinct realms of knowledge production and social order, bringing the core message of *Leviathan and the Air Pump* to life: "Solutions to the problem of knowledge are solutions to the problem of social order."[62] Anna Simmons and John Christie explore production in the context of urban manufacturing sites, affording an understanding of production sites as complex points of intersection between local and global translations involving the interaction of humans and materials with layered regimes of governance and production processes. Finally, Robert Anderson investigates interactions between academic chemists and those directly engaged in chemical industry to reflect on how we ought to understand the historical relationship between material and knowledge production. Here again we find general claims giving way to the specificities of local situations.

Though the period investigated in this volume is sometimes referred to as 'the age of revolution', debate continues regarding its 'revolutionary' nature.[63] Definitions of the Industrial and Chemical Revolutions have changed with generational regularity.[64] And while no one doubts that political revolution took place in France in 1789, discussions continue regarding its cause and character, including its relationship with Enlightenment ideas, industrialization and scientific developments.[65] The links between science and society in revolutionary France have traditionally been discussed either by chronicling how the state recruited scientists to perform specific tasks and reform productive sectors or as an aspect of intellectual history.[66] In her analysis of pedagogical

61 Roberts, "Producing" (see note 60).

62 Steven Shapin and Simon Schaffer, *Leviathan and the Air Pump: Hobbes, Boyle, and the experimental life* (Princeton, NJ: Princeton University Press, 1985), 332.

63 Hobsbawm, *Age of Revolution* (see note 1).

64 David Cannadine, "The Present and the Past in the English Industrial Revolution, 1880-1980," *Past and Present* 103 (1984): 131-172; McEvoy, *Historiography* (see note 9).

65 Robert Darnton, *The Forbidden Bestsellers of Pre-Revolutionary France* (New York: W.W. Norton and Co., 1996); idem., *Mesmerism and the End of the Enlightenment in France* (Cambridge, MA: Harvard University Press, 1968); Jeff Horn, *The Path Not Taken: French industrialization in the age of revolution, 1750-1830* (Cambridge, MA: MIT Press, 2006).

66 Gillispie, *Science and Polity* (see note 40); W.R. Albury, "The Order of Ideas: Condillac's method of analysis as a political instrument in the French Revolution," John Schuster and

reform at the École normale, Bensaude-Vincent looks instead at the kinds of mediation proposed to realize translations between "the problem of knowledge" and "the social order."

The establishment of the *École normale* in 1794, she reminds us, was part of a broader program to 'normalize' France. Alongside training teachers to teach a standard curriculum across France, a unified system of weights and measures was meant to normalize market exchanges throughout the country – itself administratively normalized through its division into *departements*.[67] The pedagogical structure that informed teacher education at the *école* built on Condillac's analytical approach, an instrument for standardizing understanding by eradicating lapses of logic and knowledge. Lavoisier had drawn on Condillac in his *Traité élémentaire de chimie*.[68] Disciplining chemistry, as advocated by Lavoisier, was seen by the school's organizers as a model for disciplining education more generally and, through that, disciplining the minds and bodies of normalized French citizens.

In fact, the ideal of standardization proved historically problematic. The metric system's introduction met so many local challenges that Napoleon withdrew it; it was finally instituted in 1837. Professors at the *École normale* resisted standardization, generally choosing instead to stress the details of their own disciplines. Charged to teach chemistry, Lavoisier's close associate Berthollet went further, opposing the approach he was supposed to defend. Where Lavoisier had claimed a seamless relation between properly disciplined understanding and the order of nature, Berthollet professed a complex topography of local circumstances and exceptions to rules that could only be governed through careful attention. Teaching (how to teach) chemistry was thereby not a question of normalization, but of training students to produce and apply knowledge in particular situations. In place of the unatainable dream of revolutionary transformation, he argued, an attentive and informed citizenry could thus work to reform the production of knowledge and society in more nuanced ways.

Berthollet was certainly not alone in criticizing the dream of organizing science and society through the unmediated application of abstract principles. But it is one thing to recognize a mismatch between revolutionary ideals and

R.R. Yeo, eds., *The Politics and Rhetoric of Scientific Method: Historical studies* (Dordrecht: Reidel Publishing Co., 1986), 203-225; Keith Michael Baker, *Condorcet: From natural philosophy to social mathematics* (Chicago: University of Chicago Press, 1974).

67 Ken Alder, "A Revolution to Measure: The Political Economy of the Metric System in France," M. Norton Wise, ed., *The Values of Precision* (Princeton, NJ: Princeton University Press, 1995), 39-71.

68 Antoine Laurent Lavoisier, "Discours préliminaire," *Traité élémentaire de chimie* (Paris: Chez Cuchet, 1789), v-xxi.

the complex variations of local circumstances, and quite another to foresee how governance regimes would actually take shape. As stressed throughout this volume, tensions amongst various interests engaged in the production of knowledge, goods and social order often led to situations that emphasized divisions rather than the applicability of universal principles. Disciplinary programs fed both the professionalization of and distinctions between individual sciences, while attempts to mediate amongst various social, political and economic interests from the standpoint of scientific knowledge often proved illusory – leading to hung juries and recriminations against (members of) the scientific community.[69] Current debates surrounding human and environmental health show how the challenges of governing translations between 'problems of knowledge' and 'social order' remain a key concern.[70]

While the revolutionary context of French attempts to engineer a clean fit between knowledge production and social order through specific governance practices set the question of their relations in sharp relief, examining the situation in Great Britain invites us to consider cases of this historical process in a less dramatic context. Given the growing importance and visibility of laboratories as the nineteenth century progressed, James' contribution to this volume, which follows the chemist Humphry Davy's tenure at two institutionally-based research laboratories in the late 1790s and early 1800s, provides a telling introduction to what followed. James' approach answers historian Graeme Gooday's call to consider laboratories in relation to the broader social worlds they inhabited: as situated between those who provided funding and institutional support, and the public, which varyingly acknowledged their status as sites of knowledge production and authority.[71] In fact, this was a fluid situation, manifesting as many variations as did laboratories themselves. A laboratory, for example, might double as a kitchen whose domestic situation complicated the gendered nature of the relation between science and the social order.[72] So too were laboratories attached to enterprises and institutions whose briefs tied knowledge production to specific purposes, such as the advancement of manufacturing, treating patients or public education.

69 In this volume, see essays by Thomas Le Roux, José Ramón Bertomeu Sánchez, and Lissa Roberts and Joppe van Driel.

70 Naomi Oreskes and Erik Conway, *Merchants of Doubt: How a handful of scientists obscured the truth on issues from tobacco smoke to global warming* (London: Bloomsbury Press, 2010).

71 Graeme Gooday, "Placing or Replacing the Laboratory in the History of Science?" *Isis* 99 (2008): 783-795.

72 Anita Guerrini, "The Ghastly Kitchen," *History of Science* 54 (2016): 71-97; Simon Werrett, "Household Oeconomy," this volume.

As indicated by Sacha Tomic in this volume, laboratories that relied on external funding had also to reckon with an accompanying regime of financial governance. This did not mean, however, that researchers were unable to exercise independence, either in terms of their research or of maneouvering between their sponsors' desires and public approbation. James presents Davy as an entrepreneurial figure who forged his career by creatively courting both patrons and the public's politer echelons, using financial and institutional support to gain a measure of autonomy for his research, which he presented publicly in ways that reinforced his and his findings' authority. That he anchored his efforts in laboratory settings is significant. Unlike modern definitions that emphasize laboratories' scientific character, the word 'laboratory' traditionally indicated a place where raw materials were 'elaborated' – worked upon to produce medicines, chemical substances, or other substances that could be put to use. Like many of his contemporaries, Davy applied the term to nature itself.[73] And in his public lectures at the Royal Institution, he drew on the claimed continuity between nature and his laboratory work to argue for public acceptance of a particular social order. Unlike his mentor Beddoes and outspoken figures such as Joseph Priestley, who championed the egalitarian ideals of radical politics and knowledge-based progress, Davy recruited nature and its investigation in a bid to privilege social stability over equality.[74]

Bringing James' thesis together with historian Jan Golinski's analysis of science as public culture in early nineteenth-century Great Britain, underscores the coincidence of an evolving regime of research autonomy – a professed characteristic of modern science – and the rise of a modern culture of 'public science' in which 'the public' was configured as a passive audience that bowed before scientific authority.[75] We might see this as part of a larger and longer transition in which science and industry became increasingly professionalized. Whatever the realities of domestic participation and use, members of

73 Gooday, "Placing or Replacing," p. 788 (see note 71); Humphry Davy, *Elements of Agricultural Chemistry, or a course of lectures for the Board of Agriculture* (London: Longman, Hurst, Rees, Orme and Brown, 1813), 14; David Gooding, "'In Nature's School': Faraday as an experimentalist," David Gooding and Frank James, eds., *Faraday Rediscovered; Essays on the life and work of Michael Faraday, 1791-1867* (London: Macmillan, 1985), 105-136.

74 Jan Golinski, *Science as Public Culture: Chemistry and Enlightenment in Britain, 1760-1820* (Cambridge: Cambridge University Press, 1992), 197.

75 Ibid.; Thomas F. Gieryn, "Boundary-Work and the Demarcation of Science from Non-Science: Strains and interests in professional ideologies of scientists," *American Sociological Review* 48 (1983): 781-795; Roger Cooter and Stephen Pumfrey, "Separate Spheres and Public Places: Reflections on the history of science popularization and science in popular culture," *History of Science* 32 (1994): 237-267.

the public were being configured as passive consumers of both knowledge and material goods made by others.[76]

The essays in this section highlight the need to recognize multiple geographies, overlapping jurisdictions and evolving identities as intrinsic to historical development. How do we reconcile narratives that stress the importance of locally available materials such as coal or wool with those that follow the movements of substances such as copper, barilla and mercury across the globe or between productive sectors? We often read that English coal powered an eclipse of Indian cotton and other foreign goods in the context of receding government intervention.[77] But other stories can be told that stress the locally heterogeneous character of industrialization.[78] Simmons' examination of pharmaceutical manufacturing in London illustrates the far-flung routes along which pharmaceutically relevant substances traveled. She simultaneously emphasizes the evolutionary character of local production processes that combined incoming substances with locally available resources under corporate and governmental oversight. Governance did more than guide behavior; it played an active role in shaping production, its components and outcomes. As demonstrated by William Ashworth, governance included identifying substances, testing for their 'purity' and composition, and making them serve purposes ranging from revenue enhancement and international competition to public health and welfare.[79]

Historical surveys of industrial production during this period generally highlight mechanization, innovation and the introduction of steam power. But London – like other industrial centers – also housed productive sectors whose sites and methods depended on more and other things than the revolutionary introduction of path-breaking machines. The pharmaceutical sector is a telling example, as it encompassed international trading companies, large-scale wholesale manufacturers and smaller-scale apothecaries and druggists, globally sourced substances and the equipment, knowledge and skill needed to produce, store and sell its products. Its market brought suppliers together with users ranging from individual customers to the mammoth British Navy and East India Company. In turn, its manufacturing sector responded to the inter-

76 John Brewer and Roy Porter, *Consumption and the World of Goods* (New York: Routledge, 1993).

77 E.A. Wrigley, *Energy and the English Industrial Revolution* (Cambridge: Cambridge University Press, 2010).

78 Thomas Misa, *From Leonardo to the Internet: Technology and culture from the Renaissance to the present* (Baltimore: Johns Hopkins University Press, 2004), 59-96.

79 Ashworth, "'Between the Trader" (see note 24).

active dynamics within and between supply and demand, mediated by the governance of firms and institutions, as well as regulations imposed by the state as its agents sought to generate revenues, oversee trade, monitor production and protect consumers.

Two resulting trends are especially worth noting. The first has to do with increasing reliance on keeping production and account records, akin to what Ursula Klein calls "paper tools".[80] As much as the chemical formulae on which Klein focuses, bookkeeping records formulized a means to manage humans, substances, laboratory hardware and the processes in which they were mutually engaged with productive effect. This growing reliance on governance through paper was especially welcome in conjunction with a second trend whereby chemical manufacturers met growing demand: expanding premises and upscaling production techniques. Because such moves required increased capitalization, producers who enjoyed privileges and prestige, or whose pharmaceutical skills and connections were matched by business savvy, were at an advantage.

Upscaling also relied on hard-won knowledge, know-how and adaptive hardware, as well as negotiations with government regulators and neighbors who faced increased nuisances. Because chemical production relied on the 'governance of fire', for example, knowledge of heat and its regulation at different scales were key components of this process.[81] Along with having to construct larger furnaces that provided constant and manageable heat, it was necessary to revamp instrumentation to allow continued access to substances while responding to problems and risks that emerged in large scale production. At this level, fumes that were slightly bothersome in small concentrations, for example, manifested a poisonous presence requiring containment.[82] Material production was thus generative of problems of chemical governance. But upscaling also afforded opportunities that deserve more coordinated investigation. Increasing production yielded both sellable products and material remains, which stimulated the utilization of industrial leftovers to achieve

80 Ursula Klein, *Experiments, Models, Paper Tools: Cultures of organic chemistry in the nineteenth century* (Palo Alto: Stanford University Press, 2002); Simon Schaffer, "'The Charter'd Thames': Naval architecture and experimental spaces in Georgian Britain," Lissa Roberts, Simon Schaffer and Peter Dear, eds., *The Mindful Hand: Inquiry and invention from the late Renaissance to early industrialisation* (Amsterdam: Royal Netherlands Academy of Arts and Sciences, 2007), 279-305.

81 See Marie Thébaud-Sorger's essay in this volume.

82 Carleton Perrin, "Of Theory Shifts and Industrial Innovations: The relations of J.A.C. Chaptal and A.L. Lavoisier," *Annals of Science* 43 (1986): 511-542, 530.

product diversification.[83] Examining individual examples provides clues for how to write a longer term history of circular economies that brings technical details together with the various types of material, entrepreneurial and governance practices that stimulated, managed and opposed them.[84]

The picture this all paints might appear path breaking and progress oriented when selectively viewed from the present, but its local ambiguities deserve closer scrutiny. John Christie answers this call in his study of entrepreneurship in industrial Glasgow. Historians variously treat 'the entrepreneur' as an important figure during our period, generally casting entrepreneurship as involving the pursuit of innovation and powering "creative destruction".[85] Historical actors, however, defined the word 'entrepreneur' differently. They were silent about the link between entrepreneurs and innovation and justifiably ambivalent about whether innovation was necessarily a good thing.[86] As Christie shows, the road to success was not always paved by novelty and innovation.

This is not to say that chemical industry witnessed no innovation between 1760 and 1840, but that the contexts and processes that marked its development were too heterogeneous to fit under a single rubric of revolutionary change. It is by delving into the biographies of Glasgow's foremost industrial entrepreneurs that Christie brings industrialization's historical variegation to the fore. Chemical industry between 1760 and 1840 included a highly diverse set of enterprises. Increasingly typical of industrializing cities at this time, Glasgow housed businesses ranging from producers of chemical substances used in other industries to those whose manufacturing processes depended on chemical practices. These businesses also varied in terms of their overall

83 Timothy Cooper, "Peter Lund Simmonds and the Political Ecology of Waste Utilization in Victorian Britain," *Technology and Culture* 52 (2011): 21-44. For examples, see John Graham Smith, *The Origins and Early Development of the Heavy Chemical Industry in France* (Oxford: Clarendon Press, 1979).

84 Telling examples include the Javel factory, south of Paris, and the chemical factory run by Watse Gerritsma in the north Dutch province of Friesland. For details, see Smith, *The Origins* (see note 83); Driel and Roberts, "Circulating Salt" (see note 24).

85 Joseph Schumpeter, *Capitalism, Socialism and Democracy* (New York: Harper, 2008), 83; Joel Mokyr, ""Entrepreneurship in the Industrial Revolution," David Landes, Joel Mokyr and William Baumol, eds., *The Invention of Enterprise* (Princeton, NJ: Princeton University Press, 2012), 183-210; Oliver Mallett, "Contesting the History of Enterprise and Entrepreneurship," *Work, Employment and Society* 29 (2015): 177-182.

86 *Dictionnaire universel de commerce* (Paris: Estienne et Fils, 1748), vol. II, 1051. Jean Baptiste Say, *A Treatise on Political Economy; Or the production, distribution and consumption of wealth*, trans. with notes by C.R. Prinsep (Philadelphia: Grigg and Elliot, 1834), 82, note 1; Barbara Cassin, ed., *Dictionary of Untranslatables: A philosophical lexicon* (Princeton, NJ: Princeton University Press, 2014), 265-268.

production strategies and the processes and products found within each. The range of strategies a single firm might employ can be mapped according to five categories: adaptive maintenance of traditional techniques, tools and 'furniture'; upscaling; introduction of new chemical techniques; mechanization; product diversification. In such complex environments, innovation emerges as a contextually bound and relative term – indicative, perhaps, of a change of scale, a revamped instrument, the adaptive introduction of a process or substance used elsewhere, or the use of material leftovers to produce other goods. It was never clear from the start which strategy would work. What is certain is that innovation guaranteed nothing. Neither did knowledge and experience guarantee success. It could take years before a novel machine or process was sufficiently stabilized to be effective, while ambient conditions including government regulations, competition for materials, fluctuating demand and international conflict could derail the best laid plans.

One thing that many industrial endeavors did share was pollution, though its effects were not experienced evenly. Wealthy elites inhabited the greener quarters of urban areas and could more easily escape the city's chemically laden, foul atmosphere. The poor had fewer choices.[87] Chemical industry took place, then, in a context marked by sociomaterial hybridity and inequality. Still true today, chemical production came with greater cost to some and greater profit for others.[88]

A final issue that needs addressing is the historical relationship between material and knowledge production. A longstanding concern amongst historians, the question has been especially highlighted by economic historian Joel Mokyr. Interested to account for British and Western European economic trends and their 'great divergence' with China since the the late eighteenth century, Mokyr dismisses explanations based on the availability of coal and colonies in favor of a 'cultural' argument that emphasizes what he sees as a coincidence between political liberalization and the increasingly pervasive production and application of 'useful knowledge'.[89] While discussions of the

87 Allan Potofsky, "Recycling the City: Paris, 1760s-1800," Ariane Fennetaux, Amélie Junqua and Sophie Vasset, eds., *The Afterlife of Used Things: Recycling in the long eighteenth century* (New York: Routledge, 2015), 71-88, on 75.

88 Andreas Malm and Alf Hornborg, "The Geology of Mankind? A critique of the Anthropocene narrative," *The Anthropocene Review* 1 (2014): 62-69.

89 Mokyr, *Gifts of Athena* (see note 58); idem., *The Enlightened Economy: An economic history of Britain, 1700-1850* (New Haven: Yale University Press, 2009); Kenneth Pomeranz, *The Great Divergence: China, Europe and the making of the modern world economy* (Princeton, NJ: Princeton University Press, 2000).

continued role of state intervention and governance regimes in Great Britain and the continent, as well as of the unmissable contribution of colonial resources and slave labor to European wealth and wellbeing, should be sufficient to put his more contentious claims to rest, the relationship between material and knowledge production to which he draws attention remains crucial.[90] As argued in this volume, focusing on the history of chemistry provides an especially revelatory lens through which to capture their historical relationship.

The primary reason is that hybrid engagement with material and knowledge production was often standard chemical practice. Ursula Klein argues that this justifies considering eighteenth-century chemistry as a "technoscience *avant la lettre*".[91] In his essay for this volume, Robert Anderson asks whether her claim should be generalized for all of Europe and answers by demonstrating that the relationship between science and industry was as much a question of social identities as of knowledge content.

Scotland provides a particularly interesting setting for Anderson's inquiry. Its universities were more open and flexible than many other European institutions during the eighteenth century, which swelled student attendance. By the 1740s the medical faculty at University of Edinburgh, where chemistry was taught, became the most popular in Europe. Simultaneously, as outlined by Christie in this volume, Glasgow, Dundee and other areas became centers for Scotland's burgeoning chemical industry. Chemistry professors found themselves teaching both medical students and those oriented toward manufacture, agriculture and law. Extended acquaintance with applicable knowledge and networks of former students who asked for advice as manufacturers, landowners, policy makers and administrators, meant that academic chemists had to consider their public identities. Joseph Black, chemistry professor at Glasgow and Edinburgh Universities for over forty years, engaged in consulting throughout his career, but insisted on keeping chemistry's social identity as a science distinct from the industrial realm. Contrariwise, Andrew Ure, professor at the Andersonian Institution (1804-1830), happily opted for a hybrid public identity, increasingly supplementing his lectures with publicized work as a popular author and industrial consultant. Quite apart from the question of anachronism, Anderson concludes, not all chemists would have accepted the identity of technoscientists.

---

90 Brewer, *Sinews of Power* (see note 20); William Ashworth, "The Ghost of Rostow: Science, culture and the British Industrial Revolution," *History of Science* 46 (2008): 250-274.

91 Ursula Klein, "Technoscience *avant la lettre*," *Perspectives on Science* 13 (2005): 226-266.

In sum, focusing on chemistry, materials and governance offers fresh perspectives on material and knowledge production in the period 1760 to 1840. A picture emerges of gradual, open-ended transformations in which various local circumstances and practices contributed to change. Alongside the new and shiny, which often disappointed, more traditional and mundane processes could make all the difference. Diverse strategies shaped production, from upscaling, administration with "paper tools" and diversification, to the maintenance and adaptation of existing practices. Chemical practitioners engaged with these strategies in various ways, overseeing global circulations of chemical goods, scaling up manufacturing processes, exploiting byproducts, generating and regulating waste and pollution. No doubt aspects of these activities were innovative, but they were integrated in processes of gradual change and long-term development, the results of which were never fully predictable.

Situating Chemistry

As a collaborative effort, the geographical focus of this volume might not be as broad as one would wish. Both Sweden and the German lands, for example, housed important centers of chemical education and activity whose content and contours were closely tied to ambient political, economic and sociomaterial conditions.[92] Chemistry, in fact, played a variety of important roles around the world between 1760 – 1840, whether informing agricultural improvement in the young American republic, producing textiles and luxury goods in Asia or engaging with pharmaceutical substances and metals in Latin America. Examining developments in all these regions through the approach of this volume would add greatly to our understanding of what is, after all, a transnational subject.

But such a huge undertaking has to begin somewhere, and the narrower focus of this volume makes sense, given that France and Great Britain (our essays' primary locations) have long been the dominant foci of attention in connection with the revolutions (the French, Industrial and Chemical) with which this period is most commonly identified. The cases presented here,

92 Anders Lundgren, "The New Chemistry in Sweden: The debate that wasn't," *Osiris* 4 (1988): 146-168; Ernst Homburg, "The Rise of Analytical Chemistry and its Consequences for the Development of the German Chemical Profession (1780-1860)," *Ambix* 46 (1999): 1-32; Hans Erich Bödeker, "Economic Societies in Germany, 1760-1820: Organization, social structures and fields of activities, " Koen Stapelbroek and Jani Marjanen, eds., *The Rise of Economic Societies in the Eighteenth Century* (London: Palgrave MacMillan, 2012), 182-211.

some of which underline the global networks in which chemical substances and practices were enmeshed, invite us to direct our attention away from the monumental, essentially distinct breaks implied by organizing historical analysis according to separate political, industrial and scientific revolutions. Instead, by considering often quotidian histories of material and knowledge production in concert with human interactions with material substances and objects and governance practices that managed productivity and social order, we point to a more unified fabric of historical development: one in which intertwined activities inductively manifested themselves in terms of a package deal of longer-term political, social, industrial and environmental changes.

This hybrid approach fits with current historiographical trends, but is rooted in the very history this volume presents. If we return to the early 1840s, we find the chemist Justus Liebig reviewing chemistry's situation in similar terms to those discussed here.[93] Among other things, he discussed the chain of events whereby the introduction in Great Britain of new bleaching techniques (themselves based on using a previously ignored by-product of soda manufacture) rendered unnecessary the employment of large tracts of rural land as bleaching fields.[94] This freed up both industrial capital for investment elsewhere and land for agricultural exploitation, the latter of which could be enriched with fertilizers that – again, thanks to productive chemical practices – combined urban and rural based waste products.

Liebig described this "as affording an excellent illustration of the dependence of the various branches of human industry and commerce upon each other, and their relation to chemistry."[95] But he and those who read his work did not simply celebrate this as unequivocal evidence of innovation-based progress. Karl Marx, for example, drew on Liebig's discussion of such hybrid chains of knowledge and material production to connect the past with concerns for the future. The sociomaterial 'metabolic system' in which these urban-rural links evolved, Marx recognized Liebig to say, was being unbalanced by capitalist production on an increasingly global scale. In place of previous oeconomic visions of sociomaterial improvement, Marx and others described a chemically mediated process of growing alienation between the

93 Justus Liebig, "Der Zustand der Chemie in Österreich," *Annalen der Pharmacie* 25 (1838): 339-347; idem., "Der Zustand der Chemie in Preussen," *Annalen der Chemie und der Pharmacie* 34 (1840):97-136; idem., *Die organische Chemie in ihrer Anwendung auf Agricultur und Physiologie* (Braunschweig: Vieweg, 1840); idem., *Familiar Letters on Chemistry* (London: Taylor and Walton, 1843).

94 Liebig, *Familiar Letters*, pp. 37-38 (see note 93).

95 Ibid., p. 31.

operative sources of production and those who saw to their managed exploitation. It was in such a context of growing antagonism that talk of 'revolution' – industrial or otherwise – gained increasing parlance.[96] A closer acquaintance with the history of chemistry that preceded such utterances can help us reconnect with the continuities and evolutionary developments overshadowed by declarations of revolution.

96 Karl Marx, *Das Kapital* (Hamburg: Otto Meisner, 1867), vol. 1, part 4, chapter 13, pp. 527-530; Victor Hugo, *Les misérables*, authorized English translation, vol. 3 (London: Hurst and Blackett, 1862), 231-233; John Bellamy Foster, "Marx's Theory of Metabolic Rift: Classical foundations for environmental sociology," *American Journal of Sociology* 105 (1999): 366-405.

PART 1

Materials and Material Objects

CHAPTER 1

Household Oeconomy and Chemical Inquiry

Simon Werrett

The history of the chemical laboratory was until recently quite obscure. But a number of studies have begun to reveal the material conditions and spatial configurations of chemical practice in a variety of settings in the period 1760 to 1840. Peter Morris's recent book *The Matter Factory* examines the laboratories of Lavoisier, Faraday and Liebig in this period, while a recent volume of *Ambix* considered eighteenth-century laboratories dedicated to chemical inquiry in, among other places, a porcelain manufactory, mining academy and assaying office.[1] The focus of these studies has been purpose-built laboratories dedicated to chemical practice, and it has been suggested that chemistry could only take place in laboratories constructed for the purpose, since they needed to contain a furnace.[2] While historians have clearly widened the repertoire of laboratories being studied from famous research institutions to military, industrial and academic sites, this chapter proposes that many sites of chemistry were not originally *dedicated* to chemical labors, and some were not laboratories at all.

Alix Cooper and Steven Shapin have noted that a great deal of early modern experimentation took place in people's homes.[3] Cooper identifies the home as a key site of scientific inquiry and the family as the central unit in domestic knowledge-making. Cooper, Shapin and others have made social relations the focus of analysis for exploring the nature of knowledge-making in the home. Cooper considers how family life shaped early modern scholarly life, while Shapin demonstrated how expectations of gentlemanly conduct in the home

1 Peter Morris, *The Matter Factory: A history of the chemical laboratory* (London: Reaktion, 2015), esp. 19-20; John Perkins, ed. *Sites of Chemistry in the Eighteenth Century*, special issue of *Ambix* 60, no. 2 (May 2013).

2 Ursula Klein, "The Laboratory Challenge: Some revisions of the standard view of early modern experimentation," *Isis* 99 (2008): 769-782, on 772-3.

3 Alix Cooper, "Homes and Households," Katharine Park and Lorraine Daston, eds., *The Cambridge History of Science*, Vol. 3: *Early Modern Science* (Cambridge: Cambridge Univ. Press, 2006), 224-237; Steven Shapin, "The House of Experiment in 17th-Century England," *Isis* 79 (1988): 373-404.

DOI 10.1163/9789004325562_003

shaped experimental etiquette.[4] This chapter also proposes that the home, among a variety of adapted spaces, continued to be an important site for chemical experimentation in the period 1760 to 1840. It shows how chemical practices were shaped by the social order of the home, and particularly ideas of *oeconomy*, a body of knowledge and practice concerning the proper management of the household (and by extension, the state or even the universe).[5] The focus here will be on Britain, and perhaps further research will reveal if similar relations to oeconomy existed elsewhere. Importantly, a fundamental focus of chemistry and oeconomy in Britain was the management of *materials*, and it is the material aspects of chemical activity in adapted spaces such as the home on which this chapter concentrates. Domestic chemistry and oeconomy were equally social and material practices, and this chapter might be seen as an exploration of *sociomateriality*, a term that reminds us that these arenas were always linked together in a rich variety of ways.[6]

Like the home itself, the material culture of chemical inquiry in this period could also be adapted, and might be said to have been in a constant state of flux, what the sociologist Karin Knorr-Cetina refers to as the "incompleteness" of objects.[7] Chemical practitioners certainly purchased or made apparatus serving some specific chemical end from an instrument-maker, but they also turned a diverse array of household goods into apparatus for their experiments. The material form and uses of a household object or instrument unfolded over time. Even dedicated instruments were not static objects, but underwent alterations and repairs. Rather than overlook this as simple expe-

4 Gadi Algazi, "Scholars in Households: Refiguring the learned habitus, 1480-1550," *Science in Context* 16 (2003): 9-42; Frances Harris, "Living in the Neighbourhood of Science: Mary Evelyn, Margaret Cavendish and the Greshamites," Lynette Hunter and Sarah Hutton, eds., *Women, Science, and Medicine 1500-1700: Mothers and sisters of the Royal Society* (Stroud: Sutton Publishing, 1997), 198-217; Deborah E. Harkness, "Managing an Experimental Household: The Dees of Mortlake and the practice of natural philosophy," *Isis* 88 (1997): 247-262.

5 On oeconomy, see Lissa Roberts, ed., "Practicing Oeconomy in the Late Eighteenth Century," special issue of *History and Technology* 30 (2014); Keith Tribe, "Oeconomic History: An essay review," *Studies in the History and Philosophy of Science* 36 (2005): 586-597; Both Roberts and Tribe offer criticism of Margaret Schabas and Neil Di Marchi, eds., *Oeconomies in the Age of Newton*, Annual Supplement to *History of Political Economy* 35 (2003).

6 For discussion of the term sociomateriality, see Wanda J. Orlikowski, "Sociomaterial Practices: Exploring technology at work," *Organization Studies* 28 (2007): 1435-1448; It has of course long been an assumption of social studies of science and technology that the social and material are fundamentally linked.

7 See Karin Knorr-Cetina, "Objectual Practice," Theodore R. Schatzki, Karin Knorr-Cetina, Eike von Savigny, eds., *The Practice Turn in Contemporary Theory* (London: Routledge, 2001), 175-188, on 181-184.

diency, this chapter suggests that practices of adaptation reflected the values of oeconomy. Oeconomy proposed that householders should care for material culture in the home, balancing the use of old goods with the purchase of new, and stewarding possessions through care, maintenance, and repairs which would ensure both the saving of time and money and the good order and harmony of the household and the environment within which it was embedded. What emerges then is a picture of chemical experimentation in the period 1760 to 1840 in which the re-use of old things was as significant a part of chemistry as the invention or consumption of new.[8]

A variety of people undertook chemical practices in the home, ranging from husbands and wives to servants and diverse networks and communities with whom the household interacted. Much of this work could be experimental and chemical, for example distilling and the preparation of medicines. This diversity will be reflected in the use of the term chemical practitioners, rather than chemists. "Material culture" was also diverse. It consisted of *substances*, the raw materials manipulated by chemical practitioners, *instruments*, the tools used to manipulate substances, and *objects*, the things chemists used besides instruments. The essay begins by examining concepts of oeconomy in the period before considering how the home was often a site for shared oeconomical and chemical practices. Householders strove to make the most of the material objects in their possession, and took care of those possessions to avoid them being damaged. Oeconomic and chemical literature offered diverse recipes for maintaining and repairing material possessions, and encouraged the re-use of broken objects and waste. The chapter concludes by considering how such an approach to materials was extended from the home to other sites such as the city and manufactories in the eighteenth and nineteenth centuries. Such extensions, which entailed rationalizations of labor and changes in scale had unintended consequences, contributing to pollution and a paradox about the value of re-using old materials which remains to this day.

The Practice of Oeconomy: Managing the Household

Early modern practices of oeconomy converged on the art of governance and could apply to the management of the family and household, the state or even the universe and divine "oeconomy of nature." There was no agreed-upon definition of oeconomy, and the many books on oeconomy that appeared in the

8 Chemistry fitted broader trends in the sciences, discussed in Simon Werrett, "Recycling in Early Modern Science," *British Journal for the History of Science* 46 (2013): 627-646.

period 1760 to 1840 included an array of advice on cleaning, cooking, and maintenance, medicine and health, gardening, husbandry and agriculture together with the management of the family and servants. Genres of oeconomic literature were often divided according to gender, with women presented as being responsible for the internal management of the home and men for husbandry outside. Numerous books written by female authors gave advice to housewives on cookery, cleaning and medicine. But as Karen Harvey has argued, "any gendered division of tasks [in the home] was unstable, even in those books specifically intended for either male or female readers."[9] Both men and women engaged in developing complex networks of exchange of recipes and practices related to chemical inquiry in the early modern period, and both men and women communicated practice orally, through letters or by keeping manuscript books.[10] However, in this period the convention was for experimental inquiries into chemistry to be published by men, while women wrote and sometimes published recipes and hints for the preparation of medicines, food, and materials related to household maintenance.[11] Bringing these apparently distinct literatures of household management and experimental inquiry together helps to make clear their interconnections in this period.

Not that oeconomy was even restricted to men and women. Oeconomy was not equivalent to economy. Although no strict contrast should be made between the two terms, oeconomy was not an abstract system of demand and supply or accounting of profit and loss, but a body of advice and examples relating to the prudent management of people and things. Both oeconomic and economic ideas related material culture, morals and social order, but economy came to do this in narrower terms than oeconomy during the nineteenth century. The subjects of oeconomy were typically humans, but numerous writers described the oeconomies of animals, birds and insects. Oeconomic thinking stressed the interrelatedness of all parts of the oeconomy of nature. "By the oeconomy of nature we understand the all-wise disposition of the creator in

9 Karen Harvey, *The Little Republic: Masculinity and domestic authority in eighteenth-century Britain* (Oxford: Oxford University Press, 2012), 27-28.

10 Elaine Leong, "Collecting Knowledge for the Family: Recipes, gender and practical knowledge in the early modern English household," *Centaurus* 55 (2013): 81-103; Elaine Leong, "Making Medicines in the Early Modern Household," *Bulletin of the History of Medicine* 82 (2008): 145-168; Sara Pennell and Elaine Leong, "Recipe Collections and the Currency of Medical Knowledge in the Early Modern 'Medical Marketplace'," Mark Jenner and Patrick Wallis, eds., *Medicine and the Market in England and its Colonies, c.1450-c.1850* (Basingstoke: Palgrave Macmillan, 2007), 133-152.

11 Michael McKeon, *The Secret History of Domesticity: Public, private, and the division of knowledge* (Baltimore, MD: Johns-Hopkins University Press, 2009).

relation to natural things, by which they are fitted to produce general ends, and reciprocal uses."[12]

In the period 1760 to 1840, many chemical practitioners appear to have shared in the values of oeconomy through a prudent stewardship of materials. Since oeconomy often focused on the proper management of the home it is perhaps inevitable that its goals of frugality and care were reflected in chemical practices that often took place in people's houses. The domestic situation of chemical activities was itself in part an expression of such concerns. Certainly the need for a furnace meant some chemical laboratories required dedicated buildings, but in many cases instead of building new laboratories, chemical practitioners preferred to make use of existing and convenient sites, converting cellars, kitchens, parlors and outbuildings into experimental spaces. As the English chemist and publisher John Joseph Griffin wrote as late as 1834,

> The notion, that a laboratory fitted up with furnaces and expensive and complicated instruments, is an absolute requisite for the proper performance of chemical experiments, is exceedingly erroneous. In fact, the truth is quite opposed to this opinion. "For general and ordinary chemical purposes," says Dr Henry, "and even for the prosecution of new and important inquiries, very simple means are sufficient: some of the most interesting facts of the science may be exhibited and ascertained with the aid merely of Florence flasks, of common phials, and of wine glasses. In converting these to the purposes of apparatus, a considerable saving of expense will accrue to the experimentalist; and he will avoid the encumbrance of various instruments, the value of which consists in show rather than real utility." It is a curious and instructive fact, that some of the most important discoveries in chemistry were made by persons who, either from choice, or motives of economy, used utensils of the very simplest character. The laboratory of the great Priestley cost a mere trifle; and it is well known how savingly Franklin went to work.[13]

---

12 Isaac J. Biberg, "The Oeconomy of Nature," Anon., *Miscellaneous Tracts Relating to Natural History, Husbandry and Physick*, trans. Benjamin Stillingfleet (London: R. and J. Dodsley, 1759), 31-108, on 31.

13 John Joseph Griffin, *Chemical Recreations: A series of amusing and instructive experiments* [...] *to which are prefixed First Lines of Chemistry*, seventh edition (Glasgow: R. Griffin and Co., 1834), 1; on Griffin see Brian Gee and William H. Brock, "The Case of John Joseph Griffin: From artisan-chemist and author-instructor to business-leader," *Ambix* 38 (1991): 29-62.

The contents of Priestley's laboratories hardly cost a trifle, but his laboratories were all rooms in houses belonging to Priestley or his patrons.[14] An engraved plate from Priestley's *Experiments and Observations on Air* (1774) showed the corner of his laboratory set up inside the orangery of Robert Adams' wing of Bowood House at Calne in Wiltshire, home to Priestley and his patron and employer the Earl of Shelburne in the 1770s. The fireplace, a round three-legged and a square table have all been put into experimental service, providing a source of heat and support for chemical instruments. Domestic and chemical functions overlapped in these spaces. Even the furnace, supposedly the marker of a dedicated chemical space, could double up as home heating. Priestley's inventory of the laboratory in his Birmingham home in the 1780s included a furnace, "containing a large Copper Vessel also iron Tubes [...] adapted to it in order to warm the Laboratory."[15] Chemical sites were also converted back into domestic space. A visitor to Priestley's Philadelphia residence in the nineteenth century noted: "His laboratory is now converted into a house for garden-tools! The furnaces pulled down, the shelves unoccupied! – the floor covered with Indian corn! A stranger might be inclined to say, "Sic transit gloria philosophiae" [Thus passes the glory of philosophy]."[16]

Sites of chemistry were thus routinely in flux, transformed between domestic, scholarly, and other uses. Cambridge colleges converted variously a cellar, a shed and an idle printing house to create new laboratories.[17] Edinburgh professor of chemistry Joseph Black occupied the Library Range of the University's Old College from 1770 to 1781, before moving to a purpose-built chemistry block. Archeology indicates that when both buildings were demolished in 1820, the larger stones were extracted from the rubble for re-use.[18] From 1790

14 Priestley himself estimated the value of instruments and books in his Birmingham laboratory to be more than four thousand pounds. See Douglas McKie, "Priestley's Laboratory and Library and Other of his Effects," *Notes and Records of the Royal Society* 12 (1956): 114-136.

15 Ibid., pp. 121-122.

16 John Finch, *Travels in the United States of America and Canada, containing some Account of their Scientific Institutions* (London: Longman, Rees, Orme, Brown, Green, and Longman, 1833), 316.

17 Kevin C. Knox, "'The Deplorable Frenzy': The slow legitimisation of chemical practice at Cambridge University," Mary D. Archer, Christopher D. Haley, eds., *The 1702 Chair of Chemistry at Cambridge: Transformation and change* (Cambridge: Cambridge University Press, 2005), 1-30, on 9.

18 Tom Addyman, "Materia Chemica: Excavation of the early chemistry stores at Old College, University of Edinburgh." Typescript, thanks to Robert Anderson for providing this essay; see also Morris, *Matter Factory*, 70-72 (see note 1).

James Watt experimented on pneumatic medicine, pottery and sculpture in a garret workshop at his home in Heathfield, Birmingham, a site which was later installed in London's Science Museum.[19] Mixtures of domestic and experimental space prompted sociomaterial puzzles. Engagement with chemical commodities operated according to social divisions of private and public space and access in these sites. In 1794, Irish physician and chemist Bryan Higgins advertised a "Society for Philosophical Experiments" at his home at 13, Greek Street, London, where participants would be given access to his "extensive Apparatus" while "the best books on subjects under examination" would "always be ready for their perusal in the drawing-room."[20] London chemist Frederick Accum "kept a considerable variety of apparatus" and a "working laboratory" in his house near Soho Square, which linked him to networks of instrument-makers and students.[21] Male inquiry depended on female goodwill. In North America, Benjamin Silliman and Robert Hare maintained a chemical laboratory in the cellar-kitchen of their lodgings at 46, Walnut Street, Philadelphia, "conceded to us by the indulgence of our hostess, Mrs. Smith."[22]

Making Use of the Home in Chemistry

The kitchen could double up as a chemical laboratory, and as the oeconomy of the household overlapped with sites of chemistry, so oeconomical practices shaped or coincided with chemical practices.[23] This might apply to shared techniques, but also to shared approaches to material culture. Chemical and oeconomic literature both described practices such as the distillation of spirits and the preparation of medications. Chemical practitioners applied to chemical instruments the same kinds of stress on stewardship and thrift that appeared in oeconomic literature on domestic management. Robert Dossie, author of a work on *Agriculture and Other Oeconomical Arts*, advised readers of his treatise on chemical arts *The Elaboratory Laid Open* of 1758 that proper maintenance of chemical instruments ensured, "considerable savings in

19 Ben Russell, *James Watt: Making the world anew* (London: Reaktion, 2014), 224-233.

20 Anon., *Minutes of the Society for Philosophical Experiments and Conversations* (London: T. Cadell Jr, and W. Davies, 1795), 6, 12.

21 George Park Fisher, *Life of Benjamin Silliman M.D., LL.D.*, 2 vols. (New York: Charles Scribner & Co., 1866), vol. 1, 142.

22 Ibid., p. 103; Chandos Michael Brown, *Benjamin Silliman: A life in the young Republic* (Princeton, NJ: Princeton University Press, 2014), 108.

23 The kitchen was also an important space for the development of anatomy and dissection according to Anita Guerrini, "The Ghastly Kitchen," *History of Science* 54 (2016): 71-97.

labour, fewel, and frequently the produce of the operation."[24] This oeconomy of materials incorporated a diverse array of techniques, ranging from care over the human body to the maintenance, repair and re-use of various vessels and goods for chemical purposes. In some cases this might correspond directly with widely available oeconomic advice and in others chemical practice raised unique problems that demanded distinctive solutions.

One feature of oeconomic advice was to highlight the diverse uses to which materials could be put in the home, on estates and on farms. In his translation of Noel Chomel's *Dictionaire oeconomique, or Family Dictionary*, the Cambridge botanist and oeconomist Richard Bradley noted of the beech tree, "It's useful for many things" before listing the various uses of beech for making "Dishes, Trays, Rimbs for Buckets, Trenchers [...] Chairs, Stools, Shovels and Spade-Grafts."[25] There was value in the capacity of things to be converted to a wide array of uses. Certainly this was an era of expanding markets for luxury items.[26] But oeconomy encouraged householders to balance the purchase of new goods with making good use of the old. An essay in the *Gentleman's Magazine* of 1731 explained that oeconomy meant "Wisdom applied to the Practice of private Life; it is situated betwixt Profuseness and Avarice, and consists in a just Medium of Concern, as to exterior Goods, between being over Careful and having no Care at all."[27]

Chemical practitioners faced with the execution of novel experiments might purchase or make new instruments for themselves but they also often adapted household items to chemical ends. The material culture of chemistry was a mixture of dedicated instruments purchased from specialist makers and a variety of household vessels and commodities such as teacups, saucers, bowls, dishes, wine glasses and furniture. Archaeological evidence supports this. Excavation in 2011 of the location where Joseph Black's chemical apparatus was stored in Edinburgh University uncovered "at least two mid-late 18th century black glass wine bottles" and,

---

24 Robert Dossie, *The Elaboratory Laid Open, or, the Secrets of Modern Chemistry and Pharmacy Revealed* (London: J. Nourse, 1758), 1; Robert Dossie, *Memoirs of Agriculture, and Other Oeconomical Arts* (London: J. Nourse, 1768).

25 "Beech-Tree," Noel Chomel, *Dictionaire oeconomique; or, the family dictionary*, trans. Richard Bradley, 2 vols. (Dublin: L. Finn, 1758), vol. 1, n.p.

26 See e.g. Sara Pennell, "'Pots and Pans History': The material culture of the kitchen in early modern England," *Journal of Design History* 11 (1998): 201-216; Amanda Vickery, *Behind Closed Doors: At home in Georgian England* (New Haven & London: Yale University Press, 2009).

27 Anon., [poss. Richard Burridge] "Oeconomie and Extravagance," *The Gentleman's Magazine* 1, no. 11 (November 1731), 489.

> [...] miscellaneous vessels of a [...] domestic nature, such as blue and white transfer-printed pearlware bowls and creamware jugs, that on the basis of residues present had evidently been employed in the preparation of chemical materials.[28]

Any number of such items are mentioned in chemical texts of the period. Less familiar household items also served chemistry. Firearms were a commonly adapted household item. Many early modern households contained muskets, pistols, or rifles, attested to by court records of firearm offenses and a 1541 statute which governed firearm ownership in England and encouraged a variety of subjects, "to have and keep in every of their houses any such hand-gun or hand-guns, of the length of one whole yard."[29] Joseph Priestley's Birmingham inventory included a brace of pistols and a "Gun with a Bayonet."[30] Gun barrels could then be adapted for use as electrical conductors or as vessels for heating. John White Webster proposed liberating gases from substances placed "in a gun barrel, the touch-hole of which has been accurately closed by an iron pin."[31] Robert Dossie proposed making an alembic for distilling mercury with a covered copper or iron pan soldered to a gun barrel which sloped down into a "common water pail" filled with water.[32] The Earl of Dundonald used gun-barrels to convey coal-gas for illumination in experiments at Culross Abbey near Dunfermline.[33] Chemical apparatus was a bricolage of material elements, some old and some new, some dedicated and some adapted, an oeconomical mixture of the specialized and the re-purposed. As much as this period saw the construction of new instruments like the ice calorimeter, it also saw the thrifty re-use of many old ones.

---

28 Addyman, "Materia Chemica" n.p. (see note 18).

29 Quoted in Joyce Lee Malcolm, *Guns and Violence: The English experience* (Cambridge, MA: Harvard University Press, 2009), 50.

30 Priestley, quoted in McKie, "Priestley's Laboratory," p. 128 (see note 14).

31 John White Webster, *A Manual of Chemistry: Containing the principal facts of the science*, third edition (Boston: Marsh, Capen, Lyon and Webb, 1839), 106.

32 Robert Dossie, *Institutes of Experimental Chemistry: Being an essay towards reducing that branch of natural philosophy to a regular system*, 2 vols. (London: J. Nourse, 1759), vol. 1, 87-88.

33 Thomas, Tenth Earl of Dundonald, *The Autobiography of a Seaman*, 2 vols., second edition (London: Richard Bentley & Son, 1861), vol. 1, 39.

Avoiding Damage

The care of materials was equally valued in domestic oeconomy and chemical inquiry. It was important to avoid damage to goods and the expense and trouble of repairing or replacing them. Householders could repair damaged household goods themselves or give them to street traders to fix. In her 1835 *Modern Domestic Cookery and Useful Receipt Book* Elizabeth Hammond explained how to mend broken iron pots, glass vessels and china, while recipes for repairs appeared in various works on chemistry.[34] Homes were embedded in complex networks of artisans and waste traders who circulated between city streets and the countryside repairing or disposing of materials for a fee. Street traders included tinkers and chair menders, the former described as wearing an apron and broad-brimmed hat, carrying saucepans, a hammer and crying "pots to mend!"

Damage during chemical experiments could be of various sorts. Instruments might be "deranged" or broken, substances could be corrupted or polluted through unwanted mixing, and practitioners' bodies might be hurt or wounded by corrosive or explosive reactions. Much chemical practice centered on the avoidance of such problems. Good design with appropriate materials and careful storage, cleaning and maintenance helped to ensure the integrity and longevity of instruments.

Practitioners reckoned making instruments durable and sturdy was critical. This applied first and foremost, according to Dossie, to the principal instrument of the chemical laboratory, the furnace,

> [...] they should be well designed, and judiciously executed; otherwise their defects greatly enhance the expense, and frustrate the intention, of the operations they are to perform; besides their being extremely liable to become, in a very short time, out of repair and uselessly ruinous.[35]

Dossie identified types of furnaces that were liable to damage or difficult to repair and warned against installing them.[36] "Damage" for Dossie was a matter of both financial loss and social disorder. Damage to flour-grinding mills or

34 Elizabeth Hammond, *Modern Domestic Cookery and Useful Receipt Book adapted for Families in the Middling and Genteel Ranks of Life* (London: Dean and Co., 1835), 246-7.

35 Dossie, *Elaboratory Laid Open*, pp. 3-4 (see note 24).

36 Ibid., pp. 9-10.

that caused by vermin or disease in sheep were among "the greatest evils [...] which the public sustains."[37] Derangement was moral and material.

In domestic and chemical practices, glass provided the principal material for ensuring substances were kept free from pollution and corruption. In 1780 *The Farmer's Wife; Or complete country housewife* explained that, "Stone or glass jars are the most proper vessels in which to make and keep pickles; for common earthen vessels are soon penetrated by the vinegar and salt."[38] As Marie Thébaud Sorger reminds us, glass vessels, including various jars and tubes, were used in chemical experiments to collect and contain substances such as airs, managed with glass stirring rods, thermometers, and stoppers.[39] Vessels should be made to minimize the risk of damage. Dossie proposed that glass receivers be made larger than those in common use.

> A greater quantity of condensing surface [renders] the operation both more profitable and safe, as it prevents the forcing of the lute, and the escape of the vapour; as well as the hazard of bursting the vessels, on the raising the fire too high.[40]

Chemical practice routinely threatened derangement. Chemistry by its very nature involved unpredictable reactions which were liable to break or damage instruments. To avoid cracking vessels heated to high temperatures they could be placed in baths. Thomas Garnett explained in his chemical lectures of the 1790s,

> Chemical vessels may be plunged [...] in a pot placed over the furnace, containing sand, water, or other matter capable of sustaining heat. These substances interposed between the vessel and the fire, compose [...] a bath, and are very helpful in imparting an uniform heat [...] Without this contrivance, glass vessels would often fly and crack.[41]

---

37 Ibid., p. 177.

38 Anon., *The Farmer's Wife; Or complete country housewife* (London: Alex Hogg, c.1780), 57.

39 See Chapter 3. Robert Harrington, *A Treatise on Air, containing New Experiments and Thoughts on Combustion; Being a full investigation of Mr. Lavoisier's system* (London: T. Evans, 1791), 194; William Nicholson, *The First Principles of Chemistry* (London: G.G.J. and J. Robinson, 1792), 48.

40 Dossie, *Elaboratory Laid Open*, p. 29 (see note 24).

41 Thomas Garnett, *Outlines of a Course of Lectures on Chemistry* (Liverpool: J. M'Creery, 1797), 47.

Protecting instruments with cases, boxes, and crates for storage and transport was another solution to such problems. Chemical goods routinely circulated between homes in the hands of potentially unreliable couriers and post-men.[42] Joseph Priestley shared many chemical practitioners' anxieties over glassware sent to him by mail coach. In 1781, he wrote to Josiah Wedgwood from Birmingham about two boxes of retorts that Wedgewood had sent to him, "The cover of the larger box was quite off, and ten of the retorts broke, most of them so as to be of no use at all."[43] Things got even worse when Priestley moved to North America.[44]

Managing damage was "sociomaterial", involving both human and artificial bodies. Chemical instruments were fragile and had to be carefully looked after. The same was true of practitioners' bodies, which they equally sought to preserve from damage. Oeconomy concerned health and safety as much as the saving of expense. Worcester surgeon William Sandford described his collection of medical advice on "the oeconomy of health" as an effort "to enable the uninformed in medical knowledge, to understand in some degree, upon what principles life is sustained, and how it may probably be prolonged, with ease and comfort to ourselves, and benefit to our posterity."[45] The bodies of chemical practitioners also required sustaining. Explosions, broken glass, electric shocks, and corrosive chemicals all threatened their integrity. To protect his eyes during experiments James Watt adapted ordinary spectacles with flat glass lenses.[46] Michael Faraday recommended wearing "glass masks, goggles, &c." when making experiments with carbonic acid.[47] Chemists invoked theory when determining safety techniques. Kings College London professor of chem-

42 Kenneth Ellis, *The Post Office in the Eighteenth Century: A study in administrative history* (Oxford: Oxford University Press, 1958).

43 Joseph Priestley to Josiah Wedgwood, August 8, 1781, in Joseph Priestley, *Scientific Correspondence of Joseph Priestley. Ninety-seven letters addressed to Josiah Wedgwood, Sir Joseph Banks* [...] *Dr. Benjamin Rush, and others*, ed. Henry Carrington Bolton (New York: privately printed, 1892), 29-30, on 29.

44 See e.g. Joseph Priestley to John Vaughan, March 21, 1799. Joseph Priestley Papers, American Philosophical Society, B P931.

45 William Sandford, *A Few Practical Remarks on the Medicinal Effects of Wine and Spirits; With observations on the oeconomy of health* (Worcester: J. Tymbs, 1799), vii; see also e.g. Andrew Harper, *The Oeconomy of Health, or, a Medical Essay: Containing new and familiar instructions for the attainment of health, happiness and longevity* (London: C. Stalker, c. 1785).

46 Watt's safety goggles, Watt's Workshop Collection, Science Museum, London, Inventory number 1926-1075/440.

47 Michael Faraday, *Experimental Researches in Chemistry and Physics* (London: R. Taylor and W. Francis, 1859), 92.

istry John Frederic Daniell described the way glass halted the passage of radiant or "dark" heat and explained how, as a result, "This property of glass is sometimes usefully employed where it is desirable to see the light of a fire without being incommoded by the heat; [so] glass screens are used to protect the eyes when it is necessary to inspect the action of a hot furnace."[48]

Chemistry could cause severe damage to the hands. In 1837 a young Matthew Arnold was taken away from Winchester College after severely burning his hand in a chemistry class (he nearly lost two fingers).[49] Hands were protected with gloves. Since phosphorus was "sometimes thrown out of the mortar" during its preparation, American surgeon John Lee Comstock warned chemistry students that "it is therefore advisable to protect the hand with a glove, and keep the face out of the way."[50] But there were no cheap, disposable gloves available at the turn of the nineteenth century, and gloves might hinder dexterity. Experimenters thought twice before exposing gloves to damage. In 1816 Richard Davenport experimented on the communication of heat in boiling tar. Although he was convinced that his gloved hand would not be "dreadfully burnt" if he plunged it into boiling tar, Davenport hesitated. "Not choosing to sacrifice a pair of gloves to the trial of an effect I had no belief in, I wrapped a newspaper double about my hand, and plunged it in up to the wrist. I retained my hand in the tar longer than I could when naked without feeling any pain."[51]

Maintenance

Chemical practice was a trade-off between damage, durability, and discovery, where by necessity the circulation of materials, whether in the laboratory or across countries or continents, could and did break things. In addition to these labors to keep things from derangement, chemical practitioners went to great lengths to repair things if they did break. Repair constituted a major part of enlightened artisanry and many "makers" spent more time cleaning and repairing goods than making them. As David Edgerton has argued, repair and

48 John Frederic Daniell, *An Introduction to the Study of Chemical Philosophy: Being a preparatory view of the forces which concur to the production of chemical phenomena* (London: J.W. Parker, 1839), 188.

49 My thanks to Geoffrey Day of Winchester College who provided the letter from Arnold to his parents, dated April 7, 1837.

50 J.L. Comstock, *A Grammar of Chemistry, adapted to the Use of Schools and Private Students*, second edition (Hartford: S.G. Goodrich, 1825), 30-31.

51 Richard Davenport, "Curious Experiments on Boiling Tar," *Annals of Philosophy; or, Magazine of Chemistry* 9 (Jan-Jun 1817): 111-114, on 114.

maintenance "are the most widespread form of technical expertise" because the number of users typically far outweighs the number of producers of technical artifacts.[52] Oeconomical treatises discussed the cleaning and repair of buildings and household goods.[53] Repair work was also widespread in chemistry both because practitioners needed to repair their own vessels, and because they offered techniques and recipes to others for making repairs.

Adhesives, for example, fascinated oeconomic writers and chemical practitioners. Bradley's translation of Chomel's *Dictionaire oeconomique* included numerous glue recipes and advice on repairing broken glass with a mixture of liquid fat and varnish.[54] Mrs. Fisher explained in the *Prudent Housewife* how to make cements and glues for sticking stone and glass using a paste of flint powder combined with melted white rosin.[55] Robert Dossie also investigated various cements, pastes, lutes, sizes and glues, and published new recipes in works on industrial arts and chemistry.[56] Dossie's recipes were intended to allow repairs of common household commodities like china and porcelain, or specialized chemical glassware and instruments. For the chemist, he provided a recipe,

> [...] for the repairing the cracks, and replacing the broken pieces, of receivers, or other glass vessels, which admit of being used after they are in that condition; and this, judiciously applied, in an elaboratory, where many such vessels are used, will make a considerable saving.[57]

Filling a crack with a linen rag soaked in a mixture of grated Suffolk cheese, powdered quicklime and milk, "will make the part equally strong, and sound, with the rest of the vessel."[58] Other recipes ranged from the simplest paste made with flour and water to cements for fixing chemical vessels using linseed meal, whiting, gum senegal, Windsor loom and Sturbridge clay, or mixtures of

---

52 David Edgerton, *The Shock of the Old: Technology and global history since 1900* (Oxford; New York: Oxford University Press, 2007), 80.

53 See e.g. John Mordant, *The Complete Steward: Or, the duty of a steward to his lord*, 2 vols. (London: W. Sandby, 1761), vol. 1, 389.

54 See "Glass," and "Glue" in Chomel, *Dictionaire oeconomique*, vol. 1, n.p. (see note 25).

55 Mrs. Fisher of Richmond, *The Prudent Housewife; Or, complete English cook, for town and country*, fourth edition (London: T. Sabine, 1788), 80.

56 e.g. Dossie, *Elaboratory Laid Open*, pp. 49-52 (see note 24); Robert Dossie, *The Handmaid to the Arts*, 2 vols. (London, 1758), vol. 2, 21-31.

57 Dossie, *Elaboratory Laid Open*, p. 52 (see note 24).

58 Ibid.

quicklime and egg whites.[59] Cements were applied with a stick or an old tobacco pipe.[60]

Chemical practitioners avoided damage, and repaired things when this was unavoidable. And if instruments could not be repaired, they could still be made serviceable. In the inventory of electrical instruments in his Birmingham laboratory Priestley included "About forty square feet of coated Jars which had been cracked by Explosions but were of some use."[61] When no longer useable, broken items could be converted to some other use. Lavoisier explained in *Elements of Chemistry* that to contain liquids for distillation, "The best utensils for this purpose are made of the bottoms of glass retorts and matrasses."[62] A heated iron ring connected to a wooden handle could be placed around the broken vessel to make it useable.[63] This practice of using old vessels even after they were broken had a long tradition, extending well before and after Lavoisier's time. The archaeology of chemical remains related to Joseph Black at Edinburgh revealed "assorted bottle bases containing residues."[64] In 1830, Michael Faraday wrote that "very useful glass dishes and capsules are made out of old retorts, receivers, and flasks."[65] Harvard professor of Chemistry John White Webster's 1839 *Manual of Chemistry* explained that earthenware vessels could be used to liberate gases from substances if they were coated and luted before heating to prevent cracking,

> [...] horse dung, chopped hay, horse hair, and tow cut short may be incorporated with the lute. The addition of sand, renders the lute more fusible, and is not applicable when very high temperatures are to be sustained. In such cases fragments of broken glass pots, or of broken crucibles, may be used, being first well pulverized.[66]

Here broken items were used to prevent further items breaking. Fragments served as a material for making new instruments, for lutes, and also to provide

59 Dossie, *Handmaid to the Arts*, vol. 2, p. 26 (see note 56).
60 Ibid., vol. 2, pp. 26-28.
61 Priestley, quoted in McKie, "Priestley's Laboratory," p. 117 (see note 14).
62 Antoine-Laurent Lavoisier, *Elements of Chemistry*, trans. Robert Kerr (Edinburgh: William Creech, 1790), 377.
63 Ibid.
64 Addyman, "Materia Chemica" n.p. (see note 18).
65 Michael Faraday, *Chemical Manipulation: Being instructions to students in chemistry on the methods of performing experiments of demonstration*, second edition (London: John Murray, 1830), 168.
66 Webster, *A Manual of Chemistry*, p. 106 (see note 31).

physical support. Samuel Parkes, London chemical manufacturer and author of a *Chemical Catechism*, used broken glass to produce phosphoric acid by exposing sticks of phosphorus to the air inside a glass funnel. "Two or three pieces of broken glass placed in the neck of the funnel to support the phosphorus, and a small quantity of distilled water put into the receiving bottle, complete this simple apparatus."[67]

Waste

Materials and instruments that really could not be used again might be discarded as waste and swept out of the house as dust. At that point they returned to wider circulations of used materials that characterized early modern states. Few materials at this time were not re-used. Many people made a living trading in old and discarded goods in the early modern period. Since the fourteenth century in many European cities, a community of scavengers were employed by municipal authorities to roam the streets with carts collecting materials. Coal ashes, bones, metals, rags, cinders, night-soil, metals and shells were all gathered and refashioned into new products.[68] Chemical practitioners engaged with these trades. In October 1768, Joseph Black told James Watt that he had asked Ninian Hill, owner of a laboratory in Glasgow, to send Watt some things to be packed and sent by carriers to Black in Edinburgh.

> [...] among these are an absurd sort of still and a tall head to it both of copper – the tall head you may sell as old Copper – the Still or Body, I wish to have opened above by taking off the top of it which is soldered on only with soft solder and that top is also to be sold as old Copper – the rest of the Body will serve me as a boiler and may be sent packed full of the other things.[69]

Nothing was wasted as Black cannibalized the still and sold the head.

67 Samuel Parkes, *The Chemical Catechism: With notes, illustrations, and experiments*, fifth edition (London: Lackington, Allen and Co., 1812), 190.

68 On early modern waste disposal, see Emily Cockayne, *Hubbub: Filth, noise and stench in England, 1600-1770* (New Haven: Yale University Press, 2007), 183-91.

69 Joseph Black to James Watt, Edinburgh, 31 October 1768, in Eric Robinson and Douglas McKie, eds., *Partners in Science: Letters of James Watt and Joseph Black* (London: Constable, 1970), 15.

Chemical practitioners also made it their business to encourage others to make the most of waste, promoting the same oeconomic principles that guided their day-to-day interactions with materials. Practitioners argued that the application of chemistry reduced waste and improved arts. In 1800, for example, the Scots periodical writer Robert Heron praised Lavoisier's chemistry for improving the art of bleaching, which was, he said, "till lately, extremely tedious, and required long labour, a great waste of time, and an extraordinary consumption of expensive materials. The waste of lime, alkali, acids, fewel, was very great." Heron went on to argue that cookery, a prominent oeconomic practice, if subjected to more chemical science, could equally become less wasteful, "how often are the best pieces of animal food, at present, wasted? [...] much improvement might be made by a due application of chemical skill."[70]

Chemical authors also promoted the idea that chemistry could reveal new uses for byproducts and waste materials. The re-use of old materials was of course common in many chemical trades, and often international in scope. As Joppe Van Driel has documented, in the Netherlands ashes and urban filth were transported to rural areas to provide fertilizer for new raw materials for urban factories. In the name of 'sustainability' (*duurzaamheid*), administrators, entrepreneurs and chemists collaborated to steward materials.[71] In Britain, paper was made from old rags imported from Spain, Russia, and elsewhere; tanning used dogs' dung picked up off the streets by female "pure-finders"; copperas or green vitriol production mixed old iron with a liquor derived from pyrites to produce copperas crystals. Old plaster was used to manufacture niter; old bones were used to make glue.[72] Chemical authors articulated the practices of these manufactures and claimed to be able to improve them. Robert Dossie described how to make spirit of hartshorn (the horns of a male deer), used as a detergent for removing stains, by distillation in his explicitly titled *Elaboratory Laid Open*. Place pieces of hartshorn in a still and gradually

70 Robert Heron, *Elements of Chemistry: Comprehending all the most important facts and principles in the works of Fourcroy and Chaptal* (London: T.M. Longman and O. Rees, 1800), 548, 586.

71 Joppe van Driel, "Ashes to Ashes: The stewardship of waste and oeconomic cycles of agricultural and industrial improvement, 1750-1800," *History and Technology* 30 (2014): 177-206; see also Joppe van Driel and Lissa Roberts, "Circulating Salts: Chemical governance and the bifurcation of 'nature' and 'society'," *Eighteenth-Century Studies* 49 (2016): 233-63.

72 Anon., *The Art of Tanning and Currying Leather: With an account of all the different processes made use of in Europe and Asia* (London: J. Nourse, 1780), 142, 209-10; Tim Allen, Mike Cotterill, Geoffrey Pike, "The Kentish Copperas Industry," *Archeologia Cantiana* 122 (2002), 319-334; on niter, see Antoine Baumé, *A Manual of Chemistry; Or a brief account of the operations of chemistry and their products* (Warrington: W. Eyres, 1778), 269.

increase the heat from a fire. An oil, salt and spirit would result, of which "let the spirit, and salt, be mixt together again; and distilled with a gentle heat, and they will both rise purer. If this operation be several times cautiously repeated [...] the spirit will become limpid as water, and have a grateful smell."[73]

Others identified further waste substances which could be used for oeconomic ventures such as agriculture. Chemical writers criticized the poor practices of traditional managers of waste. In the *Edinburgh Magazine* of 1796 Thomas Butterworth Bayley, penal reformer and founder of the Manchester Literary and Philosophical Society, advocated to the Manchester Agricultural Society the use of waste materials for manure. Bayley lambasted farmers,

> [...] of all ranks, carrying on their lands, at a great expense of labour, time, and money, vast quantities of stable dung, whilst at home they overlook and neglect the most easy, plain, and cheap methods of accumulating manures, and enriching their farms.[74]

Bayley then identified wastes that could be turned into valuable commodities, including mud, mixed with lime, street sweepings and ashes, night soil, bones, sweepings of cotton and woolen mills, sea-weed, sea shells, river weeds, spent tanner's bark, decayed vegetables, water from steepings of flax and hemp, bleachers' ashes, soap suds and ley.[75] Also in the 1790s, following the Scottish proverb that "muck is the mother of the meal chest", Archibald Cochrane, Earl of Dundonald, explored the potential of various wastes as fertilizers in *A Treatise, Shewing the Intimate Connection that Subsists Between Agriculture and Chemistry*. Chemical substances might be turned to good use, "Muriat of Magnesia [...] may be procured in great quantities from the bitter refuse liquor which at present runs to waste at the salt works."[76] Coal-gas, another product of interest to Cochrane, was a byproduct of the distillation of coal into coke, and remained a waste product of that process, liberated into the air, through the eighteenth century. In the early nineteenth century, however, the German inventor Frederick Winsor competed with Scots engineer William Murdoch to

73 Dossie, *Elaboratory Laid Open*, p. 1 (see note 24).

74 T.B. Bailey [Bayley], "Thoughts on Collecting Substances for Manure," *Edinburgh Magazine, or Literary Miscellany* (Oct 1796): 291-293, on 291.

75 Ibid., pp. 291-293.

76 Archibald Dundonald, Earl of Cochrane, *A Treatise, Shewing the Intimate Connection That Subsists Between Agriculture and Chemistry* (London: J. Murray and S. Highley, 1795), 73, 90.

use coal gas as an illuminant. Both Winsor and Murdoch sought to give credibility to their enterprises by evoking scientific principles and backing.[77]

Such enterprises thus saw chemists seeking to co-opt traditional areas of artisanry, agriculture, and manufactures in the name of the same oeconomic goals of managing materials, reducing waste, and saving money that marked household approaches to materials and experimentation. Chemical practitioners frequently moved between manufacturing and domestic contexts in pursuit of these enterprises. Dundonald spent considerable wealth scaling up chemical manufacturing enterprises on his estates in Scotland, and passed regularly between his large-scale concerns and smaller home laboratories like that of James Watt in Birmingham.[78] Dundonald ruined his finances on this enterprise, suggestive of the way oeconomic motives might differ from the simply "economic".[79] Watt meanwhile constructed prototypes of inventions such as the separate condenser from left-over pipes and syringes, before deploying them in steam manufactures.[80] Gas-lighting schemes also moved between the home and larger urban sites. Winsor first offered customers novel stoves producing light and heat with gas, fitted inside a single house, which were eventually connected through a network of pipes and supplied by gasometers to produce a network on an industrial scale.[81]

In one case at least, claims that chemistry might improve the management of waste went alongside changes from *oeconomic* to *economic* language.[82] Oeconomy and economy shared some features but the rational management characteristic of economy focused on quantitative measures of financial profit and loss, numerical accounting, and the supposedly rational principles of the unfettered market. The mathematician Charles Babbage famously reckoned a

77 Simon Werrett, "From the Grand Whim to the Gasworks: Philosophical fireworks in Georgian England," Lissa Roberts, Simon Schaffer, and Peter Dear, eds., *The Mindful Hand: Inquiry and invention from the late Renaissance to early industrialisation* (Amsterdam and Chicago: Edita; University of Chicago Press, 2007), 325-48.

78 Thomas Barnes Cochrane Dundonald and Henry Richard Fox Bourne, *The Life of Thomas, Lord Cochrane, Tenth Earl of Dundonald, completing "The Autobiography of a Seaman"*, 2 vols. (London: R. Bentley, 1869), vol. 2, 223.

79 The theme is elaborated in John Christie's contribution to this volume.

80 Russell, *James Watt* (see note 19).

81 Leslie Tomory, *Progressive Enlightenment: The origins of the gaslight industry, 1780-1820* (London; Cambridge, MA: MIT Press, 2012).

82 For a fuller discussion of the shift from oeconomy to economy, see Joppe Van Driel, "The Filthy and the Fat: Oeconomy, chemistry and resource management in the age of revolutions," (PhD Thesis, University of Twente 2016); and Simon Werrett, *Thrifty Science* (Chicago: University of Chicago Press, forthcoming).

more careful accounting of the value of waste was necessary to improve manufactures. In an appendix to his 1832 *Economy of Machines and Manufactures*, Babbage gave precise accounts of the "profitable conversion of substances apparently of little value" at a horse-slaughtering yard in Montfaucon near Paris. Reckoning the value of hair, skin, blood, hoofs, fat, flesh, tendons, and bones converted into animal food, manure, combs, lamp fuel, and other products, Babbage demonstrated that a "dead horse [...] which can be purchased at from 8.s. 6.d. to 12 s., produces from £2. 9.s. to £4. 14.s."[83] In contrast with Babbage's optimism over the economics of waste, others were more cautious. Economists questioned whether the efficient use of materials was actually beneficial. In 1865, University College London economist William Stanley Jevons argued in *The Coal Question* that the creation of increasingly efficient engines had not led to a reduction in the consumption of coal but on the contrary to a great increase, because increased efficiency lowered the cost of engines and encouraged their consumption. As Jevons concluded, "It is wholly a confusion of ideas to suppose that the economical use of fuel is equivalent to a diminished consumption. The very contrary is the truth."[84] The efficient use of waste might generate profits and growth, but it might equally create more problems than it solved.

Conclusion

A common narrative identifies a "revolution in chemistry" in the period 1760 to 1840 and identifies this revolution with radical innovation. The rhetoric of contemporaries and historians often equated this innovation with a throwing out of the old. John Joseph Griffin explained to readers of his *Chemical Recreations* in 1834, "To reform the nomenclature of chemistry, is to cleanse the Augean stable, where rubbish has been accumulating for forty years."[85] Georges Cuvier reckoned Henry Cavendish and Torben Bergman had "cleared that Augean stable, still overspread with the rubbish of hermetical philosophy."[86] Historians follow the same lead. "Lavoisier" wrote J.D. Bernal in his 1965 *Science in History*

83 Appendix to Charles Babbage, *The Economy of Machines & Manufactures*, third edition (London: J. Murray, 1846), 393-96.

84 William Stanley Jevon, *The Coal Question* (London: MacMillan, 1865)

85 Griffin, *Chemical Recreations*, p. iv (see note 13).

86 Georges Cuvier, "Biographical Memoir of Henry Cavendish," *Edinburgh New Philosophical Journal* 9-10 (1828): 209-222, on 217.

"made a clean sweep of all the old time-hallowed chemical nomenclature."[87] "The Chemical Revolution," Mi Gyung Kim writes more recently, "was successful in cleaning the attic and thereby putting the house in order."[88]

The household and its maintenance remain potent locations for making sense of chemical practice. However, chemical inquiry understood from a sociomaterial perspective in the period 1760 to 1840 was by no means about throwing out the old. Historians focused on the dedicated culture of chemistry, on materials and locations which were designed and built exclusively for chemical inquiry, have overlooked a widespread practice of adapting and converting spaces and goods to chemical ends during these decades. Early moderns, like contemporary sociologists, viewed much material culture as "incomplete", not dedicated to a single end but pregnant with possible uses. Kitchens, cellars, wine bottles, broken saucers, gun barrels, and the hints and recipes that cleaned, cared for and repaired them were as much a part of chemical experimentation as specialized apparatus like the furnace or crucible and even the latter could be altered or repurposed according to need.

Making the household and other adapted spaces the focus in this essay has revealed a set of practices and approaches to material culture that were intimately connected with and often corresponded to practices of oeconomic household management. Oeconomic concerns, shared among family members, sought to balance the consumption and use of new goods with the careful stewardship of the old, maintaining and making good use of existing possessions where possible. This led to savings of money and time, but beyond narrow conceptions of profit and loss, it established good order and provided a model for the management of other arenas of life. Applied to chemistry, oeconomics regulated a diverse body of spaces and materials that were the location of much experimentation.

Acknowledgments

My thanks to Lissa Roberts, John Perkins, and the members of the 'Situating Chemistry' project for their support and constructive commentary throughout

87 J.D. Bernal, *Science in History*, 4 vols. (Harmondsworth: C.A. Watts, 1965), vol. 3, 451; "Lavoisier soon realized that the confusion of alchemical terminology had to be swept away along with the phlogiston theory." Sarah Regal Riedman, *Antoine Lavoisier: Scientist and citizen* (London: Abelard-Schuman, 1967), 141.

88 Mi Gyung Kim, *Affinity: That elusive dream* (Cambridge, MA: MIT Press, 2003), 393.

the development of this essay. Part of this essay was written as a fellow at the Beckman Center for the History of Chemistry at the Chemical Heritage Foundation in Philadelphia, and with the support of the Rachel Carson Center for Environment and Society in Munich for which support I am most grateful.

CHAPTER 2

The Case of Coal

Lissa L. Roberts and Joppe van Driel

Coal plays a key role in current debates regarding both the 'Anthropocene' and 'Great Divergence'. Long identified as having fueled the Industrial Revolution, coal has been celebrated and condemned for spurring material progress and productivity, global inequality and environmental degradation.[1] But what is coal? While the answer might seem straightforward, recognizing that coal's identity as a chemical substance and material resource actually evolved over time, rather than having been *a priori* essential, can help us better understand the history which has both shaped and been shaped by it. That is to say that the historical identity of coal evolved through a fluid amalgam of material characteristics and applications, knowledge claims, technological capabilities, market transactions and political decisions.[2] By uncoupling our understanding of the past from an acceptance that materials have an essential identity, we realize that coal-powered industrialization was not historically inevitable; rather it was a complex matter of choice. This recognition, in turn, accentuates the fact that our collective future is also an open matter of choice.

A partial model for considering what this rethinking entails can be found in Timothy Mitchell, *Carbon Democracy: Political power in the age of oil*.[3] In the first chapter, Mitchell contrasts the "socio-technical agencies" of coal and oil, which did so much to shape politics since the nineteenth century. Briefly, coal's extraction, transport and use depended on the workers who operated coal-mines, ran the railroads and stoked coal-fueled fires. With so many workers concentrated together in locations that were crucial to the growth of industrial

1 See especially E.A. Wrigley, *Energy and the English Industrial Revolution* (Cambridge: Cambridge University Press, 2010); Alfred Chandler, "Anthracite Coal and the Beginnings of the Industrial Revolution in the United States," *Business History Review* 46 (1972): 141-181; Kenneth Pomeranz, *The Great Divergence: China, Europe and the making of the modern world economy* (Princeton, NJ: Princeton University Press, 2000); Will Steffen, Jacques Grinevald, Paul Crutzen and John McNeill, "The Anthropocene: Conceptual and historical perspectives," *Philosophical Transactions of the Royal Society A* 369 (2011): 842-867.

2 Compare with the treatment of uranium in Gabrielle Hecht, "Africa and the Nuclear World: Labor, occupational health, and the transnational prodution of uranium," *Comparative Studies in Society and History* 51 (2009): 896-926.

3 Timothy Mitchell, *Carbon Democracy: Political power in the age of oil* (London: Verso, 2011).

 | DOI 10.1163/9789004325562_004

society, organization and political clout were bound to follow. What finally broke their power, was not only the cooperation between captains of industry and colluding politicians, however, but society's growing dependence on oil. Because of their relative geographical and practical isolation, oil's workforces – those who manned the oilrigs, built the pipelines and crewed on oil tankers – never organized as coalminers and railway workers had. The global nexus of political and economic power thereby shifted and carbon democracy took on the characteristics that continue to haunt us today.

Mitchell's analysis allows him to emphasize how it is that histories of production, distribution and use are inevitably also social, economic and political histories. Missing, however, is a consideration of whether the identities of the substances with which he begins his analysis also have a history. What does it mean, in other words, to speak of 'coal'? Answering this question takes us back to the long eighteenth century, when fossil substances were being increasingly mined and exploited across significant parts of Europe. As demonstrated in this essay, coal's identity and uses were open matters at the time. As the philosopher Annemarie Mol writes, "[*O*]*ntology* is not given in the order of things [...] instead *ontologies* are brought into being, sustained, or allowed to wither away in common, day-to-day, sociomaterial practices."[4] Chemists, natural historians, encyclopedists, scientific societies, mine operators, landowners, investors, inventors, policy-makers and law courts all contributed to shaping coal's identity, classification and use. As they deliberated, they could not know that the sum of their efforts would fuel historical development in the way that has been retrospectively explained by binding coal's 'essential' identity to industrialization.

In what follows, the initial openness of coal's identity is examined. The first section examines the categorizations through which encyclopedists, natural historians and chemists sought to define and situate coal in the realm of nature. The second section zooms in on a series of British cases in which coal was regarded as a 'political' substance whose identity could only be resolved through legal deliberation regarding its use. In the third section we shift our attention to France and examine the evolving identity of coal as a 'natural resource'. We do so by considering those for whom the opening and governed exploitation of coal mines was integrated with simultaneous efforts to improve the soil and society. In other words, French chemists, entrepreneurs and offi-

4 Annemarie Mol, *The Body Multiple: Ontology in medical practice* (Raleigh, NC: Duke University Press, 2002), 6, quoted in Lissa Roberts, "Exploring Global History Through the Lens of History of Chemistry: Materials, identities and governance," *History of Science* 54 (2016): 335-361, on 347.

cials regarded coal mining as an element of oeconomic circulation of materials, meant to stimulate agricultural, industrial and social improvement. Coal's identity was, thus, not written in stone from the beginning of time. Rather it evolved during the period under investigation here in a field of historically open possibilities regarding understanding, use and socio-environmental amelioration.

Tentative Taxonomies

> "[T]here is no standard coal [...] the very word itself is a popular term, which has entered into science."[5]

Even today, coal's identity is ambiguous. The product of decayed organic material, it is generally considered together with inorganic minerals and referred to as having fueled a "mineral based energy economy."[6] It is variously classified scientifically in terms of the sort of plant remains that compose it (humic and sapropelic), its heating value and carbon content level (lignite, sub-bituminous, bituminous and anthracite), and its chemical composition (this varies in virtually every sample because of local conditions). Classification according to chemical composition, unsurprisingly, has changed along with broader developments in analytic chemistry and its instrumentation. Coal's quadripartite division was only adopted internationally in a standardized way in the late 1930s, bringing a degree of stability after centuries of multiple systems and suggestions.

Part of this is traceable to eighteenth-century translations of the Latin term '*regnum lapideum*'. While generally translated as 'mineral kingdom' (or similar cognates in other languages), Linnaeus defined this realm to include '*petrae*' (simple stones), '*minerae*' (composite/heterogenous stones) and '*fossilia*' (aggregate mixts containing both decayed animal and vegetable and substances classed as '*minerae*'). In the early editions of his *Systema Naturae*, the class of *minerae* was further sub-divided into salts and sulfurs, the latter characterized

5 E.A. Newell Arber, *The Natural History of Coal* (Cambridge: Cambridge University Press, 1911), 6.

6 The United States government, for example, only began publishing separate production statistics for minerals and fossil fuels, including coal in 1977, following the 1973 oil crisis. See also Cornelia Klein and Anthony Philpotts, *Earth Materials: Introduction to mineralogy and petrology* (Cambridge: Cambridge University Press, 2013); E.A. Wrigley, *Continuity, Chance and Change: The character of the Industrial Revolution in England* (Cambridge: Cambridge University Press, 1988).

by its combustibility and odoriferous quality. Sulfur, in turn, contained four genera, including *bitumen*, within which the solids *asphaltum*, *gagates*, and *lithantrax* were situated. The thirteenth edition of this widely referenced book included an alternative system developed by the Swedish physician Magnus Bromelius in 1730. This system divided *Sulphura* into five groups: *sulphur, bitumen, petroleum, succinum* and *lithanthrax*, which were translated into English as sulfur, bitumen, petroleum, amber and coal.[7]

But lack of coherence was not confined to natural history systems. Encyclopedias, iconic vehicles of enlightenment, illustrate this. Both Diderot and d'Alembert's *Encyclopédie* and its 'Protestant' counterpart edited by Fortuné Barthélemy de Félice (two of the Enlightenment's most successful publishing ventures) spread discussions out among a number of articles, including those dedicated to *charbon* and its subdivision focusing on *charbon de terre*, in addition to *fossile*, and *houille*.[8] Neither could turning to chemistry textbooks provide clarification. Until the late 1780s, as Ursula Klein and Wolfgang Lefèvre point out, chemistry textbooks largely organized their classification systems according to perceptible properties (appearance, smell and taste, solubility, combustibility, etc.) and variously divided the 'mineral kingdom' into competing sets of sub-categories. Wilhelm Homberg's taxonomy, for example, employed the categories metals, salts, stones and earths, while Boerhaave classed the same set of substances as metals, salts, sulfur, stones, earths and semi-metals. Johann Christian Wiegleb divided this realm into earthy bodies, metals, salts, waters and inflammable bodies.[9] Establishing a unitary identity for 'coal' was bound to be a challenge in such a context.

Contemporary authorities nonetheless managed to turn the characteristics of the chemistry of their age to advantageous use. The French surgeon and *Académie des sciences* librarian Jean François Clément Morand, for example,

---

7 Carolus Linnaeus, *Systema Naturae* (Stockholm: Gottfried Kiesewetter, 1740), 6; Carolus Linnaeus, *Systema Naturae*, thirteenth edition, ed. Johann F. Gmelin (Leipzig: Georg Emanuel Beer, 1793), vol. 3, 17; in english, see Sir Charles Linné, *A General System of Nature* [...] *translated from Gmelin, Fabricus, Willdenow, etc.* [...] *by William Turton, M.D.* (London: Lackington Allen and Co., 1806), 11.

8 Denis Diderot and Jean le Rond d'Alembert, eds., *Encyclopédie ou dictionnaire raisonné des sciences, des arts et des métiers* (Paris: André le Breton, Michel-Antoine David, Laurent Durand, Antoine-Claude Briasson, 1751-1772); Fortuné Barthélemy de Félice, ed., *Encyclopédie ou Dictionnaire universel raisonné des connoissances humaine* (Yverdon, 1770-1780); Robert Darnton, *The Business of Enlightenment* (Cambridge, MA: Harvard University Press, 1979), especially 19ff.

9 Ursula Klein and Wolfgang Lefèvre, *Materials in Eighteenth-Century Science: A historical ontology* (Cambridge, MA: MIT Press, 2007), 163-164.

contrasted natural history's 'superficial' multiplication of names and categories with the analytical insights afforded by chemistry in his oft-cited work *l'Art d'exploiter les mines de charbon de terre.*

> [O]f all the productions of nature's three realms, it is this fossil which presents the most singularities and analytical difficulties [...] In order to acquire a just idea of the constituent parts of *charbon de terre*, it is indispensible to submit samples from as many different lands as possible to chemical analysis.[10]

Sensible and qualitatively oriented chemistry, still closely related to medicine but also increasingly allied to the improvement of both agriculture and manufacture, provided the analytical tools for linking experimentally-revealed characteristics with a growing understanding of material composition and application. Textures, smells, colors and combustibility, uncovered in the laboratory, indicated whether a given sample of *charbon de terre* was sulfuric, pyritic, acid or alkaline, in addition to the relative amount of phlogiston it contained. In turn, this helped determine the nature of health risks to local miners and possible medical applications. It further indicated which samples represented sources usable for forge work, domestic heating or fertilizing the soil.[11]

Disciplinary skirmishes continued at least through the 1830s when William Hutton, agent for the Norwich Fire Insurance Company and amateur natural investigator, noted that every variety of this "fossil" he subjected to microscopic examination revealed the presence of "vegetable remains."[12] Together with the University of London's botany professor John Lindley, he set out his findings and argued for more attention to what he called fossil botany. Like Morand, Hutton and Lindley justified their approach in opposition to natural history.

> [N]either a barren nomenclature, destitute of all attempts at determining the relations that former species bore to those of our own era, nor supposed identifications of species by vague external characters, nor hasty determinations by analogies by means of partial views of structure,

10 Jean François Clément Morand, *L'Art d'exploiter les mines de charbon de terre*, 4 vols. (Paris: Saillant et Nyon, 1768-1779), vol. 2, 1117, vol. 4, 1384.

11 Ibid., see e.g. vol. 2, pp. 980, 1143, 1115, 1150-1159, vol. 4, p. 1384.

12 The vegetable origins of coal had certainly been noted before this. See e.g. John Playfair, *Illustrations of the Huttonian Theory of the Earth* (Edinburgh: William Creech, 1802), 148-150, where he discusses the views of Buffon, Richard Kirwan and others.

> are sufficient [...] nothing short of a most rigorous examination is likely to serve the ends of science.[13]

The 'science' these authors wished to serve was directed toward understanding the past. The fossil remains of plants found in various subterranean strata provided them with an as-yet underexplored "glimpse of the early history and condition of our Planet, and of the successive races of organized bodies which have existed upon it."[14] Chemical inquiry insinuated itself differently in relation to the passage of time. Similarly able to assemble clues about the Earth's past, it was simultaneously poised to suggest possible futures, whether through its application to the promotion of health, the production of new commodities, the improvement of crop yields, or a more general, manipulable understanding of composition and decomposition.

The successful movement of chemical knowledge and processes in and out of the laboratory depended on practical exactitude, but this was not enough to provide a definitive answer to coal's identity. Tentative taxonomies continue to this day and, as a recent study intriguingly argues, it is the element of time that is responsible for coal's ambiguous identity as a chemical substance. Longstanding tests "afford no information whatsoever [...] about the nature of coal," while examining its source history emphasizes locally-situated particularities of character and composition. Perhaps it would be preferable simply, then, to "look to coal as a natural product that is subject to local and regional variations."[15]

As the following section indicates, coal's identity remained a pressing issue from the late eighteenth until the mid-nineteenth century, even if – or especially because – the authority of science was insufficient to resolve the question. But if chemistry suggested trajectories of use that might shape coal's otherwise open future, might arguing back from these projected uses provide coal with a clear definition? Crucially this required more than just moving back and forth in time. It also entailed moving from the laboratory to legal courts and legislatures, where evidence was collected and weighed with different measures and identities were decided in a manner that went beyond 'laws of nature' to include those situated at the intersection of the state, society and the market.

---

13 John Lindley and William Hutton, *Fossil Flora of Great Britain* (London: James Ridgway, 1831-1833), v-vi.

14 Ibid.

15 James G. Speight, *The Chemistry and Technology of Coal* (Boca Raton: Taylor and Francis, 2013), 97.

Adjudicating Identities

The biography of James Hutton (1726-1797) illustrates how these various threads came together in the late eighteenth century. Educated as a physician, Hutton engaged in a broad array of activities including farming, chemical manufacture, canal construction and management, and geology. Coal played a role in many of his endeavors, including his most famous publication *The Theory of the Earth*. Hutton defined coal as both the product of dead vegetable matter and as a source of subterranean heat, which was responsible for inciting geo-chemical change; it thereby provided active testimony for the cycles that lent nature's oeconomy its dynamic character.[16] This invited human stewardship of a global system whose ongoing fertility depended on maintaining its dynamic balance. It also invited analysis of coal's own nature and identity, which Hutton viewed as evolving over time.[17]

Hutton's chemical investigations and related reflections led him to believe that coal was composed of phlogiston and either a simple carbonic substance or oily compounds produced by plants. "[T]hough found in every intermediate state," he distinguished two distinct sorts, related to level of "exposure to higher degrees of heat, and to other circumstances favourable to the dissipation of their more volatile and fluid parts."[18]

> Of the one kind is that fossil coal which melts or becomes fluid upon receiving heat; of the other, is that species of coal, found both in Wales and Scotland, which is perfectly infusible in the fire, and burns like coaks, without flame or smoak [sic]. The one species abounds in oily matter, the other has been distilled by heat, until it has become a *caput mortuum*, or perfect coal.[19]

As this passage indicates, the binary nature of 'coaliness' only revealed itself through time. The first coaly substance was fusible and burned with a flame, losing its phlogiston when heated. The second, more "perfect" coal was the residue of longstanding heat and fusion, which left behind a carbonic and combustible sort of phlogiston.

---

16 For multiple meanings of the word 'oeconomy', see the introduction to this volume.

17 Douglas Allchin, "James Hutton and Coal," *Cadernos IG/UNICAMP* 7 (1997): 167-183.

18 James Hutton, *The Theory of the Earth* (*Transactions of the Royal Society of Edinburgh*, vol. I, Part II, 1788), 209-304, on 240.

19 Ibid., p. 241.

Significantly, Hutton's attention had previously been drawn to both the question of maintaining the earth's fertility and the identity of coal in circumstances that were other than purely intellectual. As an owner of farmland, Hutton held active experimental and financial interests in matters of fertility and fertilization, which led him to test the use of coal ash for improving crop yields.[20] As an investor and manager of the Forth and Clyde Canal in the Scottish lowlands, the financial success of which depended on being able to maximize the amount of goods that passed through it and minimize the duties that had to be paid on them, Hutton also became actively engaged in the question of coal's identity long before he penned his famous *Theory*.

With the passage of parliamentary legislation that set a higher duty for shipping excavated coal than for its close – but equally ill-defined – relative "culm", Hutton was anxious to set narrow boundaries around the former's identity and thereby lower the charges assessed on the fossil material that passed through the Forth and Clyde Canal. As he wrote in his 1777 pamphlet *Considerations on the Nature, Quality and Distinctions of Coal and Culm*, the lack of a legally recognized definition made it impossible to distinguish culm – purportedly the predominant fossil substance to pass through the canal – from coal. Neither did recourse to natural distinctions offer an easy solution.

> It cannot be in the nature of the fossile [sic] substance that the distinction of coal and culm consists; for in many places of the kingdom, the same seam, stratum, or mine, produces what is esteemed either coal or culm, according as it is in large pieces or broken small; therefore so far as a judgment should be formed in this way, the distinction of coal and culm would appear to consist in nothing but great and small.
>
> On the other hand, it cannot be in the size alone, that culm differs from coal, because the smallest dust of a certain species of coal always pays the duty proper to coal [...] It is therefore evident, that something else [...] must be required in order to distinguish culm from coal; and it will appear reasonable to look for this in the purposes to which those several commodities may be strictly applicable.[21]

20 James Hutton, "Elements of Agriculture," (unpublished manuscript), 117; cited in Jean Jones, "James Hutton's Agricultural Research and his Life as a Farmer," *Annals of Science* 42 (1985): 573-601, on 589.

21 James Hutton, *Considerations on the Nature, Quality and Distinctions of Coal and Culm* (Edinburgh: C. Elliot, and Richardson and Urquhart, 1777), 2-3.

Of note in this passage is that it moves from speaking of the "nature" of a *substance* to the applicability of a *commodity* without any indication that a boundary exists between the two. Hutton was sure that the use to which a commodity could be put "depends truly on the nature of the substance in question," but argued that this was best revealed "from observations that may be made in the actual application of the commodity."[22] While material identities were thus ascribed to nature by this logic, they were best sought – according to the same reasoning – through an investigation of marketable goods.

Hutton's memoir was matched by ten months of lobbying the Treasury and Board of Customs in London and answered by at least one angry counter-pamphlet.[23] In the end, Hutton suggested that fraud could easily be prevented and government revenues protected by a simple test that any revenue agent could perform. When the question arose of whether a barge shipment contained culm or coal, the attending agent had only to place a small sample in a crucible and attempt to light it; culm's fusibility would prevent it from sustaining the fire.[24] The outcome, enshrined in the passage of Parliamentary legislation in December 1777, practically set the identities of culm and coal and prevented the erosion of profits that would have resulted from a refusal to distinguish culm from coal and grant a lower rate of duty for its transport.

As industrial developments created greater possibilities for the exploitation of materials, legislation and litigation were bound to follow. Questions of ownership, transportation, safety, and revenues were aired and answered in legislative assemblies, administrative offices and law courts. As seen in the case just discussed, establishing material identities was a crucial part of the process. And while this expanded the market for chemist-consultants, their involvement neither guaranteed a solution to the problem nor a trustworthy, scientific reputation for them.[25] Consultants often disagreed with each other

22 Ibid.

23 *Remarks on "Considerations on the Nature, &c. of Coal and Culm, &c." By a Friend to the Revenue. Addressed to the Commissioners for managing his Majesty's Customs* (London, 1777), cited in *The Monthly Review or Literary Journal* 58 (1778): 482; For lobbying by Hutton's colleague, see Jean Jones, "James Hutton and the Forth and Clyde Canal," *Annals of Science* 39 (1982): 255-263, on 263.

24 Hutton, *Considerations*, pp. 12-13 (see note 21).

25 Christopher Hamlin, "The City as a Chemical System? The Chemist as Urban Environmental Professional in France and Britain, 1780-1880," *Journal of Urban History* 33 (2007): 702-28; Joppe van Driel and Lissa Roberts, "Circulating Salts: Chemical governance and the bifurcation of "nature" and "society"," *Eighteenth-Century Studies* 49 (2016): 233-63, esp. 249; J.Z. Fullmer, 'Technology, Chemistry, and the Law in Early 19th-century England," *Technology and Culture* 21 (1980): 1-28; Paul Lucier, "Court and Controversy: Patenting

for reasons that spanned experimental evidence, training and interested involvement in the case at hand. Science at the bar exposed its practitioners and their disciplines to the fickle hybridity of a world at once material and social. When "nature" was deemed unable to provide acceptable guidelines, recourse was made to precedents of law and use, generously salted with the power of courtroom or parliamentary persuasion.[26]

A celebrated court case, heard in Edinburgh in 1853, brought all these elements together. A landowning couple, Mr & Mrs William Gillespie of Torbanehill, had leased a parcel of land to James Russel & Son, coalmasters. The lease granted the Russels the right to exploit "the whole coal, ironstone, iron-ore, limestone and fireclay (but not to comprehend copper, or any other minerals whatsoever, except those specified) in the lands of Torbanehill."[27] They leased the land, which adjoined a parcel they were already working, because they assumed that the veins they had uncovered of what came to be called 'Cannel', 'Boghead' or 'Turbanite' coal continued across the properties' boundary. Since it was a key ingredient in the newly burgeoning business of manufacturing paraffin, lubricants and lamp oil, the Russels stood to make a profit. The Gillespies, however, argued that this substance was not a kind of coal, but a mineral – hence, not covered by the lease. If the Russels wanted to exploit it, they should have to sign a new lease with a higher rate that reflected this added value.

More was at stake than the price of land, though. One of the major manufacturers who depended on a reasonably priced supply of this substance was James Young, who owed his position to a patent that explicitly named the ingredient in his process as 'coal'.[28] Should the Gillespies win their case, he stood to lose his dominant industrial position; small wonder that he came to the Russels' aid. As the court case unfolded, twenty-eight geologists, mineralogists, chemists and engineers gave testimony on behalf of the Gillespies; forty-one spoke on behalf of the Russels. The judge finally charged the jury to ignore the conflicting scientific evidence in favor of whether Gillespie had included this disputed mineral as coal in the original lease. "[Y]ou are to determine whether it is coal or is not coal," he pronounced, "in the language spoken

---

science in the nineteenth century," *British Journal for the History of Science* 29 (1996): 139-154.

26 Compare with José Ramon Bertomeu's essay in this volume.

27 Alexander Watson Lyell, *A Full Report of the Trial Before the Lord Justice-General and a Special Jury of the Issues in the Action at the Instance of Mr. and Mrs. Gillespie of Torbanehill* (Edinburgh: Bell and Bradfute, 1853), 2.

28 John Butt, *James 'Paraffin' Young: Founder of the mineral oil industry* (Edinburgh: Scotland's Cultural Heritage, 1983).

in the missive […] not […] in the language of geologists."[29] After only ten minutes of deliberation, the jury found for the Russels and, implicitly, for Young's continued dominance in this coal's industrial use.

The results were mixed. The Gillespies went away feeling cheated and continued to seek redress through the courts. Individual scientific reputations were publicly attacked, which emphasized the fragility of science's claim to objectivity. Young's victory was shaky and short-lived. The advertised value of this commodity – recently identified through litigation as coal – drew inventive competitors like moths to a flame; and, by the time his patent ran out in 1860, its value began to be eclipsed by the rise of shale-oil extraction. But the domain of coal's identity had grown, determined by the interpretation of a legal document rather than the authority of science.

A Fertile Fossil

The connections between coal and industrialization, along with the social and environmental inequalities and degradation they brought in their combined wake, have been seared into our cultural consciousness by novels such as D.H. Lawrence's *Sons and Lovers* and *Women in Love* and Emile Zola's *Germinal*. Coal, in these literary monuments, warmed the homes and lined the financial portfolios of owners as it blackened the short and miserable lives of workers whose families had to fight even for the right to glean the dusty leftovers of shipments sent to stoke the fires of industry.[30] These processes have been traced forward from the second half of the nineteenth century with great effect, but carrying them back in time risks papering over important developments. Generally overlooked in histories that return to the eighteenth century is a combination of contemporary recognition of coal's various identities and the presence of *oeconomic* initiatives that incorporated the reclamation of coal fields into visions of socio-environmental improvement.[31] Focusing on these

29 Watson Lyell, *Trial*, pp. 236-237 (see note 27).

30 Giulia Pissarello, "Industrialism as "Tragedy of Ugliness": D.H. Lawrence's ecological consciousness," *Griseldaonline* 10 (2011): 31-42; Sara B. Pritchard, "Mining Land and Labor," *Environmental History*, 10 (2005): 731-733; Agnes Kneitz, "'As if the River was not Meat and Drink to You': Social novels as a means of framing nineteenth-century environmental justice," *Interdisciplinary Studies in Literature and Environment* 1 (2015): 1-16.

31 Lissa Roberts, ed., *Practicing Oeconomy in the Late Eighteenth Century* (Special issue of *History and Technology* 30 (2014): 133-279). For recognition of various identities, see "Catalogue alphabétique des differents charbons de terre," Morand, *L'art d'exploiter les mines*, vol. 1, pp. 181-195 (see note 10).

developments is not to deny coal's involvement in industrialization. It is to recall that the historical trajectory along which this involvement unfolded was neither necessary nor inevitable; alternative routes forward were acknowledged and explored.

The *Compagnie des Mines d'Anzin*, one of France's largest industrial enterprises in the nineteenth century and setting for key scenes in *Germinal*, illustrates this. Established in 1757, its incorporation document is couched in Enlightenment terms that describe serving the public good as a primary goal. Beyond rhetoric, the company stood out during its first several decades for providing its workers with housing, medical care, education and retirement pensions.[32] With this in mind, the rest of this section attends to the ways in which coal's multiple identities were intertwined with the projection of alternative historical futures in France during the second half of the long eighteenth century.

Historians usually situate eighteenth-century French coal prospecting between a deforesting past and an industrializing future. However, the fact that other – often more abundant and easier to access- energy sources were available, reminds us that presenting coal as *the* driver of industrialization imports a teleological perspective into historical interpretation.[33] Throughout the eighteenth century wood and peat remained the most commonly used fuels in French households and factories, while the energy supplied by animals, humans, wind- and waterpower continued to top the share of fuel in

32 William Henry Hurlbert, *France and the Republic* (Charleston: BiblioLife, 2007 [1890]), 328; Richard Barker, "French Entrepreneurship During the Restoration: The record of a single firm, the Anzin Mining Company," *The Journal of Economic History* 21 (1961): 161-178, on 164.

33 Michael Williams, *Deforesting the Earth: From prehistory to global crisis* (Chicago: University of Chicago Press, 2003), 171; Andrée Corvol, trans. Richard C. Holbrook, "The Forest," Pierre Nora and David P. Jordan, eds., *Rethinking France: Les lieux de mémoire*, vol. 2, *Space* (Chicago: University of Chicago Press, 2006), 109-110; Ian D. Rotherham and David McCallam, "Peat Bogs, Marshes and Fen as Disputed Landscapes in Late Eighteenth-Century France and England," Louise Lyle and David McCallam, eds., *Histoires de la Terre: Earth sciences and French culture 1740-1940* (Amsterdam: Rodopi, 2008), 75-88, on 87; Ivan T. Berend, *An Economic History of Nineteenth-Century Europe: Diversity and industrialization* (Cambridge: Cambridge University Press, 2013), 80, n. 15; The traditional economic narrative nonetheless remains persistent. Prasannan Parthasarathi, *Why Europe Grew Rich and Asia Did Not: Global economic divergence, 1600-1850* (Cambridge: Cambridge University Press, 2011), 160; Charles Coulston Gillispie, *Science and Polity in France at the End of the Old Regime* (New Jersey: Princeton University Press, 1980), 427; Michael Stephen Smith, *The Emergence of Modern Business Enterprise in France* (Cambridge: Harvard University Press, 2005), 162.

manufacturing.[34] The use of coal for fuel did indeed slowly spread in French glassworks, potteries, brickworks and breweries, favored in these manufactures because the fossil burned at higher temperatures than wood or charcoal.[35] At the same time, eighteenth-century coal enthusiasts also promoted non-industrial uses. In his previously mentioned *L'Art d'exploiter les mines de charbon de terre*, Morand drew on an international array of authors to discuss "the different ways of employing coal in manufactories, workshops and households."[36] Monitored by chemists, Morand explained, the nature and uses of coal proliferated as it was transformed into various states. Using the fossil as fuel in manufactures, in what he referred to as "the governance of fire," was only one option.[37] When oily, physicians employed it to combat ringworms, abscesses or sexually transmitted diseases, while craftsmen converted it into varnish. When smoky, it attacked scurvy and measles. Painters used coal-impregnated water to produce black and red pencils. Artisans processed hard and spongy coal to plaster vaulted ceilings. In a powdery state of ash, it found employment in cement, dyestuff and glassworks, or as fertilizer.[38]

At mid-eighteenth century, French nobles were especially prominent among those who mobilized their assets to capitalize on the newly discovered coal fields in northern France.[39] And here, the promise of mining coal for its fertile ash was attractive – accessing the subsoil to bolster soil fertility. Recognizing an analogy with Dutch successes using peat ash for fertilizer, many nobles invested in coal mining to seek local substitutes that would free them from dependence on imported coal ash from Hainault in the Southern Netherlands and wood and peat ash from the Netherlands.[40] Their efforts were partly trig-

---

34 Rondo Cameron, "A New View of European Industrialization," *The Economic History Review* 38 (1985): 1-23.

35 Fernand Braudel, *The Wheels of Commerce*, trans. Siân Reynolds (Berkeley and Los Angeles: University of California Press, 1992), 328-29.

36 Morand, *L'Art d'exploiter*, vol. 2, no. 4, pp. 739-1356 (see note 10).

37 Morand, *L'Art d'exploiter*, vol. 2, p. 1195 (see note 10).

38 Roland de la Platière, "Abrégé historique de l'usage des cendres de tourbes superficielles & souterreines pour fertiliser les terres en Hainaut & dans la haute Picardie," *Art du Tourbier*, in J.E. Bertrand, *Descriptions des arts et métiers, faites ou approuvées par Messieurs de l'Académie royale des sciences, nouvelle édition*, vol. XIX (Neuchatel: L'Imprimerie de la Société Typographique, 1783), 472-565, on 546.

39 Marc Rouff, *Les mines de charbon en France au XVIIIe siècle* (Paris: Rieder et Cie, 1922); Reed Geiger, *The Anzin Coal Company: Big business in the early stages of the French Industrial Revolution, 1800-1833* (Newark: University of Delaware Press, 1974).

40 Roland de la Platière, "Abrégé historique," pp. 545-546 (see note 38); On Dutch practices, Joppe van Driel, "Ashes to Ashes: The stewardship of waste and oeconomic cycles of agri-

gered by recent geo-political developments. The War of the Spanish Succession ended in 1713 with new borders that delegated part of Hainault to the Austrian Hapsburgs, including the coal mining areas of Mons and Charleroi. Inhabitants on the French side of the border, long accustomed to securing supplies of wood-, peat- and coal ash via these areas, now faced fees that could be raised at will by foreign rulers.[41]

Noble French landowners, involved in managing vast tracts of arable fields, had a direct interest in relieving this insecure supply of fertilizers.[42] This led them to invest in coalmines for two reasons. First, coalmines granted access to coal that, when burned for fuel, left fertile ashes – the use of which as fertilizer for arable agriculture was widely documented in agricultural handbooks going back to the late sixteenth century.[43] Second, in some cases it was possible to mine specially targeted fossils that could be directly manufactured into fertilizers.

This last strategy gained import in 1753, when coal prospectors in Picardy found shallow deposits of a substance that spontaneously combusted. In a report on their discovery, the *intendant* of the local administrative center (at Soisson, near Laon) noted that "cultivators and laborers" had found a way to control this spontaneous combustion to produce blackish or reddish ashes that "contained salts specific to vegetation."[44] The report soon circulated widely. The *intendant* communicated the findings to the *Inspecteur-général des manufactures de Picardie,* who was charged with consulting the central Council of Commerce on new policies.[45] The *Inspecteur* appealed to cultivators and land managers across the country, to harness and experiment with the fertile fossils. As such, the report was taken up in Diderot's *Encyclopédie* in 1766 and included by the agriculturalist, inspector general of the marine and *Académie*

cultural and industrial improvement, 1750-1800," *History and Technology* 30 (2014): 177-206.

41 Roland de la Platière, "Abrégé historique," pp. 545-546 (see note 38); P.M. Jones, *The Peasantry in the French Revolution* (Cambridge: Cambridge University Press, 1988), 15.

42 Ibid., p. 15.

43 See e.g. Johann Coler, *Oeconomia oder Hausbuch*, 6 vols. (Wittenberg, 1593-1606); Gervase Markham, *Markhams Farwell to Husbandry or, The inriching of all sorts of Barren and Sterile grounds in our Kingdome, to be as fruitfull in all manner of Graine, Pulse, and Grasse, as the best grounds whatsoever* (London: Roger Iackson, 1625); Pieter van Ængelen, *De Verstandige Hovenier* (Doornick; Marcus Willemsz., 1659), Duhamel du Monceau, *Éléments d'agriculture*, 2 vols. (Paris: Guerin and Delatour, 1762).

44 Charles-Blaise de Méliand, "Houille," Diderot and d'Alembert, *Encyclopédie*, vol. 8 (1766), pp. 265-68 (see note 8).

45 Gillispie, *End of the Old Regime*, pp. 425-26 (see note 33).

des sciences member Henri Louis Duhamel du Monceau in his popular *Descriptions des arts et métiers* – reaching a wide audience of aristocrats and governors.[46]

By this time, more than fifty new coalmines, the explicit purpose of which was to mine for these self-combustible fossils, had been listed in the region. (see Table 2.1.) The fertilizers manufactured from their yields circulated as "*cendres de charbon de terre*", "*cendres d'engrais*" or "*houille d'engrais*" (the term *houille* was the word for coal used in Lorraine and around Liège). Farmers eagerly bought and used them as fertilizer in arable agriculture. To govern these and other coal mining activities, France's royal administration took an intermediary role, reconciling the drive toward 'improvement' with nobles' assets and the know-how of practitioners. As landowners, the noblemen of northern France were entitled to the land's surface, but the crown could lay claim to everything underneath. By a new decree of 1744 the government assumed control over the distribution of mining concessions, reaffirming that a royal permit was required to open a mine.[47] Officials prioritized noble requests, but only if they kept production going and hired experts to do the job. In this role, the administration stimulated the exploitation of Picardy's coalfield as a natural resource for the domestic production of fertilizers. For example, one nobleman acquired a concession to quarry all the land between the villages of Ham and Laon, including, as the official document read, the property of others who were currently "unable to undertake the exploitation," provided that he would mine for *cendres de charbon de terre*, consult other ash mining companies and hire proper "gens de l'art" from Flanders.[48]

The strategy worked. Already in 1766, administrative reports cited over 400 farmers using coal-based fertilizers in the region of Aisne in Picardy alone.[49] Subsequent surveys continued to cite their frequent use throughout northern France until well into the second half of the nineteenth century.[50] Stabili-

46 Roland de la Platière, "Abrégé historique" (see note 38); Charles-Blaise de Méliand, "Houille," pp. 265-68 (see note 44).

47 *Arrest du conseil d'etat du roy,* PORTANT *Règlement pour l'exploitation des Mines de Houille ou Charbon de terre. Du 14 Janvier 1744* (Lyon: P. Valfray Fils, 1744).

48 Germain Martin, *La grande industrie en France sous le règne de Louis XV* (Paris: A. Fontemoing, 1900), 158; Rouff, *Mines de charbon,* p. 245 (see note 39).

49 The following is based on the empirical information provided in M. Lenglen, "Etude de quelques particularités relatives à l'Histoire des Engrais," *Bulletin des Engrais* 10 (1937): 222-24, 257-59, 305-08, 318-20, 353-55, 380-82, 403-06, 449-53.

50 Christophe Dieudonné *Statistique du département du Nord,* vol. 1 (Douai: Marlier, 1804): 407-12; J.I. Pierre, *Chimie Agricole ou l'agriculture considérée dans ses rapports principaux avec la chimie,* fifth edition (Paris: Libraire Agricole, 1863), 489-90.

zation of mining, production and use of *cendres de charbon de terre*, however, was not straightforward. Its difficult history provides further evidence, of widespread engagement with mining coal for fertilizers, of the involvement of chemistry in these endeavors, and of the changing identity of coal in the period around 1800.

To begin, the new fossils were initially surrounded by much publicity, as newspaper announcements and advertising posters boasted of the fertilizing quality of what was mined – often without regard to the detailed rules for how to convert these fossils into fertilizers.[51] High promises could backfire, as some farmers soon started reporting on diminishing returns, voicing fears of long-term soil depletion. In his previously mentioned report, the *intendant* also acknowledged the occasional hesitations, citing local concerns that *cendres de charbon de terre* might harm the soil by their burning properties, that they would give a bad taste to crops and keep fodders green for too long, against the taste of the cattle.[52] One prominent physician joined the skeptics, namely Joseph Raulin (1708-1784), royal pensionary and inspector-general of mineral waters. Raulin argued in a published treatise that the self-combustible fossil, when turned into ash, developed a high concentration of vitriolic acid that was harmful to skin and eyes. When used as fertilizer in fodder cultivation, these properties might be communicated to the crops, causing cattle diseases. The mined substances were surely useful, he stressed, but precautions and strict guidelines were required.[53] Merchants who traded in Dutch wood- and peat ashes, operating close to the border with the Southern Netherlands (near Cambrai), mobilized such arguments to put pressure on local authorities. These, in turn, started circulating warnings on the potential dangers of mined coal ash (as opposed to the ashes retrieved as a byproduct from burning fuel). Meanwhile, at least one northern French municipality communicated a citizen petition to their provincial council with complaints about increased fire risks, recent disease outbreaks and sulfurous fumes damaging fruit trees, all linked to the nearby *cendrières* where the self-combustible fossils were transformed into fertilizers. Some agricultural societies tried to dismiss these protests on the ground that they were initiated by landowners with a vested interest in obstructing coal excavation on their lands. Others argued that bad experiences were caused by the fraudulent circulation of forged species.

51 Lenglen, "Etude" (see note 49).

52 Méliand 1766, "Houille," pp. 267-68 (see note 44).

53 Joseph Raulin, *Examen de la houille, considérée comme engrais des terres* (Paris: Vincent, 1775).

Meanwhile, chemically trained officials tried to steer ongoing practices in a good direction. They did so by classifying the mined fossils and formulating strict rules for identification and usage, tied to explicated improvement goals. Practicing chemists at the *Académie des sciences*, such as Jean Hellot and Gabriel Jars, took the lead in developing the administrative organization of both mining and the use of mined substances.[54] Combining chemical inquiry with diplomacy and political management, both traveled abroad to examine a wide array of mined substances and their organized extraction. In this context, Hellot experimented with *cendres de charbon de terre* in the small-scale setting of his garden.[55]

Another member of the *Académie*, Duhamel du Monceau, followed up on these experiments. After chemically analyzing excavated ashy coal substances in his own laboratory retort, Duhamel discussed their simultaneously natural *and* political attributes in his widely translated Éléments d'Agriculture (1762).[56] As such, he formulated detailed guidelines for finding, extracting and using *cendre de charbon de terre*. This particular "caustic earth", he stressed, "announces itself" by leaving an oily film covering any surface waters in its surroundings.[57] To convert it into fertilizer, it should be watered slightly, mashed and kneaded into cakes of seven inches in diameter, arranged to form conical structures, allowing well-ventilated, slow combustion that should last for up to three days, before being spread over the land in early spring, in a ratio of sixty to eighty pounds per acre.

Similarly, the *intendant* adopted sensuous chemical classifications in his original report, to summarize ongoing practices and extract rules for ascertaining quality and proper use of mined self-combustible fossils.[58] He classified the mined fossils as a salty and bituminous type of "fossil coal", verifiable by its sulfurous odor and the way it lit up when placed in the smoldering remains of a fire. Once the odor faded away, it could be used to cultivate grains, vegetables, vines and fodders, or to maintain meadows.

---

54 Gillispie, *End of the Old Regime*, pp. 425-429 (see note 33).

55 Ibid, pp. 427-433; Doru Todericu, "Les mines du pays de Liège dans les papiers du savant Français Jean Hellot (1685-1766)," *Technologia* 6 (1983): 61-68; Méliand, "Houille," p. 266 (see note 44).

56 Duhamel du Monceau, *Éléments d'agriculture*, vol. 1, pp. 182-186 (see note 43); Idem., trans. Philip Miller, *The Elements of Agriculture*, 2 vols (London: Vaillant, Durham and Baldwin, 1764), vol. 1,163-166.

57 Duhamel du Monceau, "Cette terre est caustique" [...] "s'annonce", *Éléments d'agriculture*, vol. 1, p. 183 (see note 43).

58 Lissa Roberts, "The Death of the Sensuous Chemist: The 'new' chemistry and the transformation of sensuous technology," *History and Philosophy of Science* 26 (1995): 503-529.

The written reports of chemically trained officials further circulated through a network of agricultural societies. This way, their descriptions of local practices and their chemical assessments were attached to articulated improvement goals. Tellingly, the initiative to establish these societies – some fifteen founded in 1761-1763 – came jointly from the noblemen Louis-François-Henri de Menon, marquis de Turbilly and the controller-general of finance Henri Bertin. The texts that these societies produced suggest that the initiatives to enroll coalmines in soil fertility management should not be understood solely as a response to international politics. Rather, they extend beyond the quest for import substitution to include broader concerns with oeconomy. The largely noble coal entrepreneurs sought to bridge tradition and change, tying productive investment to promotion of publicly beneficial arts and sciences.[59] On one hand, they sought to maintain feudal customs of landownership and land cultivation in the face of competitors keen to exploit subterranean resources. On the other, they joined an emerging network of self-proclaimed improvers, involving themselves in the formulation of oeconomic norms, values and goals.

For example, in a series of published treatises on *cendres de charbon de terre*, the baron Léon de Perthuis de Laillevault (1757-1818), member of the *Société d'agriculture de Meaux*, emphasized that efforts to promote their use first sprang from "the study of morality."[60] His writings exemplify the prevailing stance among improvers to present the maintenance of both soil fertility and society-wide improvement as a morally informed project, made possible by organized material exchange and chemical inquiry. Such exchange and inquiry were needed to build self-supporting moral communities, they believed, resilient to the vicissitudes of war and international competition.[61] Having visited several coalmines and farms while garrisoned in northern France as an army engineer, the baron decided to do his part by communicating his experiences, noting the confusion arising from the various terms indiscriminately used by other writers. He distinguished three types of mined earths: *peat*, brownish, light, mixed with plant materials and extractable from marshes; *houille*, a pyritic substance, darker, heavier, purer and extractable from strata as deep as

59 Lissa Roberts, "Geographies of Steam: Mapping the entrepreneurial activities of steam engineers in France during the second half of the eighteenth century," *History and Technology*, 27 (2011): 417-39, 421-423; Geiger, *The Anzin Coal Company* pp. 14-29 (see note 39); Rouff, *Les mines de charbon* (see note 39).

60 "L'étude de la Morale", Léon de Perthuis de Laillevault, *Expériences et nouvelles observations sur les houilles d'engrais* (Paris: Clouiser et Jombert, 1780), 1.

61 Ibid., pp. 1-17, 21, 37.

forty feet; and *charbon*, the darkest, heaviest, purest and deepest.[62] The baron described various applications of the ashes of *houille* that he had witnessed, including, beyond fertilization, the treatment of diseased cows with the fumes and waters flowing from *cendrières*.[63]

Duhamel also linked his chemical assessment of fertile fossils to a more inclusive political program of mixed social and material resource husbandry, centering around agriculture. As the historian Etienne Stockland has argued, his politics should *not* be seen as deriving from physiocracy, understood as a school of thought that theorized agriculture as the backbone of (national) surplus production. Rather, it was bound to a practice-based form of political *oeconomy*.[64] Duhamel promoted collaboration together with a hierarchical division of labor, in which landowners would enable trials; magistrates would protect participating tenants; naturalists would generalize local practices; clergymen would instruct their subjects based on published communication; and working farmers would develop insights through practice.[65] Defining agriculture as both a "science" and a "branch of government," he viewed such collaborative work as forming, "the true basis of commerce" *and* moral well-being, as it "estranges [inhabitants] from Vice" and spreads "sentiments of probity."[66]

Agricultural historians mostly remember Duhamel for propagating the work of Jethro Tull (1674-1741), who sought to circumvent manuring by introducing new sowing and ploughing devices. Yet in both his *Traité de la culture des terres suivant les principes de M. Tull* (6 vols. 1750-1761) and in his *Éléments d'agriculture*, Duhamel embraced fertilizers as central to his project and advertized mined coal ashes as part of an extensive list of fertilizer production. Tying ash-based fertilization to the promotion of discarded or previously neglected "litter", he called on every citizen to search for similar "hidden treasures."[67] Duhamel organized the search by distinguishing between fertilizers from the mineral kingdom (including mined *cendre de charbon de terre*, but also "the ashes of the fossil-coal burnt in glass-houses, brew-houses and other manu-

62 Léon de Perthuis de Laillevault, *Observations critiques sur un ouvrage intitulé "Examen de la houille, considérée comme Engrais des terres"* (Meaux: Charlie, 1777), 5-9.

63 "Lettre de M. le Marquis de Flavigny", Perthuis de Laillevault, *Expériences et nouvelles observations*, pp. 134-138 (see note 62); Idem., *Instruction familière, adressée aux grand et petits cultivateurs, Sur l'usage des Houilles d'engrais, des Tourbes & de leurs cendres* (Paris: Jombert et Clousier, 1781), 34.

64 Etienne Stockland, "'La Guerre aux Insectes': Pest control and agricultural reform in the French Enlightenment," *Annals of Science* 70 (2013): 435-460.

65 Duhamel du Monceau, *Elements of Agriculture*, vol. 1, pp. viii, xi-xii (see note 43).

66 Ibid., pp. vi, xiv.

67 Ibid., p. 180.

factories", "old rubbish of mortar", "rubbish of old walls", etc.) the vegetable kingdom (including "soot of Chimney-sweepers", "saw-dust", "sea-wreck", etc.) and the animal kingdom (including "the offals and cleansings of slaughter-houses", discarded leather "cuttings", "night-soil", etc.).[68]

Assembling a wealth of such discarded materials might have been straightforward, but they had to be carefully processed to take part in the enactment of the desired socio-material order. Duhamel provided rich information on how to carry this out in practice. Organizing his discussion along the two standard modes of contemporary chemical production – the wet and dry way, he noted that "sensible farmers" accumulated their litter in watered holes, then laid it to rot, while the extracted juices in the holes further "served to enrich and rot the fresh litter."[69] Alternatively, farmers stuffed decaying materials into kilns that were specially constructed to allow for slow burning, yielding ashes impregnated with fertile salts. (See Figure 1.) Such discussion was not uncommon among French *amateurs*. The marquis de Turbilly, patron of the provincial agricultural societies, for example, likewise acknowledged that when preparing "artificial manure [...] if we lack water, we must turn to fire."[70]

As valued by these writers, the mined fossils thus gained political attributes. When carefully selected and prepared, they could join other resources to maintain self-supporting moral communities. This vision, in turn, translated into expressions of what constituted good 'oeconomy', connected to values of diligence, thrift and avoidance of waste. Duhamel concluded that proper collaborative management prevented the resources embodied in materials, soils and people from being squandered. "Thus, the best advice I can give to good oeconomists, is first to get their plowed lands in proper order, before they think of breaking up wastes."[71]

While numerous aristocratic enthusiasts experimented with communicated techniques on their own estates, the oeconomic vision of mutually attuned social classes, materials and landscapes to which promoters attached these techniques obviously failed to materialize. But even if a durable interaction between agricultural societies, mostly occupied by local administrators and aristocrats, and practicing farmers was not achieved, the continued engagements with mining and using fertile fossils in northern France shows that individual successes were booked.[72]

---

68 Ibid., pp. 144, 151, 153, 166, 171-174.

69 Ibid., p. 182.

70 Marquis de Turbilly, *Memoire sur les défrichemens* (Paris: D'Houry, 1760), 122.

71 Duhamel du Monceau, *Elements of Agriculture*, vol. 1, p. 100 (see note 43).

72 Gillispie, *Science and Polity*, p. 370 (see note 33).

In the long run, however, these activities became disconnected from the Western-European coal business. During the first half of the nineteenth century, the regional involvement of coal mines in soil fertility management was increasingly overshadowed by the contemporary growth of mechanizing industry. As mining of coal for fuel increased, dramatically altering the socio-material landscape as it brought improvement for some, these once innovative practices faded into what was perceived as the homogenous fabric of rural traditionalism, gradually narrowing coal's recognized identity.

The historical trajectories that helped shape this development are vast and complex, as the daily use of coal, the political organization of mining and the classification of matter continued to be deeply entangled. Very briefly, a new series of coal-fired steam engines of the early nineteenth century began to allow for more efficient use of coal as fuel, making it an increasingly attractive power source in places where capital investment and access to coal converged – this last extended by expanding transport systems. This went hand in hand with increased consumption of coal as fuel in European industry, reinforced by contemporary industrial upscaling. In this context, the *Compagnie des Mines d'Anzin* quickly monopolized the resources to mine and sell French coal with the financial aid of Parisian private bankers and sustained governmental tariff protection.[73] As the uses of French coal were mounting, coal's recognized identity was both narrowed and generalized as fossil fuel. Meanwhile, at mid-nineteenth century, the fertile ashes once known as *cendres de charbon de terre* came to be referred to in purchase agreements and scientific textbooks as the very particular "*cendres noires de Picardie*".[74]

Conclusion

What, then, is coal? This essay provides at least two lessons that help answer this question in a historically meaningful way. The first takes us back to Timothy Mitchell's discussion of what he calls coal and oil's "socio-technical agencies."[75] As powerfully insightful as his analysis is, he begins by classifying these substances as carbon fuels, thereby black-boxing their material identities – even

73 Geiger, *Anzin Coal Company* (see note 39).

74 Pierre, *Chimie Agricole*, pp. 489-490 (see note 50); M. Andraud et al., *Dictionnaire du commerce et des marchandises*, vol. 1 (Paris: Guillaumin, 1837), 495; A. Payen et A. Richard, *D'Agriculture théorie et pratique a l'usage des écoles d'agriculture, des propriétaires et des fermiers*, vol. 1 (Paris: Hachette, 1851), 38-39.

75 Timothy Mitchell, *Carbon Democracy*, pp. 12-42 (see note 3).

as he integrates them in his analysis. By including the material character of coal in the historically evolving mix out of which its identity flows, this essay suggests the interpretive fecundity of recognizing that agency is at once social, technical and material. By mutually informing each other through time, these three aspects constitute an entity's historically evolving identity. Coal was not always already a carbon fuel, leading ineluctably to the socio-technical promises and threats of the industrial world. Rather, its identity evolved over time through interaction with and use of its embodied properties (themselves quite situational in character) in the field, the laboratory and the workshop, on the printed page, in the marketplace, courts and legislative assemblies.

This first lesson thus responds to a broad tendency to consider a substance's identity as providing a source of historical explanation, based on its purportedly ahistorical character. The argument made here is that this is precisely what needs to be investigated because the entwinement of the material with the social and the technical renders identity innately historical. The second lesson relates to the historical consideration of substances as commodities. This essay illustrates the struggles entailed in transforming coal from an inchoate field of confusedly classified material substances into more clearly identified and regulated substances liable to profitable excavation and exploitation. So too does it document contestations and alternatives. In doing so, it reminds us that considering substances such as coal as 'resources' should not be done in a narrow economic sense.[76] Neither is it warranted to focus narrowly on the historical links between material 'resources' such as coal and the technological contrivances with which they were processed and put to use. In both these cases, a further step often involves speaking of 'human capital' whereby people, along with their knowledge and skill are viewed as one more set of resources that contributed to economically measured productivity. This essay has instead shown how human actors and the socio-material networks in which they were engaged sometimes worked toward goals other than economic progress for its own sake, *oeconomically* seeking to steward human and material 'resources' in ways that benefited both nature and society. Given that concerns with the combined economic and environmental crises we currently face have led to paying increased attention to the era covered by this essay and volume of which it is a part – viewing it as the birthplace of modern industrialization, environmental decline and socio-geographical inequalities – recognizing the past existence of such alternatives might prove salutary as we plan for our future.

76 Lissa Roberts, "Producing (in) Europe and Asia, 1750-1850," *Isis* 106 (2015): 857-865.

TABLE 2.1 *Coal mines exploited for fertilizers ("des mines de terre de houille")*

5	Location	Opened in*	Region	Source
1	Beaurain	1753	Picardy, Aisne	A, p. 265
				B, p. 546
				E, p. 44
				J, p. 15
				K, p. 395
				N, p. 95
				O, p. 524-25
				P, p. 185
2	Suzy, Faucoucourt, Cessieres and Lizy	1756	Picardy, Aisne	A, p. 265
				B, p. 546
				E, p. 44, 55
				I, p. 68
				P, p. 185
				R, p. 28
				T, p. 374
3	Armay	1750s	Picardy, Aisne	B, p. 545
4	Annoy and Rumigny	1760	Picardy, Somme	A, p. 265
				B, p. 545
				E, p. 44
				I, p. 67
				J, p. 15
				P, p. 185
5	Jussy	1760	Picardy, Aisne	I, p. 67
				J, p. 15
				O, p. 524-25
				P, p. 185
				R, p. 28
6	Hinacourt	1760	Picardy, Aisne	I, p. 67
				J, p. 15
7	Sissonne	1761	Picardy, Aisne	I, p. 65-66
8	Anvi and Montigny	1753-1761	Picardy, Amiens	C, p. 186
				D, p. 195.
9	Villé and Breuil	1762	Bas-Rhin	I, p. 65
				R, p. 28

TABLE 2.1 *Coal mines exploited for fertilizers ("des mines de terre de houille") (cont.)*

5	Location	Opened in*	Region	Source
10	Itancourt	1767	Picardy, Aisne	E, p. 44 G, p. 407 I, p. 67 R, p. 28
11	Vendeuil	1769	Picardy, Aisne	E, p. 44 I, p. 67 P, p. 185 R, p. 28
12	Homblières	1771	Picardy, Aisne	E, p. 44, 69 G, p. 407 I, p. 67
13	Beaurieux	1772	Picardie, Aisne	F, p. 404 I, p. 65 T, p. 374
14	Travecy	1753-1772	Picardy, Aisne	B, p. 546 H, p. 216 R, p. 28 E, p. 44 I, p. 67 P, p. 185
15	Rheims (direction Soissons)	1753-1772	Marne	E, p. 45 Q, p. 357
16	Canly	1775	Picardy, Oise	N, p. 95 O, p. 524
17	Baurin	1753-1775	Picardy, Somme	F, p. 397 G, p. 407
18	Chavignon	1753-1776	Picardie, Aisne	E, p. 44, 69 F, p. 403
19	La Fère / Charmes	1777	Picardy, Aisne	B, p. 546 I, p. 68 P, p. 185 R, p. 28
20	Mezy-Moulins and Passy-sur-Marne	1779	Picardy, Aisne	E, p. 45, 54, 69, 72, 211 I, p. 64 R, p. 28 T, p. 374

5	Location	Opened in*	Region	Source
21	Mailly	1753-1780	Picardy, Somme	B, p. 546 E, p. 44, 69-70
22	Golancourt	1753-1780	Picardy, Oise	B, p. 546 E, p. 44 O, p. 524-25 P, p. 185
23	Liez	1753-1780	Picardie, Aisne	B, p. 546 E, p. 44 I, p. 67 P, p. 185
24	Luzancy	1753-1780	Brie, Seine-et-Marne	E, p. 43, 54, 102, 118
25	Marlemont	1753-1780	Champagne, Ardennes	E, p. 41, 45, 118
26	Aubigny	1753-1780	Picardy, Somme	E, p. 41, 45, 118
27	Ecogne	1753-1780	Champagne, Ardenne	E, p. 118
28	bois de Prie (Mésières)	1753-1780	Champagne, Ardenne	E, p. 41, 45, 118
29	Muirancourt	1753-1780	Picardy, Oise	E, p. 44, 71 K, p. 395 O, p. 524-25
30	Beuvraignes	1753-1780	Picardy, Somme	E, p. 44 O, p. 524-25
31	Benet	1753-1780		E, p. 44
32	Thiérache	1753-1780	Picardy, Aisne	E, p. 45
33	Rocroi	1753-1780	Champagne, Ardenne	E, p. 45, 55, 118, 212
34	Saint-Aude	1753-1780		E, p. 90, 104, 118,
35	Vandeuil	1753-1783	Champagne, Marne	B, p. 546
36	Bassay	1753-1783		B, p. 546
37	Hinnacourt	1753-1783	Picardy, Oise	B, p. 546
38	Lambays	1753-1783		B, p. 546
39	Le Santerre	1753-1783	Picardie, Somme	B, p. 546

TABLE 2.1 *Coal mines exploited for fertilizers ("des mines de terre de houille") (cont.)*

5	Location	Opened in*	Region	Source
40	Roye	1753-1783	Picardie, Somme	B, p. 546 M, p. 827-28
41	Chaulnes	1753-1783	Picardie, Somme	B, p. 546
42	Rollot, Montdidier	1753-1783	Picardie, Somme	B, p. 546 O, p. 524 T, p. 374
43	Verberie	1753-1783	Picardie, Oise	B, p. 546 L, p. 20 O, p. 525
44	Compiègne	1753-1783	Picardie, Oise	B, p. 546 N, p. 95 O, p. 524 T, p. 374
45	Benay	1753-1787	Picardy, Aisne	I, p. 67 P, p. 185
46	Urcel	1753-1796	Picardy, Aisne	I, p. 63-64, 66
47	Mesnil-Saint-Laurent	1753-1796	Picardy, Aisne	I, p. 67
48	Hallencourt	1753-1796	Picardy, Somme	I, p. 67
49	Cerisy	1753-1796	Picardy, Somme	I, p. 67
50	Lisfontaine	1753-1796	Picardy, Oise	I, p. 67
51	Gibercourt	1753-1796	Picardy, Oise	I, p. 67
52	Eaucourt	1753-1796	Picardy, Somme	I, p. 67
53	Cugny	1753-1796	Picardy, Aisne	I, p. 67
54	Blérancourt	1779-1796	Picardy, Aisne	I, p. 68 T, p. 374
55	Arsy	1775-1812	Picardy, Oise	N, p. 95 S, p. 108
56	Jonquières	1775-1812	Picardy, Oise	N, p. 95
57	Moyvillers	1775-1812	Picardy, Oise	N, p. 95
58	Remy, Estrées-Saint-Denis	1775-1812	Picardy, Oise	N, p. 95

* When a source mentions a specific year of opening or concession for a mine, this year is listed in the table. Otherwise, the period in which the mine opened is estimated by reference to the earliest known source that cites the mine.

Sources

A. De Méliand, Charles-Blaise. "Houille." In Diderot and d'Alembert, eds., *Encyclopédie, ou Dictionnaire raisonné des sciences, des arts et des métiers*, vol. 8, 265-68 (Paris: Briasson, 1766).

B. De la Platière, Roland. "Abrégé historique de l'usage des cendres de tourbes superficielles & souterreines pour fertiliser les terres en Hainaut & dans la haute Picardie," *Art du Tourbier*, in J.E. Bertrand, *Descriptions des arts et métiers, faites ou approuvées par Messieurs de l'Académie royale des sciences, nouvelle édition*, vol. XIX (Neuchatel: L'Imprimerie de la Société Typographique, 1783), 472-565.

C. Duhamel du Monceau, Henri-Louis. *Éléments d'agriculture* (Paris: Guerin and Delatour, 1762).

D. Duhamel du Monceau, Henri-Louis. *Traité de la culture des terres.* Vol 6. (Paris: Guerin and Delatour, 1761).

E. Perthuis de Laillevault, Léon. *Expériences et nouvelles observations sur les Houilles d'engrais* and *Recherches sur la houille d'engrais et les houilleres* (La Haye: Clousier et Jombert, 1780).

F. *Dictionnaire minéralogique et hydrologique de la France.* Vol. 4 (Paris: Brunet, 1776).

G. Review of Raulin, Joseph. "Examen de la houille, considérée comme engrais des terres", *Journal Encyclopédique.* Vol. 6 (Bouillon, 1775), 406-418.

H. *Courier du Bas-Rhin* 1772.

I. "Suite du Tableau des Mines et Usines de la France," *Journal des Mines.* No. 25 (1796): 49-73.

J. Brayev, B.L., *Statistique du département de l'Aisne* (Laon: Melleville, 1825).

K. *Journal des mines, ou recueil de mémoires sur l'exploitation des Mines, et sur les Sciences et les Arts qui s'y rapportent.* Vol. 12 (Paris: Bossange, Masson et Besson, 1801).

L. Grave, Louis. *Précis statistique sur le canton de Pont-S.[TE]-Maxence, arrondissement de Senlis (Oise)* (1843).

M. Tessier, Alexandre-Henri. *Encyclopédie méthodique: Agriculture.* Vol. 2 (Paris: Panckoucke, 1791).

N. Graves, Louis. *Précis Statistique sur le canton d'Estrées-St-Denis, arrondissement de Compiègne (Oise.)* (1832).

O. De la Platiere, Roland. *Art du Tourbier*, in J.E. Bertrand, *Descriptions des arts et métiers, faites ou approuvées par Messieurs de l'Académie royale des sciences, nouvelle édition*, vol. XIX (Neuchatel: L'Imprimerie de la Société Typographique, 1783), 471-558.

P. Monnet, Antoine-Grimoald. "Memoire Sur la Terre pyriteuse qui se trouve en Picardie & dans le Soissonnois, & sur les moyens qu'il y a d'établir des Fabriques de Vitriol avec cette matiere", in Rozier, ed. *Observations sur la physique, sur l'histoire naturelle et sur les arts*. Vol. 11 (Paris: Bureau du Journal de Physique, 1778), 183-186.

Q. *Dictionnaire minéralogique et hydrologique de la France*. Vol. 2, no. 2 (Paris: Costard, 1772).

R. *Statistique générale et particulière de la France et de ses colonies* (Paris: Buisson, 1804).

S. Graves, Louis. *Précis statistique sur le canton de Liancourt, arrondissement de Clermont (Oise)* (1837).

T. Cuvier, F.G., ed. *Dictionnaire des Sciences Naturelles,* vol. 26 (Strasbourg: Levrault, 1823).

CHAPTER 3

Capturing the Invisible: Heat, Steam and Gases in France and Great Britain, 1750-1800

Marie Thébaud-Sorger

Over the course of the eighteenth century the common perception of air, that invisible but omnipresent element of nature, experienced a profound change. This essay argues that a common field of knowledge emerged through the materialization of aerial fluids, including gases, steam and heat. This topic inspired the creativity of a hybrid milieu of practitioners, who extended the investigation of air while embedding it in public concerns. A growing culture of consumption, especially in urban contexts in France and Britain, helped nurture a number of new devices and apparatus aimed at mastering these fluids for various purposes and in everyday life. They offered the capacity to reshape the interplay between scientific results, social needs and political incentives, presenting new horizons for the public good and public health.

This essay reformulates assumptions (and raises questions) regarding the 'sites' where new approaches to air were forged. From the seventeenth century, air was closely linked to the rise of the experimental sciences. Calculating the weight and pressure of the air and understanding the vacuum, were crucial to a change of perception embodied in devices such as the air-pump. Imbuing immaterial air with a new kind of materiality fostered the emergence of "a set of practices which centered on the climate, meteorology, the atmosphere and electricity." According to Simon Schaffer, "aerial philosophy" played a major role in this change and "acted as a wider and grander theater of power and *also* as a space in which a new economy of understanding and control might operate."[1] Discoveries in the 1770s of various elastic fluids challenged and diffracted the category of "air" as a unified, natural body and, together with the identification of various gases, reframed the growing field of pneumatic chemistry.[2] However, while the role of prominent European chemical practitioners in this story is familiar, this essay explores how investigations of air engaged the creativity of a less familiar and more heterogeneous set of practitioners,

1 Simon Schaffer, "Natural Philosophy and Public Spectacle in the Eighteenth Century," *History of Science* 21 (1983): 1-43, on 16.

2 Joseph Priestley, *Experiments and Observations on Different Kinds of Air* (London: W. Bowyer and J. Nichols, 1774).

DOI 10.1163/9789004325562_005

who engendered novel communities and new audiences around their inventions and devices.

To do this entails a methodological displacement through a focus on the sociomateriality of the devices through which this broad range of practitioners contained, investigated and manipulated aerial fluids, including small-scale inventions, machines and commodities such as lamps, ventilators, gas masks, firebombs, "œconomic" stoves and furnaces. Discussion of specific cases draws on a body of printed ephemera – leaflets, advertisements, subscriptions, short essays, trade cards – and project descriptions sent to various societies, academies and societies of arts. The essay thus moves beyond tired distinctions between scientific and more commercial practices which were in fact closely interconnected. In this account the manufacture of certain technical inventions may be seen to have operated as both what Rheinburger has called an "experimental arrangement" and as a social and epistemological one.[3] Considering "air" as a boundary object helps to formulate new assumptions about the epistemic nature of the devices that materialized invisible aerial fluids for a wide range of audiences.[4] This enables a more general questioning of changes in perceptions of nature at the end of the eighteenth century, and the relationship between material and knowledge production, which entailed the co-construction of an investigative field. Mapping the intellectual and social milieux in which people engaged with materiality through the making, use and understanding of small-scale devices and the substances they contained reveals how changes occurred at the level of daily practices.

This essay centers on an understanding of aerial fluids through technical work that practically interacted with air qua matter. André Leroi-Gourhan's anthropology of techniques is an inspirational source for the essay's approach.[5] The specificity of aerial fluids (such as rarefied air, noxious air, expanded air, inflammable air) inspired specific operative work on the shape and composition of devices (such as containers and vessels) designed to capture, investigate and make use of the fluids. Reinterpretating chemistry from the angle of technology recalls the pattern of Catherine Jackson's 'glass revolution' – the way in

3 Hans-Jörgen Rheinberger, "Experiment, Différence and Writing I," *Studies in History and Philosophy of Science* 23 (1992): 305-331, on 309.

4 Susan Star, James Grisemer, "Institutional Ecology, 'Translations' and Boundary Objects: Amateurs and professionals in Berkeley's Museum of Vertebrate Zoology, 1907-39," *Social Studies of Science* 19 (1998): 387-420.

5 André Leroi-Gourhan, *L'homme et la matière* (Paris: Albin Michel, 1943).

which materially embodied processes eventually transformed the whole understanding of 'air'.[6]

The essay begins by examining how, on both sides of the English Channel, practitioners proposed a diverse array of commodities and objects that emerged from their investigations into air. These offered practical improvements in public and private everyday life through different means of mastering aerial fluids. Subsequent sections consider how the effects of these objects on air reshaped the social and material arrangements in which they were developed. Focusing on inventive practices makes clearer transfers of skills and hybridizations that occurred between entrepreneurs, amateurs and craftsmen. The careers of Ami Argand and Flavien Marie Scanegatti demonstrate this with particular clarity. While material devices provided the means to do experiments and create narratives about air, they also raised questions regarding both the social and intellectual boundaries of chemistry along with the knowledge they helped to forge.

Sites of Air's Commodification

Historians have recently attended to an "aerial sensibility" that emerged over the course of the eighteenth century, focused particularly on environmental medicine – a field that connected issues of health, a global understanding of weather and an emergent pneumatic chemistry.[7] The changing understanding of 'airs' in this period has been extensively studied from the point of view of the prominent European chemists of the period.[8] Studies of the intersection

6 Catherine M. Jackson, "The 'Wonderful Properties of Glass'," *Isis* 106 (2015): 43-68. Technology will be used here with its original meaning of *techno-logos*, "science of operations" or "science of the arts", rather than "applied science". See Eric Schatzberg, "From Art to Applied Science," *Isis* 103 (2012): 555-563.

7 Simon Schaffer, "Measuring Virtue: Eudiometry, enlightenment and pneumatic medicine," Andrew Cunningham and Roger French, eds., *The Medical Enlightenment of the Eighteenth Century* (Cambridge: Cambridge University Press, 1990), 281-318; Marco Beretta, "Pneumatic vs. 'aerial Medicine': Salubrity and respirability of air at the end of the eighteenth century," *Nuova Voltiana: Studies on Volta and his time*, Pavia, 2 (2000): 49-71; Vladimir Jankovic, *Confronting the Climate. British air and the making of environmental medicine* (New York: Palgrave Macmillan, 2010); Jan Golinski, *British Weather and the Climate of Enlightenment* (Chicago: University of Chicago Press, 2011).

8 Bernadette Bensaude-Vincent, Isabelle Stengers, *Histoire de la chimie* (Paris: La Découverte, 1993); Marco Beretta, *The Definition of Chemistry from Agricola to Lavoisier* (Canton, MA: Science History Publications, 1993).

of their practices with the social milieux of instrument-makers and artisans have shed light on dynamic collaborations but also revealed tensions and ways in which knowledge was challenged by new measurement practices.[9] However, little attention has been paid to the role of inventive practices that were supported by a growing market for new consumer commodities. Entrepreneurs, amateurs, architects, physicians and a large community of skilled artisans including instrument-makers, stove and pump manufacturers, distillers, coppersmiths and tinsmiths all engaged with aerial substances. They designed objects dedicated to mastering the elements, including a wide range of devices for cooking, heating rooms, preventing and extinguishing fires, fumes and every type of suffocation. They used steam and included small appliances dedicated to hygiene and comfort (steam baths, ventilators, lamps, portable stoves, and so on). This "commodification of air", recently noted by Vladimir Jankovic, spread in the second half of the eighteenth century.[10] It fostered public interactions in various sites: public lectures, workshop displays, repositories of the new sociabilities of improvement, demonstrations and trials occurring *in situ* – on river banks, in gardens, parks, at a cesspool. Audiences interacted at these sites with devices that have not usually been considered as tools of pneumatic chemistry investigation; nor have these audiences been recognized as participants in a new ontology of air, which emerged at this time.

Recognizing changes in the culture of consumption permits a different approach to these sites so that, rather than exploring separate spaces, it is possible to ask how evolving practices forged links between various spaces such as laboratories, workshops, factories and public squares. Diverse fields of action, such as medicine, architecture, urban safety and industry, which have until now been studied separately, are thereby seen as connected. Indeed, from this perspective, aerial fluids can be seen to have entailed a common approach. Writing in the 1940s, the anthropologist André Leroi-Gourhan proposed that fluids ought to be considered as matter if they could be decanted and poured. This entails expanding the category 'fluids' beyond liquids to include all kinds of containable materials. Leroi-Gourhan's interest in prehistoric tools in the

9 Larry Stewart, "'Ordinary' People and Philosophers in the Laboratories and Workshops of the Early Industrial Revolution," Margaret C. Jacob and Catherine Secretan, eds., *In Praise of Ordinary People: Early Modern Britain and the Dutch Republic* (New York: Palgrave Macmillan, 2013), 95-122; Lissa L. Roberts, "The Death of the Sensous Chemist: The 'new' chemistry and the transformation of sensuous technology," *Studies in History and Philosophy of Science Part* 4 (1995): 503-529; H. Otto Sibum, "Les gestes de la mesure. Joule, les pratiques de la brasserie et la science," *Annales Annales. Histoire, Sciences Sociales* 4-5 (1998): 745-774.

10 Jankovic, *Confronting the Climate*, p. 70 (see note 7).

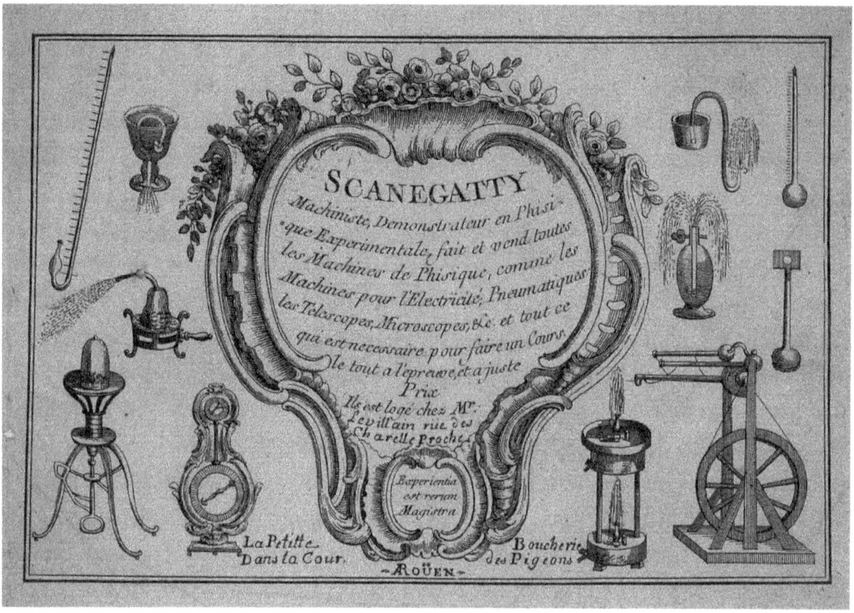

FIGURE 3.1 *Anon., Trade Card of Scanegatty, Machine Inventor and Demonstrator of Physics Apparatus, c. 1775; etching and engraving on paper; 160 × 202mm; Waddesdon, The Rothschild Collection (The National Trust) Bequest of James de Rothschild, 1957; acc. no. 3686.1.65.123. Photo: University of Central England Digital Services* © THE NATIONAL TRUST, WADDESDON MANOR.

1940s led him to single out a pattern of technological action, distinguishing social-technical human activities according to their relation to different types of matter. This approach proposed a classification of various "gestures" and "operations" that are done to matter. Consequently, Leroi-Gourhan defined fluids by the processes developed to enable their containment and circulation, that is, through the vessels in which they were contained. For certain eighteenth-century practices, diverse forms of aerial fluids, vapors and emanations might be understood in this way, through the objects and techniques used to capture, investigate and manipulate them.

The trade card of Flavien Marie Scanegatti, an Italian-born instrument-maker who settled in the French city of Rouen in the early 1760s, perfectly illustrates this perspective (Fig. 3.1).[11] It depicts the instruments that could be found in his shop, described two decades later by Arthur Young as a "considerable room furnished with mathematical and philosophical instruments and

11 Waddesdon Manor Trade Cards Collection, Acc. No. 3686.1.65.123, "Trade Card of Scanegatty, Machine Inventor and Demonstrator of Physics Apparatus," n.d., c. 1775.

models."[12] Scanegatti used these in the public physics lectures he advertised regularly in gazettes, where he performed curious experiments for local audiences with electricity, hot air and inflammable air. His trade card presented various instruments including a barometer and thermometer, and demonstration apparatus such as an air pump, an æoliopyle on a stand, an electrical machine with a wheel, a glass tube with measurements and a pot of water with a siphon. The *Journal de l'abbé Rozier* reported on other Scanegatti inventions such as an *aéromètre* or *pèse-liqueur* (scales designed to precisely measure the weight of several liquid substances), and, in 1774, new medical tools such as an improved "fumigatory apparatus" to save people from drowning. Scanegatti's activities extended beyond his shop door, taking in industrial sites for the needs of local production (textiles machines, enamel at Saint Sever); industrial chemistry, with the development of a lead chamber for sulfuric acid production; and a plaster kiln fueled by coal – an invention for which he performed experiments in front of local elites and in academic accreditation exams for the *Bureau du commerce* and the *Académie royal des sciences*.[13] He also undertook the scientific supervision of balloon experiments in 1784 on behalf of the Avignon city council, where he had been called to work on hydraulic projects.[14] There were continuities across this diverse and impressively creative activity, which has normally led to him being described as eclectic.[15] Motivated by utility and industry, a good part of the inventions he designed, despite their heterogeneous appearance, were connected by a common desire to capture fluids and enable their transformation.

Situated between what is retrospectively identifiable as philosophical curiosity and craft-based practicality, the creative activities of Scanegatti appear exemplary of an intermediary world, of interactions between local academies, artisanal elites and industrial networks. The career of the entrepreneur Ami Argand followed a similar trajectory, although Argand's reputation was not

---

12 Arthur Young, *Travels in France, During the Years, 1787, 1788, 1789* (London: George Bell and Son, 1909), 142 (5th October 1788). Thanks to John Perkins for his help on Scanegatti.

13 Archives Nationales de France, Paris (subsequently A.N.F), series F12-2380, Scanegatti, "Application du charbon de terre à la cuisson du plâtre," 1786 à 1788; Archives of Cnam, T.666, 24, p. 2 engravings.; Academie des sciences archives, Paris, "Pochette de séance, 31st January 1787: Fourneau pour cuire le plâtre de Scanegati," (Report from Vandermonde, Sage et Monge, 3rd February 1787).

14 *Courrier d'Avignon* 26 (Tuesday 30th March 1784): 108; Air and Space Museum Archives, Le Bourget, Montgolfier folders, XV-33, "Letter of the marquis de Brantes to Joseph de Montgolfier, Avignon, 7th April 1784."

15 Frédéric Morvan-Becker, "L'École gratuite de Dessin de Rouen, ou la formation des techniciens au XVIIIe siècle" (PhD Thesis, Université Paris VIII-Saint Denis, 2010), 779-790.

built up in one place, but through traveling across Europe to develop his invention of a lamp using double air currents.[16] The son of a Geneva watchmaker who was educated by Horace-Bénédict de Saussure, Argand's interests covered a wide range of connected chemical and mechanical processes, such as distillation, combustion and evaporation. He made the acquaintance of the Montgolfier brothers, and Etienne Montgolfier engaged his help with hot-air balloon experiments in Paris in autumn 1783.[17] Louis-Paul Abeille, a former factory inspector and at that time the French royal government's Secretary of Commerce, recorded his visit to their open-air workshop situated in the garden next to Jean-Baptiste Reveillon's famous wallpaper factory in the Faubourg Saint Antoine. Abeille described a diverse community of actors surrounding the balloon prototype, including state and city administrators, entrepreneurs such as Argand and Reveillon, a foreman of the factory Giroud de Villette, apothecaries such as Quinquet and Lange, and Meusnier de la Place, a mathematician and correspondent of the *Académie Royale des sciences*, not to mention neighbours and curious onlookers.[18] Leaving Paris for England, Argand sought the support of Boulton and Watt, who helped him take out a patent for his lamp that he later lost in a trial. Returning to France he obtained a *privilège* to set up a lamp manufactory in Versoix in 1787.[19] Projecting for the lamp in Birmingham and London was punctuated by other experiments – first with air-pumps and gas balloons, which were shown to George III at Windsor in November 1783, and then with the large-scale production of inflammable air for Lunardi and Blanchard's public balloon ascents made in London in 1784.[20] Argand moved through a variety of different contexts in which aerial fluids

16 Michael Schrøder, *The Argand Burner: Its origin and development in France and England, 1780-1800: an epoch in the history of science illustrated by the life and work of the physicist Ami Argand, 1750-1803* (Odense: Odense University Press, 1969).

17 Charles Coulston Gillispie, *The Montgolfier Brothers and the Invention of Aviation, 1783-1784* (Princeton, NJ: Princeton University Press, 1983); Marie Thébaud-Sorger, "Amitiés, entraides et circulations techniques: les affinités électives de l'entrepreneur Argand," Michel Cotte, ed., *Circulations techniques, en amont de l'innovation: Hommes, objets et idées en mouvement* (Montbeliard: Presses Universitaires de Franche-Comté – UT Belfort-Montbéliard, 2004), 111-128.

18 Louis-Paul Abeille, *Découverte des lampes à courant d'air et à cylindre par M. Argand* (Geneva: 1785), 13.

19 John Wolfe, *Brandy, Balloons, and Lamps: Amy Argand, 1750-1803* (Carbondale: Southern Illinois University Press, 1999).

20 Air and Space Museum Archives, Le Bourget, Montgolfier folders, XIII-39-45: letters from Argand to Etienne de Montgolfier, 1783-1785.

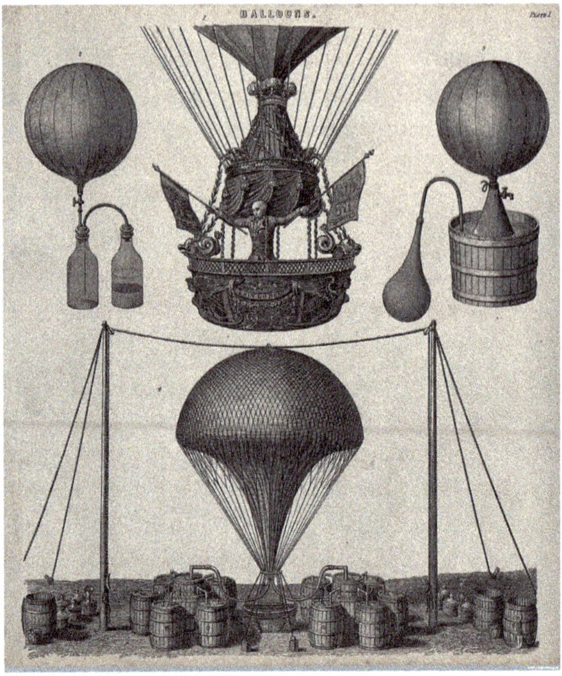

FIGURE 3.2
Balloons. Engraving. [*s.n.*] [*S.l.*]. *Wellcome Library, London, Iconographic Collections, ref.* ICV *No 41432.*

were worked on and exhibited, including factories, gardens, public squares, shops and workshops.

What was original in this period was the way a dynamic set of actors from various backgrounds, including entrepreneurs, instrument-makers, craftsmen, merchants, servants, noble amateurs and apothecaries, moved from one field of practice to another, transferring skills and knowledge as they did so. They thereby helped to establish common interests among large audiences, which fostered intersections between the pleasure of curiosity and industrial achievements, private comfort and public concerns. These intersections created a constellation of "sites" around their objects where the same processes could be staged and reinterpreted in a variety of ways. Many practitioners, such as Argand and Scanegatti, maintained complex relationships with academics, seeking out support, recognition, institutional and social credit and their further involvement in projects. Their practices were shaped by porous institutional, political and territorial boundaries and by intellectual tensions, which differed between Britain and France. In France, for instance, the Academy's pre-eminence in the evaluation of inventions was on the increase.[21] But

21 Liliane Hilaire-Pérez, "Technical Invention and Institutional Credit in France and Britain in the Eighteenth Century," *History and Technology* 16 (2000): 295-306.

while local configurations diverged in many respects between Britain and France, they were nevertheless forged by symmetrical dynamics.

In the 1760s fire hazards, noxious air, hygiene and health became prominent topics in society. Individuals engaged with these issues prompted a wide circulation of processes and forged new audiences, encountering in particular the interest of "alternative" new sites of sociability, such as the prosperous London Society of Arts, founded in 1757, and many similar societies across Europe. Scanegatti promoted a scheme to open a *Société des arts* in Rouen affiliated with the Parisian *Société d'émulation pour l'encouragement des arts* (1777-1782), an ephemeral society founded by a physiocrat, the Abbé Baudeau.[22] Embodying a public consisting of mixed audiences of aristocrats, savants, elite artisans and men who exercized the machinery of power, these societies were particularly concerned with issues related to the public good, while at the same time increasing support for patriotic and capitalist businesses.[23] Offering new spaces of legitimation, they launched contests, offered rewards and issued medals. They were sensitive to supporting all kinds of artifacts in which technological improvement and the management of global reform were at stake, such as ventilation and rescue apparatus, extinguishers and fire engines, stoves and furnaces. For instance, William White Esquire approached the special committee established by the London Society of Arts on "hand ventilators" with an "air machine" of his invention. (The committee ran for several years, organizing a prize for this category in 1791-1792.) Besides being exhibited in his workshop, White's machine was on display in the society's repository, alongside similar devices for comparison.[24] White also approached the Humane Society, which gave him several accreditations in support of the development of a patent.[25] These efforts coincided with the growth of philanthropic societies dedicated to resuscitating drowned people in Amsterdam, Hamburg,

22 A.N.F., T*160.5, *Grand registre du secrétariat de la société d'émulation de l'abbé Beaudeau depuis 1778 à 1782*, "Mémoire de Mr Raymond de St Sauveur sur une proposition de Mr Scanégatti de Rouen sur un comité provincial a établir en cette ville, Retenu au comité d'inspection, 10 fev. 1778."

23 Liliane Hilaire-Pérez, *L'Invention technique au siècle des Lumières* (Paris: Albin Michel, 2004).

24 *Diary or Woodfall Register* (1792) Tuesday 5th June.

25 London Metropolitan Archives, 4517/A/01/01/001, *Minutes of the Humane society*, meetings of the 30th January 1776 and 7th May 1783; Royal Society of Arts, Society of arts archives, Loose letters, PR.MC/101/10/512, "Letter from William White about his air machine," 30th January 1792; *Extracts from the reports of the Royal Humane Society, with certificates, letters, & c. which fully evince the utility of an air machine, or patent ventilator; invented and sold by William White* (1794).

London, Paris and Geneva.[26] Founded mostly by physicians, they dealt with issues of asphyxia and reanimation and scrutinized various inventions including gas masks, fumigation boxes and fumigation apparatus, such as that improved by Scanegatti, or conceived by the physician and alderman Philippe-Nicolas Piat in Paris in the 1770s.[27]

In England and France, a similar concern emerged regarding risk and safety, the mastering of new energies and a shared desire to discipline urban areas.[28] Identifying technical devices with the capacity for improvement enabled a new marketing dynamic based on claims to be able to transform the immediate environment. Through the commercialization of various commodities, apparatus and objects, a new material culture took root in the social and urban landscape of the eighteenth century that gradually changed people's relationship to the elements of nature. Infused with this aerial knowledge for large audiences, for whom the natural world was submitted to intensive changes, each invention intersected with these different spheres. Producing an apparatus or new commodity using 'airs' was the result of a complex reception and production process that made their achievement possible and, in return, shaped communities around their materiality.

Devices Shaping Communities

Through the design of devices an evolving sociomateriality situated the "air" in a densely populated field that highlighted the possibilities offered by aerial

26 Luke Antony Francis Davison, "Raising up Humanity: A cultural history of resuscitation and the Royal Humane Society of London, 1774-1808" (PhD Thesis, University of York, 2001).

27 Philippe-Nicolas Pia, *Détails des succès de l'Établissement que la ville de Paris a fait en faveur des personnes noyées* (Paris: S. Yves, 1773) with an engraving of his "Boîte-entrepôt"; "Boite fumigatoire de Gardanne," *Gazette du commerce, de l'agriculture et des finances* 84 (1778): 668-669 (20th October); On the disinfecting devices of Guyton de Morveau and Dumotiez, see the essay by Elena Serrano in this volume.

28 Sabine Barles, *La ville délétère. Médecins et ingénieurs dans l'espace urbain, XVIIIe- XIXe siècles* (Seyssel: Champvallon, 1999); Catherine Denys, *Police et sécurité au XVIIIe siècle dans les villes de la frontière franco-belge* (Paris: L'Harmattan, 2002); Marie Thébaud-Sorger, "Innovation and Risk Management in Late Eighteenth-Century France: The administration of inventions in French cities at the end of the *ancien régime*," Christelle Rabier, ed., *Fields of Expertise. A comparative history of expert procedures in Paris and London. 1600 to present* (Cambridge: Cambridge Scholar Press, 2007), 261-189; Thomas Le Roux, *Le laboratoire des pollutions industrielles. Paris, 1770-1830* (Paris: Albin Michel, 2011).

substances. A flourishing consumption culture encouraged many multi-functional inventions in which aerial fluids were used, fitting within a general agenda of improvement. These new apparatus appeared not only as goods ready for sale on the market but also as unfinished objects whose completion would require wider participation. Consumers, experts and administrators would be involved in such practices, which would also generate communities of hybrid actors and allow different manufacturing techniques to converge into collaborative enterprises that might lead to pioneering work on various materials.

A perspective that includes technology, rather than just thinking about gases themselves, makes it possible to reveal the inventions and materials used to capture and thereby frame them. Adapting materials articulates a way of acting upon aerial matter which must be understood in terms of operations such as wood turning, glass blowing and polishing, tinning, welding and varnishing, which required expert knowledge and much skill in order to adapt the reaction between the air enclosed and the material used. These operations, according to Leroi-Gourhan, are embodied in concrete local and historical configurations. The effectiveness of devised craft methods depended precisely on the physical specificities of each kind of fluid, which entailed appropriate constructive "processes", aimed at the performance of sealing, resistance, capture and circulation. They inspired pioneering work with various shapes and materials (valves, sheet metal, glass and metal tubes, various chimneys, varnishes, soft fabrics, leather bellows, burners) that created an "operational chain": a set of sequences that performed an effective action upon matter.[29]

Balloon workshops embodied these processes perfectly and, since the invention of hot-air balloons in 1783, opened up a new and stimulating field of research. These balloons involved complex technology – though it must be added that components thereof might be objects that were traditionally available and quite mundane.[30] This included wooden barrels in which sulfuric acid was diluted with water, and then poured onto iron filings to produce inflammable air; pipes intended to capture gases and conduct them into a soft fabric balloon, covered with a varnish that had to be elastic enough to allow a variation in pressure and prevent the escape of the volatile air inside and any risk of explosion; and tin-plate cooling systems designed to avoid the danger-

29 This concept was also part of André-George Haudricourt's approach, *La technologie, science humaine. Recherche d'histoire et d'ethnologie des techniques* (Paris: Maison des sciences de l'homme, 1987), and was related initially to French sociologist Marcel Mauss and the techniques of the body.

30 See Simon Werrett's essay in this volume.

ous heat that occurred during chemical reactions, and which also purified the carbonic gas. The processes at stake in these machines were part of a research trend associated with the industrial development of new substances, which included not only the chemistry of gases, but also sulfuric acid and waterproof varnish made from rubber: in short, chemical practices that were deployed at several levels of collaboration[31]. Every experiment with a balloon required the production of a prototype, the making of which involved a mixture of entrepreneurs, provincial amateurs, semi-learned people and manufacturers, chemists and craftsmen, often financed by public subscription.[32] Much more than a commercial practice, this procedure enabled people to engage with the processes being developed. Prototypes enhanced their inventor's credit, and were used for many other inventions, such as the air pump, ventilator and steam engine.[33]

Abeille's account of the Reveillon factory describes the energy that Argand had put into his supervision of Montgolfier's construction of the hot air balloon, which involved many impassioned exchanges with the people gathered around the machine. To all those present, the analogy between the hot air balloon and the process involved in Argand's lamp seemed obvious. The flame was stimulated by vital air (oxygen), and issued by way of the resistance of the walls of the machine. The way in which, in general, containers behaved through a transformation of temperature and composition of the air enclosed was also crucial for the lamp. Argand's lamp was also a composite object, made of glass, sheet metal, welding, varnish, wheels, wick, and oil. The tube was a crucial issue. Argand wished to replace the metal chimney located above the flame with a glass one, which might help to increase the effectiveness of the light. In order to develop a suitable material, he needed to find glass able to resist alternate heating and cooling. British knowledge of flint glass might reveal the solution, so Argand moved to England to overcome this obstacle and to find

---

31 Leslie Tomory, *Progressive Enlightenment. The origins of the gaslight industry, 1780-1820* (Cambridge, MA: MIT Press, 2012).

32 Marie Thébaud-Sorger, *L'aérostation au temps des Lumières* (Rennes: Presses Universitaires de Rennes, 2009).

33 See, among many examples, Thomas Tidd, *Considerations on the Use and Properties of the Æolus a New Invented Portable Machine for Exchanging and Refresching the Air of Rooms, &c.* (London: J. Reeves, 1755); Anon., "Pompe nouvelle et portative pour les incendies et arrosements," *Journal de littérature, des sciences et des arts par Mr. l'abbé Grosier* (1780): 358-359, referring to a subscription opened in Paris by a mechanic, Charpentier. Lissa Roberts also discusses this entrepreneurial process in "Geographies of Steam: Mapping the entrepreneurial activities of steam engineers in France during the second half of the eighteenth century," *History and Technology* 27 (2011): 417-439.

adequate support to develop his invention. Similar issues were raised and discussed widely among various audiences and through transnational networks of expertise. Argand's connections extended into French-English capitalism, networks of administrators, savants and investors who were all crucial in chemistry's evolution in the last decades of the century.

Synergies, analogies and systematization emerged in the development of aerial contraptions. These objects constituted a common taxonomy of different shapes (vessels, tubes, connecting pipes, spring, bellows, blades and so on) and a repertory of know-how that could be put to diverse uses. Coating and varnishing, critical techniques for the containment of air, emerged as prominent fields of investigation relating to fireproofing and the production of airtight seals. Every element in one invention could potentially be transferred to another. Inventions were composite artifacts, and adaptable to a wide range of applications at different scales (not to mention prices). Inventors then shaped communities around their objects employing many common forms of action such as display, advertising, seeking the approval of various societies, or engaging in other spheres of accreditation. Commercial literature played a crucial role in the dissemination of these processes: through advertisements, leaflets and calls for subscription, entrepreneurs required consumers to cultivate the ability to understand, compare and judge the relevance of a device.[34] Objects were bonded together. While drawing attention to Argand's great assistance in gas-making for Blanchard's balloon in London, an article published in the *Courier de l'Europe* advertised his new lamp in an enthusiastic way.[35] Artifacts such as lamps, stoves and ventilators were designed on the border between private and public concerns. The announcement of Whites' "air machine" claimed that it was conceived to serve in various contexts where air was confined in public buildings, ships, hospitals and mines, and also in private rooms. He targeted not only London markets, but also the colonies. This kind of device thus existed at the intersection of health and private comfort, in addition to being intended as an improvement for people at work. Yet this polysemy was based on one coherent principle: that invisible air became palpable through devices and their components (here the fan blades stirring the air) producing an obvious effect that everyone could grasp and comprehend. Many kinds of devices were designed to fight asphyxia. Inventors acted either

34 Liliane Hilaire-Pérez, Marie Thébaud-Sorger, "Les techniques dans l'espace public. Publicité des inventions et littérature d'usage au XVIIIe siècle (France, Angleterre)," *Revue de Synthèse* 127 (2006/2): 393-428; Jeffrey R. Wigelsworth, *Selling Science in the Age of Newton: Advertising and the commoditization of knowledge* (Farnham: Ashgate, 2010), 7.

35 *Courrier de l'Europe* (September 1784), 24.

on the surrounding "air" or they invented equipment for the exposed body, such as the fumigator apparatus. The description of Scanegatti's apparatus involved "flexible tubes", "cannula" and a "bellows" designed to inject tobacco smoke into the bodies of drowned people to help recover respiration after drowning.[36] The opening up of a new understanding of 'airs' addressed the issue of being able to breathe, in both a literal and a metaphorical sense, in the human and the social body.

Capturing, changing and making air drew on a wide range of actions and a new comprehension of the nature of aerial substances, which occurred in various social spaces on various scales. Bringing these together reveals a continuity, where linking one object to another in a whole system made sense. Emerging through decomposition and re-composition, the staging of inventions presented relationships between simple natural bodies in a new light, initiating a shift in sensitive and conceptual frameworks of material life. Scanegatti offered lessons in which he presented this kind of interconnection by exploiting the intrigue of striking experiments.

> Among the many items to be exhibited before the eyes of the public will be a hydraulic pendulum of very regular movement that will act as an alarm in the morning, via two small canon blasts, which light a vessel or lamp filled with spirits, which heats a small pot of broth during the time it takes to get dressed, which will be ready to drink at the end of one's *toilette*.[37]

Mastering this operative sequence on a small scale set up the wonderful power of working with fluids, by means of a curious mechanical arrangement, using "explosions", "liquor", "heating" and "evaporation" to achieve individual accomplishments like waking up and making a pot of broth. By using the possibilities offered by "aerial fluids", a new kind of relationship with materials emerged for large audiences, changing the way in which they could be understood, mastered and turned into an effective action.

Small-scale inventions, even those that were amusing and curious, encouraged a common way to explore the new materiality of air and its effect upon daily improvements, where mastering these substances played a major role "in

---

36 "Noy," Félix Vic d'Azyr, *Encyclopédie méthodique*, 8 vols. (Paris, 1783), vol. 5, 366; Joseph-Jacques Gardane, *Gazette de santé, contenant les nouvelles découvertes sur les moyens de se bien porter & de guérir quand on est malade* (Paris, 1774), 298.

37 Notice of Scanegatti's course of experimental physics beginning 1st July 1764. *Annnonces, affiches et avis divers de Normandie*(1764), 20.

the providential economy of nature."[38] Scanegatti later explored this creativity by tackling objects for urban management through his fumigation apparatus and his furnace, which dealt with the crucial oeconomic issue of the shortage of wood for production. As part of the promotion of his plaster kiln in 1786, he organized extensive trials in front of the king's representative in Rouen, several manufacturers, town aldermen and local men of science. This demonstrated further what kind of achievement the mastering of combustion could provide when embedded in a political œconomy that dealt with the management of energy and the search for fuel substitutes.[39] Seemingly technical practices would actually change the sociomaterial process of knowledge production and contribute to a transformation in the perception of these invisible substances.

'Airs' as Boundary Objects

The materiality of 'airs' could only be revealed when they interacted with other materials. As Lavoisier wrote in his first essay on elastic fluids in 1777, "They escape the sense of touch, except where their resistance to bodily movement renders them discernible and to a certain extent palpable."[40] By working with 'airs' (vaporized substances, noxious air, hot air) via their devices, practitioners tried to track this 'materiality', simultaneously producing experimental systems in which the identification of these airs' properties was put to the test. Partly because they presented prototypes, projects and inventions under development, each artifact could act as an experimental tool, similar in some ways to the "generator of surprise", to recall Rheinberger's category.[41] The degree of uncertainty inherent in the success of these apparatus and devices made their performance in various sites a source of permanent wonder. Their effectiveness depended on many material arrangements that constantly needed to be re-adjusted.

Inventors scrutinized the interaction between 'airs' and the materials of the containers they adapted. Aerial fluids, unlike a liquid with an identical density throughout, diverged because they might be made of several fluids with

38 Golinski, *British Weather*, p. 161 (see note 7).

39 See the essay by Roberts and Van Driel in this volume.

40 Antoine-Laurent de Lavoisier, "Expériences et observations sur les fluides élastiques en général et sur l'air de l'atmosphère en particulier," (1777), Jean-Baptiste Dumas and Edouard Grimaux, eds., *Œuvres de Lavoisier* (Paris: Imprimerie nationale, 1862-93), vol. 5, 271.

41 Rheinberger, "Experiment, Différence and Writing I.," p. 307 (see note 3).

different properties and densities. Therefore practical observations and work on various technical arrangements led practitioners to formulate observations according to these distinctions. Air being perceived as a single entity seems to have been diffracted by various components resulting from the heating process, such as the vaporization of various liquid substances, which created invisible fluids, and combustion, which liberated several substances including fumes and invisible components such as the hypothetical *air du feu*, equated by some with phlogiston. The relationship between aerial gases and atmospheric air was not resolved by the discovery of new gases such as vital air (oxygen), noxious air (carbonic gas), and various inflammable 'airs' (such as hydrogen and methane).[42] The idea that air was composed of several chemical elements only emerged gradually and in the meantime, ideas about air interacted with inventive practices. Observers were able to record accurately how 'airs' changed the vessels in which they were contained, through corrosion, condensation, fermentation and inflammation. In response, they sought out materials that could maintain a flame, avoid toxic emanations, reduce fumes or exploit possible motions.

These dynamic processes engendered a wide range of experiments, in surface treatments or ingenious connecting mechanisms such as flexible strips and elastic springs. However, the behavior of invisible aerial fluids in their devices raised many assumptions due in part to this feature of invisibility. Pneumatic chemists undertook a tests to define the different properties of elastic fluids, through reactions with combustion (explosion, extinction) made thanks to the possibility of isolating them in glass vessels to carry out analysis. Other practitioners used their technical devices to explore the same properties, with opposite expectations: for instance, to avoid explosions or to restore the vivacity of a living organism. They scrutinized the effects that capturing air made possible, such as the elasticity that provided motion when air was compressed. They sought to master 'aerial matter' while experimenting concretely with the behavior of invisible aerial fluids that revealed their different natures and compositions.

Rather than considering inventions as applications of scientific knowledge, we need to consider how seemingly technical practices changed the sociomaterial process of knowledge production and contributed to the understanding of these substances. Such practices – and the objects they engaged – often

42 Maurice Crosland, "'Slippery Substances.' Some practical and conceptual problems in the understanding of gases in the pre-Lavoisian era," Frederic Lawrence Holmes and Trevor Harvey Levere, eds., *Instruments and Experimentation in the History of Chemistry* (Cambridge, MA: MIT Press, 2000), 79-89.

pre-empted theory, as was the case for the understanding of heat and steam, or the nature of noxious air. Long before any chemical identification of carbonic gas, for example, the link between an understanding of 'atmospheric air' and 'health' was made through practical research into ventilators such as those proposed by Stephen Hales.[43]

Combustion processes were at the heart of chemical practices. The reaction of "air", fire and various materials and substances remained a central issue. Understanding the role of air in combustion was crucial for the design of many devices, which dealt with producing heat and fire, whether they were intended to extinguish a flame, or conversely to maintain it while circulating the heat produced. Far from being restricted to an empirical level of practice, practitioners engaged with newly generated knowledge of pneumatic chemistry. Activating fire with air was at the core of many apparatus, especially lamps, as highlighted by Argand's lamp. Argand was engaged in enthusiastic discussions with Meusnier de la Place during Montgolfier's experiments in Paris in 1783, especially concerning the "action on dilated air by fire", the vaporization of substances, and the formation of droplets condensing on the sides of the fabric envelope that particularly caught their attention. For Argand, as for Meusnier, who was working at that time with Lavoisier on the experiment of the decomposition and recomposition of water, vapor was related to water, understood as being composed of vital air and inflammable air.[44]

Elasticity was particularly complicated to comprehend because it could be and was approached from both a physical and chemical standpoint. Lavoisier, among others, moved between these approaches in the pattern he proposed regarding aerial elastic fluids and, in particular, the property of many acids vaporized under the effect of a change of temperature.[45] By the end of the century, a permanent porosity between physical and chemical ideas concerning the states and composition of matter existed in the category of fluids. As Robert Fox demonstrated in his book on the caloric theory, a chemical approach was maintained for a long time in the understanding of heat.[46] Thus while many practitioners did not identify themselves as chemists, a large community worked with the materiality of invisible aerial substances, exploring

43 Stephen Hales, *A Description of Ventilators: Whereby great quantities of fresh air may with ease be conveyed into mines, goals* [sic] *hospitals, work-houses and ships, in exchange for their noxious air, which was read before the Royal Society in May, 1741* (London: W. Innys, 1743).

44 Abeille, *Découverte des lampes à courant d'air* (see note 18).

45 Lavoisier, "Expériences et observations sur les fluides élastiques" (see note 40).

46 Robert Fox, *The Caloric Theory of Gases from Lavoisier to Regnault* (Oxford: Oxford University Press, 1971).

not just their behavior but also their nature. The chemical knowledge of an "engineer" such as James Watt, for instance, was relatively great, exemplifying the fact that we need to overcome the anachronistic separation between physics and chemistry.[47] 'Airs' were boundary objects that shaped the field of knowledge in addition to distinctions between practitioners.

Argand brought the evidence of his ability to transgress apparent boundaries between an empirical approach and chemical knowledge while performing practical work. Understanding perfectly the role of oxygen in combustion as he did, he qualified as the most appropriate person to supervise the chemistry required by ballooning projects, helping to manage balloon ascents in London by mixing the diluted sulfuric acid in each barrel with an iron stick in order to accelerate the hydrogen production process.[48] Argand's activities illustrate that the processes involved in mastering heat and the production, evaporation and manufacture of gases were linked by an exploration of the risks they entailed, such as explosion and flammability. They simultaneously point to the social dimensions of such work. While Argand took part in the Coffee House Philosophical Society in London and befriended the Lunar Society group in Birmingham, especially Priestley, his integration may have been easier in England than in the intellectual surroundings of the French *Académie des sciences* where the reform of empirical practices of chemistry was at stake.[49] However the majority of relations between established men of science and various practitioners involved close interactions, especially in provincial cities, in France and Britain alike. Provincial French academies such as those in Nancy or Rouen, which admitted Scanegatti as a fellow in 1775, offered a favorable framework of convergence for scientific and practical arts, fortunes and talents.[50] Recognized as "a fine glass blower," a skill necessary for the success of meteorological experiments that he helped to perform, Scanegatti also earned recognition among the elite of Rouen for his lectures in physics, and demonstrations of his "fumigatory apparatus", which focused particularly on the

47 David Philip Miller, *James Watt, Chemist: Understanding the origins of the steam age* (London: Pickering and Chatto, 2009).

48 *Courrier de l'Europe* (September 1784), 24.

49 Jan Golinski, "Conversations on Chemistry: Talk about phlogiston in the Coffee House Society, 1780-1787," Trevor Harvey Levere and Gerard L'Estrange Turner, eds., *Discussing Chemistry and Steam: The minutes of a coffee house philosophical society, 1780-1787* (Oxford: Oxford University Press, 2002), 191-205; Peter Jones, *Industrial Enlightenment: Science, technology and culture in Birmingham and the West Midlands 1760-1820* (Manchester: Manchester University Press, 2008).

50 John Perkins, "Creating Chemistry in Provincial France before the Revolution: The examples of Nancy and Metz," *Ambix* 51 (2004): 4375.

injection of hot air and complemented theoretical essays on "the noxious and mephitic air of cesspools" that he presented at the local academy.[51]

The fact that debates surrounding the identity of aerial fluids rested on recognizing their characteristics demonstrates that these men were no less informed or capable of tackling issues relating to 'airs' than their better-known academic contemporaries. They evolved within the same epistemic world, offering solutions to mastering steam and fire, producing hydrogen and purifying the atmosphere of carbon dioxide. Their activities helped to shape both the intellectual boundaries of 'air' and a social world through the design of artifacts. The outlines of the social milieux engaged with 'airs' shed light on new arenas of practices and exchanges where different communities interacted, creating opportunities to speculate about the concrete expectations of 'airs' aroused in the public sphere. This was particularly the case with the prevention of hazards that aimed at disciplining urban spaces and spreading social reform. In *ancien regime* Paris, prefiguring the creation of the Committee for Salubrity in 1791, public authority and the politics of regulation fostered much expertise dedicated to salubrity. Skilled chemists such as d'Arcet or Cadet de Vaux supervised a large field of research that connected hygiene and philanthropy.[52] The nature of 'airs' was not only discussed in restricted areas such as Royal Society of Medicine, but also staged through experiments in the center of the city, that linked a variety of public buildings with risk-filled places such as mines, cesspools and rivers, targeting both injured workers and drowned people.[53] They generated new expectations in the public sphere by staging "miracles of chemistry" that underlined the "real solutions" offered to the city's "suffering humanity" by a discipline that had been freed from its old formulas.[54] Mastery of air was one among many 'miracles' provided by chemistry. Prompting a large number of essays, prints and explanations, discussions

51 "Extrait d'une lettre écrite par Boin correspondant de l'académie de Rouen," in the section "Observations sur le froid de l'hiver de 1776," *Oeuvres de Lavoisier,* vol. 3, 394; *Annnonces* (see note 37); "Sur le méphitisme des fosses d'aisances (1781)," in "Liste des mémoires lus à l'Académie dans ses séances particulières et publiques, depuis 1781 jusqu'en 1793," *Précis analytique des travaux de l'Académie royale des sciences, belles-lettres et arts de Rouen, depuis sa fondation en 1744 jusqu'à l'époque de sa restauration* 5 (Rouen, 1821): 16.

52 Le Roux, *Le laboratoire des pollutions industrielles* (see note 28).

53 Thébaud-Sorger, "Innovation and Risk Management," (see note 28); Philippe-Nicolas Pia, *Avis patriotique concernant les personne suffoquées par la vapeur de charbon qui apparaissent mortes et qui ne l'étant pas, peuvent recevoir des secours pour être rappelées à la vie* (Paris: Veuve Thiboust, imprimeur du roi, 1776).

54 Louis-Sébastien Mercier, *Tableau de Paris* (Hambourg: Virchaux & co., 1781), volume 1, chapter 53 «Air vicié,» 62.

occurred in various social spaces that interacted with the development of a chemistry embedded in wider debates. A directory of practical solutions was explored by a great variety of actors, in and beyond academic circles. The interaction between air and the city, seen as a living organism, facilitated a reconciliation between and convergence of investigations into nature and social reform.

Conclusion

This essay has shown how devices used to contain 'airs' reorganized the sociomaterial world around them and reshaped the environment at the end of the eighteenth century. Inventive practitioners aspired not only to capture invisible fluids but also to modify their actions, and even to produce them artificially. Each device they created comprised a material arrangement that might perform the impossible: resuscitation, mastering fire and heat, using the power of steam for daily comfort and transforming the irrespirable atmosphere into breathable air.

Embodied in various adaptable commodities, apparatus and objects designed to perform useful actions, a new material culture brought about a concretisation of gases, heat and steam. This entailed a mixture of sounds, smells, movements, changes of temperature and changes in the size and shape of containers. On a broader scale, it brought together chemical practice and an evolving culture of consumption in ways that gradually changed society and the urban landscape. These devices were not necessarily designed to serve just one purpose but were, on the contrary, intended to be transferred and applied to a great variety of contexts: urban, rural, industrial, domestic and public. From lighting streets and theaters to preventing fires, resuscitating drowned people and overcoming gravity, these devices exhibited a wide range of possibilities that fostered the creation of a heterogeneous network of practitioners who understood and promoted their intensive development for new purposes.[55]

Many "improvers" were engaged in a synthetic approach based on comparisons, analogies, reconciliations, convergences and proportions, that engaged with some of the most recently generated knowledge on materials and substances. They thereby contributed to the further understanding of the 'airs' they sought to master. These persons were not just academic figures or typical projectors, but hybrid entrepreneurs, craftsmen, lecturers, manufacturers,

55 Leslie Tomory, *Progressive Enlightenment* (see note 31).

administrators, physicians, apothecaries, architects and clergymen. Their identities drew on a whole gallery of Enlightenment occupations. They contributed through their material devices to the creation of a new oeconomy of nature, which intersected with social expectations, the market place and entrepreneurial strategies in addition to the governance of matter.

Far from seeing this complex process in terms of the application of theoretical chemical knowledge to practice, this essay's more anthropological approach transcends the supposed opposition between know-how and empirical knowledge, on one side, and theory on the other. It refuses, that is to say, the separation of epistemic investigation from the material arrangements that made it possible. Concentrating on the materiality of devices further affords a historicisation of the distinction between physics and chemistry, and the social distinction between chemists and other practitioners.[56]

Whereas the question of the nature of 'airs' (intersecting with water and fire) was surely a concern of transnational debates, one could argue that the involvement of a much wider audience occurred through advertisements, public displays and experiments; by coming into contact with the everyday experiences of the inhabitants of European cities, this context forged a new culture of the natural elements through demonstrated knowledge of their properties. In conclusion, air stood at the boundary of a whole range of actors, linking diverse activities from industry to public lecturing, and various practices attempting to contain and manage aerial fluids. These labors had transformative sociomaterial effects, taking place in a wide range of sites, creating the grounds for new understandings of air, encouraging new audiences and public interactions, and prompting the creation of a variety of ingenious small-scale inventions.

Acknowledgments

I would like to thank Liliane Hilaire-Pérez with whom I gave the very first version of this paper in January 2012, "Les techniciens des gaz et de la vapeur au 18e siècle: savoirs, micro-inventions et industrial Enlightenment" at the Research seminar LIED, Paris Diderot Paris 7. My thanks to Delaina Haslam, Simon Werrett and Lissa Roberts for their intensive work revising my text.

56 Jackson, "The 'Wonderful Properties of Glass'," p. 68 (see note 6).

CHAPTER 4

Spreading the Revolution: Guyton's Fumigating Machine in Spain. Politics, Technology, and Material Culture (1796-1808)

Elena Serrano

Salve Morvó! [sic]
Your inquiring mind
Made a sweet breath of life [...]
Which flying into the atmosphere
Purifies and destroys at once
The corrupted germ of death.[1]

In 1806, the Spanish poet Rosa Gálvez (1768-1806) published a seven-page poem celebrating the lawyer, chemist and politician Louis-Bernard Guyton de Morveau (1737-1816). During the first years of the century, an epidemic of yellow fevers caused thousands of deaths on the Spanish coasts. Guyton had arguably fabricated a gas that destroyed the agents of contagion that stubbornly remained in the atmosphere and goods for years. This "sweet breath of life" as the poet called it, was the controversial oxy-muriatic gas.

Guyton was a champion of oxy-muriatic gas. He not only wrote about its properties, but also with the prestigious French instrument-makers the Dumotiez brothers, he developed a machine that released the gas.[2] The fumigating machine embodied two essential features of Lavoisier's system of chemistry: the theory of acids and the theory of combustion.[3] As is well known, Lavoisier believed that all acids contained oxygen (including muriatic acid,

1 Rosa Gálvez, "Oda en elogio de las fumigaciones de Morvó [sic]," *Minerva o el Revisor General* 52 (1806): 3-10, on 8. My translation.
2 The reports do not distinguish between the two brothers, Louis Joseph and Pierre François. See Maurice Daumas *Les instruments scientifiques aux XVII et XVIII siècles* (Paris: Presses Universitaires de France, 1953), 378-79.
3 In Spanish it is often referred as *Máquina fumigatoria;* in French as *Appareil de désinfection.*

DOI 10.1163/9789004325562_006

which Humphry Davy later demonstrated to be composed of hydrogen and chlorine). According to Guyton, the fumigating machine supplied a highly oxygenated compound of muriatic acid – oxy-muriatic acid– which destroyed contagious particles in a process akin to combustion.[4] The machine was intended to prevent gangrene in soldiers' wounds and to disinfect the air of poisonous sites such as jails, hospitals, theaters, churches, and ships. It was also widely used during epidemics of yellow fevers in Europe.

This essay focuses on the fumigating machine as a means to explore how beliefs and attitudes became embedded in societies and also inversely, how ways of interpreting nature, society, and politics became embedded in artifacts. It will show, first, how the machine served to spread the new French chemistry; second, how it came to embody a new relationship between citizens and the state, and third, how this artifact was imported by the Spanish absolutist state, appropriated, and used for its own propaganda. It thereby adds to this volume's general argument against simplistic narratives regarding the intellectual foundations of the chemical and industrial revolutions and argues against a "linear model" of technological development.[5] By focusing on a chemical artifact, it shows a historically more complex and significant interweaving of theory, material culture, and politics.

Simon Schaffer and Ken Adler have shown how instruments and technological artifacts are deeply political, moving beyond the view that instruments simply embody theory and visions of nature.[6] Schaffer has stressed the links of eudiometers with dissenters' political agendas, while in his classic book *Engineering the Revolution: Arms and Enlightenment in France, (1763-1815)* Alder confronts the question of the politics of revolutionary guns.[7] He argues that a

4 Ruth Ashbee, "The Discovery of Chlorine: A window to the chemical revolution," Hasok Chang and Catherine Jackson, eds., *An Element of Controversy: The life of chlorine in science, medicine, technology, and war* (London: British Society for the History of Science, 2007), 15-40; William A. Smeaton, "Guyton de Morveau, Louis Bernard," Charles C. Gillispie, ed., *Dictionary of Scientific Biography* (New York: Scribner, 1976), 600-4.

5 See the essay by John Christie in this volume for a statement of this argument.

6 Simon Schaffer, "Measuring Virtue: Eudiometry, enlightenment, and pneumatic medicine," Andrew Cunningham and Roger French, eds., *The Medical Enlightenment of the Eighteenth Century* (Cambridge: Cambridge University Press, 1990), 281-318; On the relationship of instruments and theory, Trevor H. Levere, "The Role of Instruments in the Dissemination of the Chemical Revolution," *Éndoxa: series Filosóficas* 19 (2005): 227-42; Bernadette Bensaude-Vincent, *Lavoisier: mémoires d'une revolution* (Paris: Flammarion, 1993); John Tresch, *The Romantic Machine. Utopian science and technology after Napoleon* (Chicago: University of Chicago Press, 2012).

7 Ken Alder, *Engineering the Revolution: Arms and enlightenment in France, 1763-1815* (Princeton: Princeton University Press, 1997); Charles C. Gillispie and Ken Alder, "Exchange: Engineering the revolution," *Technology and Culture* 39 (1998): 733-54.

new, intimate relationship between politics and technology was forged during this period. Alder recognizes the interaction of artifacts and politics at different levels. The most obvious concerns the way in which technologies were bound up in struggles over sovereignty, over foreign policy, and over relations with different political groupings. The relationship between the oxy-muriatic acid fumigation and the politics of the Spanish state has been analyzed in these terms. During the 1970s, Spanish scholars construed the polemic between followers and detractors of acid fumigations as an example of the impossibility of pursuing authentic science in an authoritarian political regime.[8] It may be useful to recall that at that time, Spain was moving away from Franco's dictatorship, a regime notorious for its purges of scientists and its censorship practices. Recently, José Ramón Bertomeu and Antonio García Belmar have argued for a more nuanced view, in which a broad consensus about the efficacy of fumigation was fabricated not by the Government alone, but with the cooperation of other groups that shared academic and economic interests in fumigation.[9]

Artifacts could also be "potent icons."[10] For instance, Alder identifies the pick as a symbol of the revolutionary power of the French people. But his most important contribution from the viewpoint of this analysis is his turn to politics for an explanation of the design and functioning of artifacts. Rather than using technological or social determinism to explain why particular objects take the form they do at particular times, he stresses the political dimension of choices: "the deep structural level of politics necessarily shapes the way material objects and technological knowledge are organized and directed."[11] The essay will explore how the practices of fumigation afforded changes in the relationship between the citizen and the power of the state.

We must, however, be aware of the dangers of over-emphasizing the agency of artifacts on one hand, and of considering them as "empty vessels to be filled

8 Luis García Ballester and Juan L. Carrillo, "Un ejemplo de represión de la ciencia en la España absolutista: la supresión del capitulo 15 de la 'Breve descripción de la fiebre amarilla' (1806) de J.M. Arejula," *Revista de Occidente* 134 (1974): 205-11; Juan L. Carrillo, Pedro Riera Perelló, and Ramón Gago, "La introducción en España de las hipótesis miasmática y prácticas fumigatorias. Historia de una polémica (J.M Aréjula – M.J Cabanellas)," *Medicina e historia* 67 (1977): 8-26; Luis García Ballester and Juan L. Carrillo, "The repression of Medical Science in Absolutism Spain: The case of Juan Manuel de Aréjula, 1755-1830," *Clio Medica* 9 (1974): 207-11.

9 Antonio García Belmar and José Ramón Bertomeu, "España fumigada. Consensos y silencios en torno de las fumigaciones ácido-minerales en España, 1770-1804" (in progress). The author was unable to consult this source before this essay was completed.

10 Gillispie and Alder, "Exchange," p. 745 (see note 7).

11 Ibid., p. 743.

with meanings" on the other.[12] To avoid this danger, this essay approaches the fumigator through the study of its affordances. According to Susan J. Douglas, affordance may be defined as, "what certain technologies privilege and permit that others don't."[13] The concept of affordance has a relational ontology, and thus the affordances of objects change as their historical context changes. Affordance refers to "those functional and relational aspects of technology that frame but do not determine the possibilities for action in relation to an object."[14] The analysis that follows takes into account this dynamic and relational construction of artifact-meanings. It is divided into two sections. The first deals with Guyton's fumigating machine, while the second follows the instrument's journey to Spain and the complex history of this relocation.

Guyton's Fumigating Machine

The disinfection apparatus that Guyton and the Dumotiez brothers designed basically consisted of a closed vessel that stored oxy-muriatic acid gas ready to use. Figs. 4.1 and 4.2 show the 1805 version for disinfecting big rooms. The machine ensured a controlled emission of the gas by way of an ingenious method of keeping it under moderate pressure in a glass bottle, which was housed in a wooden frame. A large screw held a wooden cap (H) that was pushed over a thick tap of glass (I) to keep the bottle closed. When the screw was loosened, the cover rose up under pressure from the gas, which escaped into the room. Notice that the piece (H) was specifically designed to engage with the columns (B), so it could easily be slipped through the columns. When one needed to refill the apparatus, the screw (E) was loosened, so that the bottle could be removed from its setting onto the surface of the board (D). Even when the apparatus required moving around the room while emitting the gas, the bottle remained safely in place.

12 Francesca Bray, *Technology, Gender and History in Imperial China: Great transformations reconsidered* (London; New York: Routledge, 2013), 8.

13 Susan J. Douglas, "Some Thoughts on the Question 'How Do New Things Happen?'," *Technology and Culture* 51 (2010): 293-304, on 293; Ian Hutchby, "Affordances and the Analysis of Technologically Mediated Interaction," *Sociology* 37 (2003): 581-89; Idem, "Technologies, Texts, and Affordances," *Sociology* (2001): 441-56.

14 From Brian Rappert criticizing Ian Hutchby, in Brian Rappert, "Technologies, Texts, and Possibilities: A reply to Hutchby," *Sociology* 37 (2003): 565-80, on 566; For an insightful discussion of the types of affordances, see Mats Frindlund, "Affording Terrorism: Idealists and materialities in the emergence of modern terrorism," Max Taylor and P.M. Currie, eds., *Terrorism and Affordance* (London: Continuum, 2012), 73-92.

FIGURE 4.1
A large disinfection apparatus for military hospitals and other public spaces. Museo Galileo's Photographic Archives: Pressure receiver (Inv.3778). COURTESY OF MUSEO GALILEO, FLORENCE.

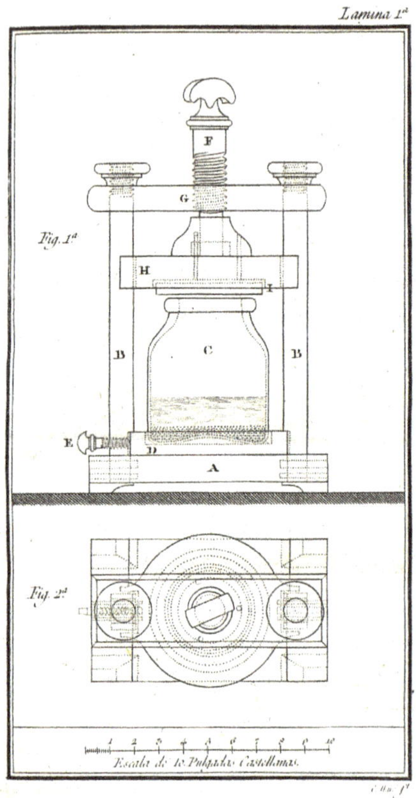

FIGURE 4.2
Sketch of the large version of Guyton's disinfection apparatus as published in the Semanario de Agricultura y Artes a los Párrocos. COURTESY OF BIBLIOTECA DEL REAL JARDÍN BOTÁNICO, CSIC, MADRID.

No metal parts were used because the acid gas would corrode them. Once the apparatus was filled with reagents, one needed only to unfasten the screw, let the fumes of the oxy-muriatic acid flow, and fasten the screw back. For disinfecting hospitals, this was to be done once or twice a day, for a period of two to six minutes, dependent upon the size and occupancy of the ward. According to the leaflet that accompanied the machine, the gas lasted six months if used daily. In addition to the large version, the Dumotiez brothers also designed machines of a smaller size (Figs. 4.3 and 4.4). This latter was designed for carrying– with the caution of keeping it upright – in order to visit the sick, attend funerals, concerts, theaters, and masses.

FIGURE 4.3
A portable version of Guyton's disinfection apparatus. COURTESY OF SCIENCE MUSEUM, SOUTH KENSINGTON, LONDON.

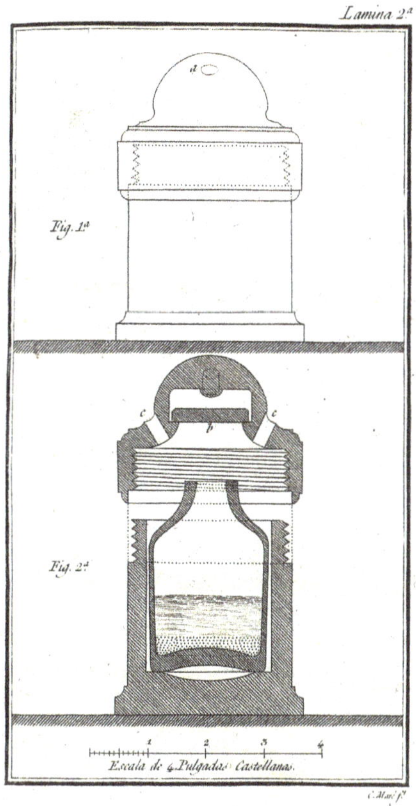

FIGURE 4.4
Sketch of the portable version of Guyton's disinfection apparatus as published in the Semanario de Agricultura y Artes a los Párrocos. COURTESY OF BIBLIOTECA DEL REAL JARDÍN BOTÁNICO, CSIC, MADRID.

Although the fumigating apparatus might look simple, its construction involved a great deal of material research.[15] The system that confined the gas with the wooden screw was decided upon after models using ground-glass stoppers failed because the gas corroded them. The bottle was made with a new technique of grinding glass developed by Guyton. It also needed to have thick borders to resist the pressure of the screw and the gas. In addition, the

15 William A. Smeaton, "Platinum and Ground Glass: Some innovations in chemical apparatus by Guyton de Morveau and others," Frederic L. Holmes and Trevor H. Levere eds., *Instruments and Experimentation in the History of Chemistry* (Cambridge, MA; London, England: MIT Press, 2000), 211-37.

glass disk that sealed the bottle needed to be flat, as did the edge of the bottle, to allow perfect contact and avoid leaks of gas. The Dumotiez brothers recommended following the same technique as in the pneumatic machine.[16] The machine required construction by skillful craftsmen. In Spain, a Guyton-style machine was made by a gifted artisan and member of the Barcelona Academy of Sciences, Pelegrín Forés y Madaula (1775-1841).[17] The fact that this was highlighted in Forés y Madula's obituary indicates the high prestige that constructing such an apparatus supposed. Indeed, one of the biggest issues faced during attempts to replicate the machine on a large scale in Spain was precisely the lack of specialized glassmakers and turners available to perform the work.[18]

Guyton's machine in its various forms was just one part of a spectrum of fumigation techniques performed with everyday gadgets and materials. Since ancient times, people had evaporated fumes of odoriferous stuff, including thyme, rosemary, juniper, wormwood, myrrh, incense, and vinegar, simply by heating pots.[19] Contemporary treatises on domestic economy included recipes for disinfecting with sulfuric, muriatic, and nitric acid, in which no special devices were needed.[20] To fumigate using sulfuric acid for instance, the Spanish agricultural magazine *El Semanario de Agricultura* suggested filling a normal clay pot with salt and putting it on a portable oven full of coal embers. The salt was stirred with a simple stick until one felt the heat, and then the sulfuric acid was carefully poured on.[21] In the fumigation of the Russian hospital ship *Union*, doctor Archibald Menzies heated sand in a clay receptacle, inserted a teacup containing sulfuric acid, and added powdered niter to produce fumes of nitric acid.[22] This raises the question of how contemporaries justified buying special-

16 Louis-Bernard Guyton de Morveau, *Traité des moyens de désinfecter l'air* (Paris: Bernard, 1805), 388.

17 Carlos Puig-Pla, "Els primers socis-artistes de la Reial Acadèmia de Ciències i Arts de Barcelona (1746-1842)," Agustí Nieto Galán and Antoni Roca Rosell, eds., *La Reial Acadèmia de Ciències i Arts de Barcelona als segles XVIII y XIX. Història, ciència i societat* (Barcelona: Institut d'Estudis Catalans, 2000), 287-310, on 300.

18 Levere, "The Role of Instruments" (see note 6).

19 Antonio Pérez de Escobar, *Avisos populares médicos y domésticos* (Madrid: D. Joachin Ibarra, 1776), 74-5; Félix Martínez López, *Reflexiones del Dr. Félix Martínez López sobre las enfermedades* (Valladolid: En la oficina de la viuda e hijos de Santander, 1788), 13; Marie Armande Jeanne Gacon-Dufour, *Moyens de Conserver la santé des habitants des campagnes* (Paris: Buisson, 1806), 165.

20 Miguel José Cabanelles, *Observaciones sobre los gases ácido- minerales* (Cartagena: Manuel Muñiz, 1802), 18-22.

21 Anon., "Medicina Doméstica," *Semanario de agricultura y artes* 1 (1797): 70-2.

22 Juan Manuel Aréjula, *Memoria sobre el modo y ocasiones de emplear los varios gases para descontagiar los sitios epidemiados* (Sevilla: Imprenta Mayor, 1800).

ized precision instrumentation to do something that could be done with pots and pans.[23]

The fumigating machine was part of an explosion of contemporary gadgets aimed at transforming the "atmosphere" of densely populated cities, which people of the last decades of the century felt were becoming alarmingly contaminated.[24] Frightening warnings about urban airs were heard almost everywhere. In Barcelona, for example, members of the Royal Academy of Medicine vividly described how Barcelona air, being "full of fetid particles, corrupted, acrid, corrosive, and poisonous", damaged health.[25] Thomas Garnett, of the Royal Institution in London, described English city air as a, "chaos of eternal smoke and volatile corruption from the dead, the dying, sick'ning, and the living world."[26] A British leaflet announced the sale of fumigating ingredients for removing the "foetid smells, stagnated and putrid air" which were "the cause of many dreadful diseases [...] which so frequently prove fatal."[27] In this climate of anxiety the fumigating machine provided an easy, quick, and handy way of disinfecting.

Oxy-muriatic gas became an object of consumption, a commodity. The fumigating machine was marketed as a reservoir of a potent disinfectant ready to use, which provided a standardized, reliable means of fumigating.[28] It offered educated urban elites a new optimistic feeling of controlling contagion. Chemists succeeded in enclosing hermetically (or almost hermetically) a powerful new gaseous substance. Designed with the latest material technology, filled up with kits of ingredients prepared by chemists or pure ingredients purchased in apothecaries the fumigating machine fostered the authority of chemical – mostly male – experts, and nourished the prestige of the new chemistry.

---

23 On the adapted use of household items, see Simon Werrett's essay in this volume.

24 Vladimir Janković, *Confronting the Climate: British airs and the making of environmental medicine* (New York: Palgrave Macmillan, 2010); Candance Ward, *Desire and Disorder: Fevers, fictions, and feelings in English Georgian culture* (Lewisburg: Bucknell University Press, 2007).

25 Academia Médico-Práctica de Barcelona, *Dictamen de la Academia Médico-Práctica* (Barcelona: Carlos Gibért y Tutó, 1784), 26.

26 Thomas Garnett, *A Lecture on the Preservation of Health* (Liverpool: J. M'Creery, 1797).

27 Gerard William Groote, *Fumigating Ingredients, to Remove Offensive Smells, Foul, Putrid and Stagnated Air* ([London], [1780?]).

28 That was probably the reason why in 1807 the Spanish Government spent 258 *Reales de Vellón* on two fumigating machines for disinfecting La Corte jail. Archivo Histórico Nacional (AHN): Consejos, 1397, folio 375.

Empowering Oxygen

In 1801, disappointed by the city of Genoa's management of the fever epidemic of 1800, Guyton published his *Traité des moyens de désinfecter l'air* (Treatise on the Means of Purifying Infected Air).[29] The *Traité* enjoyed three editions (1801, 1802, and 1805), and was translated into English, Spanish, German, and Italian.[30] Furthermore, it was widely publicized in journals, magazines, and domestic economy manuals. The way in which Guyton demonstrated in the *Traité* the superiority of the acids, and particularly, the oxy-muriatic acid, is revealing of the epistemic and moral affordances that the fumigating machine offered to contemporaries.

Guyton described his experiments for testing the disinfectant properties of different substances with a kind of chemical miasma-test that he developed. The miasma-test constructed the authority of oxy-muriatic acid as the supreme disinfectant, and so it is worth describing it in detail. Guyton made three assumptions. First, he elaborated on the relation of "fetid and pernicious" odors (supposedly caused by maladies within bodies), and concluded that they could only come from some constituent of the body.[31] Second, he logically argued that since odors were part of bodies, and a body only remained the same while it preserved all its properties, it followed that to destroy the odor was to destroy the body – a dangerous challenge.[32] Here Guyton made a crucial distinction between chemists and lay people. Lay people many times confounded "destruction" with "masking"; only the knowledgeable chemist could determine when a disinfectant was working. With these assumptions, Guyton put into practice his test. He left three samples of meat under a glass bell until it became "perfectly putrid" after six days. Then he kept the pestiferous odor in a bottle, which he connected to another one that contained the supposed disinfectant.[33] If the disinfectant destroyed the foul odor, it meant that it might also destroy the miasmas.

Guyton tried perfumed waters and mineral acids and concluded that only mineral acids had the power to destroy the odor, and therefore, the miasmas. Now, seeking for a chemical rationale to explain why acid destroyed putrid

29 Louis-Bernard Guyton de Morveau, *Traité des moyens de désinfecter l'air* (Paris: Bernard, 1801).

30 For quotations in English, I use the English translation, Louis-Bernard Guyton de Morveau, *A Treatise on the Means of Purifying Infected Air*, trans. R. Hall (London: J. & E. Hodson, 1802).

31 Guyton, *Traité*, p. 92, point 59 (see note 29).

32 Ibid., p. 93, point 60.

33 Ibid., pp. 68-9.

miasmas, Guyton argued that miasmas could not be simple bodies because simple bodies could not reproduce. Miasmas should then be organic bodies, from which it followed that miasmas could be destroyed by fire. Using a dramatic image for his readers, Guyton asked "When the clothes or furniture of a person dead of the plague are burned, does any one suspect that the virus, with which they were infected, can be found entire in the ashes?"[34] For the very same reason then, miasmas could not resist the "condensed oxygen" of the mineral acids, which produced the combustion of organic bodies, in Guyton's words "the most astonishing of combustions." Guyton celebrated the power of the oxy-muriatic acid with these words: "Such are the properties of oxygen, of super-oxigenants, of acid fumigations, and, above all, of the oxygenated muriatic acid gas."[35]

The oxygenated muriatic acid was thus construed as a product of chemical research, based, according to Guyton, on "the most exact experiments", "the application of principles the most evident," and "the consequences of observations drawn from the most authentic sources."[36] Modern chemistry discovered that fumigating with acids, especially with the oxy-muriatic acid, had the same purifying effect as fire: "Such is the grand instrument of disinfection which modern chemistry has brought to our knowledge."[37]

Moralizing Fumigations, Empowering Citizens

Four years later, Guyton went even further. In the 1805 edition of the *Treatise*, fumigating with oxy-muriatic acid acquired a moral dimension. Guyton had already received the Napoleonic *Legion of Honour*. According to the award letter (published in the *Treatise*), the reasons were not only his numerous writings that advanced chemistry, but also the discovery that fumigations with muriatic acids could stop the contagion of yellow fevers, "the rival of the plague." Moreover, the letter continued, he had invented a fumigation apparatus that was "very useful."[38] In addition to including three sketches of the fumigating machines, Guyton concluded the *Treatise* with a meaningful paragraph. After stressing that he had provided all kind of proofs of the efficacy of mineral

34 Guyton, *Treatise*, p. 218 (see note 30); Guyton, *Traité* p. 263 (see note 29).

35 Ibid., p. 268.

36 Guyton, *Treatise*, p. 221 (see note 30); Guyton, *Traité* pp. 267-8 (see note 29).

37 Guyton, *Treatise*, p. 223 (see note 30); Guyton, *Traité*, p. 266 (see note 29).

38 Guyton, *Traité*, p. viii (see note 16); The apparatus were sold in Dumotiez's shop at 12 Rue des Jardinets.

fumigations, the theoretical principles on which their action was grounded, the ways to apply them, and even the instruments for making the practise easy, Guyton stated, in upper case: "THE CONTAGION CANNOT BE BORN AND SPREAD IF NOT BY THE MOST CULPABLE NEGLIGENCE."[39] Fumigating was no longer a personal choice, but a moral responsibility. Thus, the disinfection machine embodied the conceptual and moral power of the new chemistry, warranting chemists' intervention in both the private and the public spheres for the sake of the whole of society.

The machine also embodied the changing relationship between political power and the citizenry. Epidemics expose how power is exercized and how deep social inequalities may be. The control of the contagion implied a tight control of the population. The traditional way of fighting epidemics resembled traditional forms of conquest. Troops sent by the sovereign besieged the infected city. Quarantines and Lazarettos served to isolate populations at the discretion of the authorities.[40] The city doors were closed and a tight military *cordon sanitaire* prevented traffic between the infected city and the outside.[41] No one could travel without a sanitary passport, except the rich, who as soon as the epidemic was declared, fled the city.[42] Sick poor indigents were moved to hospitals, those suspected of being sick to the lazaretto, while the dead were buried with quicklime. Prisoners were forced to conduct the carts of corpses.[43] Cannon were fired in infected neighbourhoods in the belief that this dispersed

39 Ibid., p. 596.

40 Mercedes Pascual Artiaga, "La ciudad ante el contagio: medidas políticas y administrativas dictadas en la epidemia de fiebre amarilla de 1804 en Alicante," *Asclepio* 54 (2002): 125-53, on 133-4; On the *cordon sanitaire* in Malaga see *Gaceta de Madrid* 71 (04/09/1804): 791; See also *Gaceta de Madrid* 77 (25/09/1804): 857-9. The Government forbade the people of Madrid to communicate with Málaga, Vélez, Antequera, Montilla, and Alicante under punishment of five years of exile for nobles and prison for lay people.

41 Anon., *Edicto general comprehensivo de todas las reales provisiones* (Barcelona: Manuel Texéro, 1800), 202-10; See also Capitanía General Cádiz, *Cerciorado ya de que la enfermedad que reina en Málaga* (S.l., 1803?). <http://bdh-rd.bne.es/viewer.vm?id=0000070734&page=1> Accessed 6 December 2015.

42 Juan L. Carrillo, and L. García Ballester, "El comportamiento de las clases y grupos sociales de Málaga en las epidemias de fiebre amarilla," *Cuadernos de la historia de la medicina española* 11 (1972): 88-95; See also Juan Manuel Aréjula, *Breve descripción de la fiebre amarilla padecida en Cádiz* (Madrid: Imprenta Real, 1806), Figure 6, in which he numbered the people who fled the city of Malaga.

43 Andrés Pérez Baylón, *El templo de la muerte* (Malaga: Francisco Martínez de Aguilar, 1804).

the infection.[44] Those who dared to escape from the lazarettos were shot, as were burglars of contaminated houses. Commerce was strongly controlled and smuggling was punished with the gallows. The cost of the epidemic was huge. Military and the other expenses were great, while commerce withered. Ships from infected ports were quarantined and not allowed to anchor in other ports.[45] Once the epidemic was over, goods suspected of being infected –bed-clothes, wool, furniture and books – were burnt or buried with limestone. "Only fire, gold, and gallows cure the fevers," a medical saying stated.[46]

In contrast, the fumigating machine fought the contagion in a profoundly different way. The wellbeing of the whole community was assured by fumigating with gas that, arguably, democratically disinfected the houses of poor and rich identically. Unlike military violence, this manner of stopping the contagion appealed to citizens' moral responsibility, by asserting their responsibility for their own health and that of their peers. Citizens were thereby empowered but were also culpable if infected. The discourse shifted from a state that assured control of epidemics and the population by means of force to a more subtle register, in which it assured control through its own citizens, who were now responsible for protecting themselves and their peers.

In 1804, the Spanish Prime Secretary Manuel Godoy (1767- 1851) decided to manufacture thirty thousand fumigating machines for distribution among the population of southern Spanish towns suffering from yellow fever epidemics.[47] The Spanish ambassador in Paris sent three models (large for hospitals, medium for households, and a portable version) to guide production.[48] However, it soon became apparent that massive and rapid replication was impossible. Guyton's apparatus proved too expensive and sophisticated to be produced on this scale. As other essays in this volume argue it was often a process of elaborating and extending already known procedures rather than

44 Six cannon shots were fired in Malaga over two days in 1803. AHN: Consejos, 11975. "Junta de Sanidad: Sobre las precauciones con motivo de la enfermedad de Málaga."

45 Archivo Histórico de la Univesidad Complutense (AHUC): "Reglamentos Navales" in *Providencias generales,* artículo 93.

46 Anon., *Reflexiones acerca de la epidemia que reyna* [*sic*] *en Cádiz* (Madrid: Imprenta Real, 1800), 46.

47 In 1800, the Cadiz fever epidemic killed 7,387 people (population 48,520), in Seville 14,685 (population 80,568); in Malaga 1803, 6,887 (population 51,7459); in 1804, 11,486. The 1804 epidemics extended to Alicante (2,472 dead people, population 13,212), Antequera (2,948, population 14,5779), Velez Málaga (5,245, population 12,700), Córdoba (400, population 40,000), Cádiz (28,92, population 54,899), and Cartagena (11,445, population 33,222). Aréjula, *Breve,* figures 1-6 (see note 42).

48 Anon., *Mercurio de España* (15/4/1805): 69-71.

radical innovation that solved problems posed by large-scale production.[49] The royal apothecary, entrepreneur, and chemistry professor Pedro Gutiérrez Bueno (1745-1822) responded in this case with a new model of Guyton's machine, rendering it easier to replicate.

The Fumigating Machine Travels to Spain

Gutiérrez Bueno's ability to move comfortably between artisanal, fashionable, and academic milieu was the key to establishing the Spanish fumigating machine's authority.[50] Educated as an apothecary, in 1788 Gutiérrez Bueno translated the *Méthode de nomenclature chimique* of Lavoisier, Guyton de Morveau, Fourcroy and Berthollet.[51] He taught chemistry in the surgeons' college of San Carlos (1801-1804) and was director of the Royal Laboratory of Chemistry from 1787. Situated in the center of Madrid, the laboratory hosted his popular chemical classes which were frequented by craftsmen, apothecaries, and aristocrats.[52] Gutiérrez Bueno was close to influential people, such as the poet and politician Leandro Fernández de Moratín, who called him "Petrus *Bonus*", and the editor Juan Antonio Melon, who published Gutiérrez Bueno's treatise on dyes and glass making.[53] In 1790, Gutiérrez Bueno famously accom-

49 See the essays by Christie and Simmons, this volume. Thomas P. Hughes, "The Evolution of Large Technological Systems," Wiebe E. Bijker, Thomas P. Hughes, and Trevor J. Pinch eds., *The Social Construction of Technological Systems: New directions in the sociology and history of technology* (Cambridge, MA: MIT Press, 1987), 51-82, esp. 57-58; Thomas P. Hughes, *American Genesis: A century of invention and technological enthusiasm, 1870-1970* (Chicago: University of Chicago Press, 2004), 53 ff.

50 Paula Carrasco Jarabo, "Vida y obra de Pedro Gutiérrez Bueno," *Boletín de la Sociedad Española de Historia de la Farmacia* 15 (1964): 154-69; 16 (1965): 10-24; 71-86; 101-18 and 153-77; Carrasco transcribes a document in the Archivo General del Palacio Real de Madrid (Leg. 490, Exp. 26), in which Gutiérrez Bueno listed his merits ((1965): 113-14). He prepared the acids for fumigating, the machines, and the leaflets to be sent to Granada, Cádiz, Valencia, and other villages.

51 Louis-Bernard Guyton de Morveau et al., *Método de la Nueva Nomenclatura Química* (Madrid: Don Antonio de Sancha, 1788).

52 Antonio García Belmar and José R. Bertomeu Sánchez, "Pedro Gutiérrez Bueno, los libros de texto de química, y los nuevos públicos de la química en el último tercio del siglo XVIII," *Dynamis* 2 (2001): 351-74; José R. Bertomeu Sánchez and Antonio García Belmar, "Pedro Gutiérrez Bueno (1745-1822) y las relaciones entre la química y la farmacia durante el último tercio del siglo XVIII," *Hispania* 208 (2001): 539-62.

53 Melon published the journal *Semanario de Agricultura y Artes*, in which Gutiérrez Bueno's daughter María Antonia collaborated.

panied an aristocratic female association, the *Señoras de las cárceles*, to analyze the Madrid prisons' air.[54] He was also keen to participate in public health issues. He discussed the quality of Madrid airs and waters and the proper way of coating pots for preventing certain type of fevers.[55] In addition, some of his entrepreneurial activities were closely related to the oxy-muriatic acid.[56] He directed the production of sulfuric acid in a manufactory beside the Manzanares River. In 1790, he translated Berthollet's treatise on the use of oxy-muriatic acid for bleaching, and employed Berthollet's method in the Royal Manufactory of San Idelfonso. He even designed a domestic machine, which according to him, easily bleached cloth at home with oxy-muriatic acid.[57]

It was this acquaintance with materials, devices, and large-scale production that allowed Gutiérrez Bueno to substantially cheapen Guyton's models. This he did using low-cost wood, substituting round forms for square boxes (which were easier to mold), and inventing a new system for holding the gas (see Fig. 4.5). Instead of screws, Gutiérrez Bueno used wedges. To release the gas, one opened the box by pulling up the cover and dragging out the wedge. Gutiérrez Bueno also developed a large apparatus similar to Guyton's, minus the round forms and screw, which was both simpler and easier to repair. If the wedges came loose, one only needed to add another piece. In addition, the bottles that held the gas could be cheaper, since the system of sealing the glass did not require as much pressure as the Parisian ones. Finally, the Spanish portable machines apparently lasted longer than some of the Guyton models, because the wood did not directly suffer from the corrosive effect of the gas.[58] The

---

54 Elena Serrano, "Chemistry in the City: The scientific role of female societies in late eighteenth-century Madrid," *Ambix* 60 (2013): 139-59.

55 Pedro Gutiérrez Bueno, "Informe," *Memorial Literario*, Agosto (1790): 73-8; Pedro Gutiérrez Bueno, *Analisis de las aguas de Madrid* (Madrid: Imprenta de Villalpando, 1800); Pedro Gutiérrez Bueno, *Método práctico de estañar las vasijas de cocina* (Madrid: Imprenta de Villalpando, 1803); See also Carrasco, "Vida y obra", pp. 107-114 (see note 50).

56 Pedro Gutiérrez Bueno, *Instruccion práctica para destilar las aguas fuertes* (Madrid: Don Blas Román, 1787).

57 Pedro Gutiérrez Bueno, *Memoria sobre el blanqueo del lino, algodón y otras materias* (Madrid: Don Antonio de Sancha, 1790).

58 Anon., *Memoria sobre las disposiciones tomadas por el Gobierno para introducir en España el método de fumigar* (Madrid: Imprenta Real, 1805), 8, footnote; The author explained that out of his thirty-five French portable models, the wooden cases of twenty-four were cracked. For the models sold by Gutiérrez Bueno, see *Descripcion y uso del aparato permanente para desinficcionar el ayre* [*sic*] (Madrid: Imprenta de Villalpando, 1805).

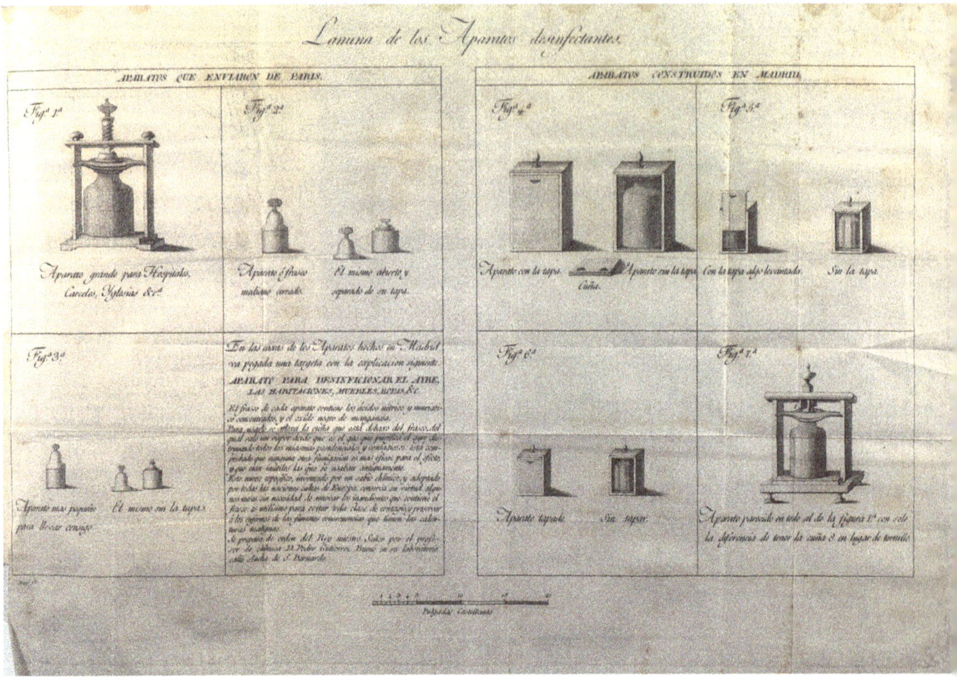

FIGURE 4.5 *Fumigating machines made in Paris alongside those made in Madrid. Anon.,* Memoria sobre las disposiciones tomadas por el Gobierno para introducir en España el método de fumigar *(Madrid, 1805).* COURTESY OF BIBLIOTECA DE CATALUNYA, BARCELONA.

Spanish machines were first distributed in 1805 for the "complete disinfection" of Cartagena.[59]

Marshalling with Enlightened Reformers

The Spanish fumigating machines were construed as material proof of the providence, enlightenment, and effective *oeconomic* policies of the Government. In her poem about Guyton's fumigations mentioned at the beginning of this essay, Rosa Gálvez not only praised Guyton, but also the Prime Secretary Manuel Godoy for popularizing fumigation in Spain. In Galvez's words, Godoy was a "beneficent hero", who destroyed envy, false piety and vile superstition (referring to clergymen who refused to fumigate churches).[60] In fact, the

59 Anon., *Memoria,* p. 9 and pp. 27- 32 (see note 58).

60 Gálvez, "Oda", p. 10 (see note 1).

Government issued substantial propaganda promoting its role in stopping the yellow fevers. In 1805, it published a two hundred-page treatise, comprised of an extensive narration and supporting documents, the *Memoria sobre las disposiciones tomadas por el Gobierno para introducir en España el método de fumigar* (*Memoire on the dispositions taken by the Government for introducing in Spain the fumigating method*).[61] These included descriptions of experiments, readers' letters from the infected cities to local newspapers, correspondence between Godoy and the Supreme Health Board (*Junta Suprema de Sanidad*) and translations of foreign documents, such as the Napoleonic regulations of military hospitals.

As illustrated in figure five, the treatise presented the Spanish machines side by side with the Parisian ones, in a way that encouraged comparison.[62] The image suggested that the king had done a great service to the Spanish population by promoting the benefits of oxy-muriatic disinfection. Moreover, on the label that accompanied the bottles, the reader learned that the apparatus was "invented by a wise chemist [...] adopted by all the educated nations of Europe" and prepared by order of the "King, our Master."[63]

At that time, Spain endured great economic and political turmoil. The weakness of the Monarchy was evident to Spaniards and foreigners alike. Alliance with France in 1797 pulled the country into conflict with Great Britain with devastating consequences for colonial commerce and state finances. The changing relationship with France seriously damaged the uneasy equilibrium between Spain's three traditional political forces, namely, the church, the so-called aristocratic party of the Count of Aranda, and the reformers. The Inquisition gained power and former members of the Government were now prosecuted. Defending the mineral acid fumigations for combating the contagion was a convenient way for Godoy and the Spanish Crown to align with the reformers, without thereby taking on additional political risks.[64]

---

61 Anon., *Memoria* (see note 58).

62 Ibid., figure 1.

63 Ibid., p. 9.

64 Emilio Laparra López, "La inestabilidad de la monarquía de Carlos IV," *Stvdia histórica. Historia Moderna* 12 (1994): 23-34; Jean René Aymes, *España y la revolución francesa* (Barcelona: Crítica, 1989); Claude Morange, "Las estructuras de poder en el tránsito del Antiguo al Nuevo Régimen," Joseph Pérez and Armando Alberola, eds., *España y América entre la Ilustración y el liberalism* (Alicante; Madrid: Casa de Velázquez- Instituto de Cultura Juan Gil-Albert, 1993), 35-7; Emilio LaParra, *La Alianza de Godoy con los revolucionarios* (Madrid: Consejo Superior de Investigaciones Científicas, 1992); Emilio LaParra, "Ilustrados e Inquisición ante la Iglesia constitucional francesa," *Revista de Historia das Ideias* 10

Disinfecting public spaces such as jails and hospitals was deeply connected with new ideas about the role of the state in public welfare. In Spain, from the 1780s onwards, societies of "friends of the country" and scientific associations translated foreign treatises and experimented on the disinfecting properties of different acids. The Royal Academy of Medicine of Madrid translated Jean Janin's *L'antiméphitique* and tested the disinfection power of vinegar. In Barcelona, Carles Gimbernat championed disinfection with nitric acid. He invented a heating lamp for evaporating the fumes of the acid, and translated Smyth's account of the disinfection of the Russian ship *La Union*.[65] The bishop of Barcelona, Pedro Díaz Valdés a cleric with sympathy for the Spanish Jansenist movement, ordered the printing and distribution of Guyton's disinfection method.[66] Valentín de Foronda, a member of the Economic Society of Vascongadas and author of numerous essays on political economy, translated Guyton's article from the *Encyclopédie Méthodique*.[67]

The Government took a very active role in promoting oxy-muriatic acid disinfection. In particular, it promoted public experiments. As scholars have shown, engaging audiences was an effective means to circulate ideas and practices, selling instruments, gaining adepts, and legitimating experts.[68] In July 1805, a commission of prestigious savants did experiments at three different sites: the pharmacy of Gutiérrez Bueno, the Real Casa Hospicio, which hosted Madrid vagabonds and poor people, and its stables.[69] The conclusion was that fumigation with oxy-muriatic acid could be safely applied to goods, people, and animals. These experiments were projected as crucial for the Spanish economy. To prove the advantages that the practice of fumigation would bring, the *Memoria* included the orders that the Supreme Board of Health (*Junta*

(1988): 359-74; Gonzalo Anes, *Economía e ilustración en la España del siglo XVIII* (Barcelona: Ariel, 1969).

65 Jean Janin, *El antimefitico* (Madrid: Imprenta Real, 1782); James Smyth, *Relación de los experimentos hechos por Mr. Menzies* (Madrid: Viuda de Ibarra, 1800).

66 Joaquim Puigvert, ed., *Bisbes, Illustració i jansenisme a la Catalunya del segle XVIII* (Vic: Biblioteca Universitaria, 2000); Juan Bada, "Don Pedro Díaz de Valdés, obispo de Barcelona (1798-1807)," *Anthologica Annua* 19 (1972): 651-74.

67 José Manuel Barrenechea, *Valentín de Foronda, Reformador y Economista Ilustrado* (Álava: Diputación Foral, 1984).

68 Bernadette Bensaude-Vincent and Christine Blondel, eds., *Science and Spectacle in the European Enlightenment* (London: Ashgate Publishing, 2005); Geoffrey Sutton, *Science for a Polite Society: Gender, culture, and the demonstration of enlightenment* (Denver: Westview Press, 1995); James van Horn Melton, *The Rise of the Public in Enlightenment Europe* (Cambridge: Cambridge University Press, 2001).

69 Anon., *Memoria* "Informe de los Facultativos," pp. 37-40 (see note 58).

Suprema de Sanidad) circulated in Malaga in 1804, and those circulated in Cartagena in 1805 for readers to compare. The Malaga orders involved older methods used to disinfect the city after the *cordon sanitaire* period ended.[70] For four months, goods could not be taken out of the city and houses that sheltered sick people had to remain closed. Mattresses, linen and clothes were burnt or buried in the waste grounds with lime. In the ports, quarantine was imposed on all goods. Textile goods were hung in the sun and air for weeks; other merchandise was kept in storehouses for months, or burnt. In contrast, the measures employed in Cartagena in 1805 used machines designed by Gutiérrez Bueno. Twice daily, neighbourhood directors carried out domestic fumigations.[71] The disinfection was completed in a month.

In a way, the acid fumigating technology revived the circulation of goods and people halted by the infection. The doctor Miguel Cabanellas designed special fumigating machines for disinfecting clothes in the lazaretto of San Joseph, Cartagena (figures 4.6 and 4.7). The first was a kind of big closed box with a grid in the middle for letting the gas pass through the textiles. The second was a kind of hut in which one sat while the fumes of the oxy-muriatic acid percolated from below, and which included a breathing-tube for persons worried about inhaling oxy-muriatic acid (item k in Fig. 4.7).

Cabanellas also designed a lazaretto that can be understood as a chemically-based plant for recycling people and goods back to normal circulation (Fig. 4.8). The lazaretto was separated from the city by ditches and walls. It was organized in individual cells where the sick were placed for recovery.[72] Special units for disinfecting clothes and objects were strategically situated. Mattresses, bedclothes, furniture, animals, and people were properly disinfected before being returned to the other side of the ditch. Fumigating allowed goods to be safely reintegrated in the ambient sociomaterial environment.[73]

There is one puzzling question that must be asked. Did the oxy-muriatic acid work? A partial answer may be found in an apparently trivial comment of Cabanellas. While fumigating the matrasses of the Cartagena lazaretto, he noted that all the bedbugs and cockroaches died.[74] The insecticide properties

70 Ibid., "Número Sexto," pp. 33-5; Artiaga, "La ciudad," (see note 40); Mariano Peset and José Luis Peset, *Muerte en España. Política y sociedad entre la peste y el cólera* (Madrid: Seminarios y Ediciones, 1977); Esteban Rodríguez Ocaña, "La cuestión del lazareto marítimo permanente en la España del siglo XVIII, de Cádiz a Mahon," *Asclepio* 40 (1988): 265-76.

71 Anon., *Memoria* "Número Quinto," pp. 27- 32 (see note 58).

72 Quim Bonastra, "Los orígenes del lazareto pabellonario. La arquitectura cuarentenaria en el cambio del setecientos al ochocientos," *Asclepio* 60 (2008): 60-61.

73 On sociomaterial environments, see Lissa Roberts and Joppe van Driel, this volume.

74 Miguel Cabanelles, *Defensa de las fumigaciones ácido-minerales* (Madrid: Repullés, 1814).

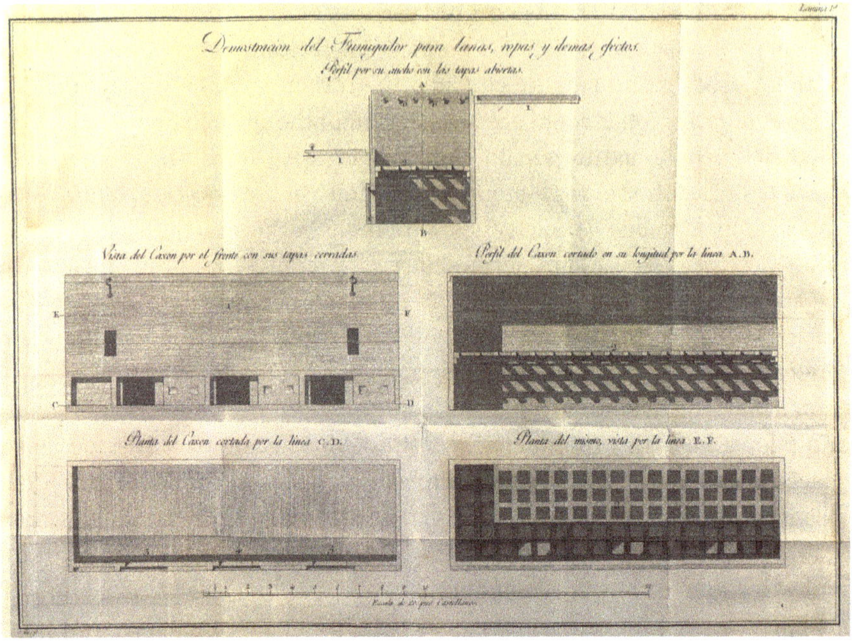

FIGURE 4.6 *Fumigating machine for objects in Anon.,* Memoria. COURTESY OF BIBLIOTECA DE CATALUNYA, BARCELONA.

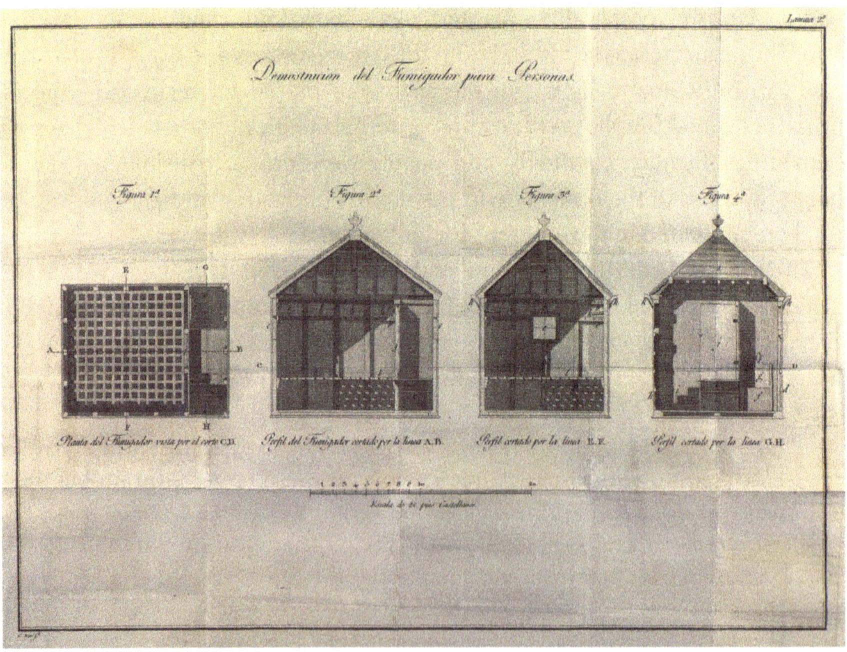

FIGURE 4.7 *Fumigating machine for people in Anon.,* Memoria. COURTESY OF BIBLIOTECA DE CATALUNYA, BARCELONA.

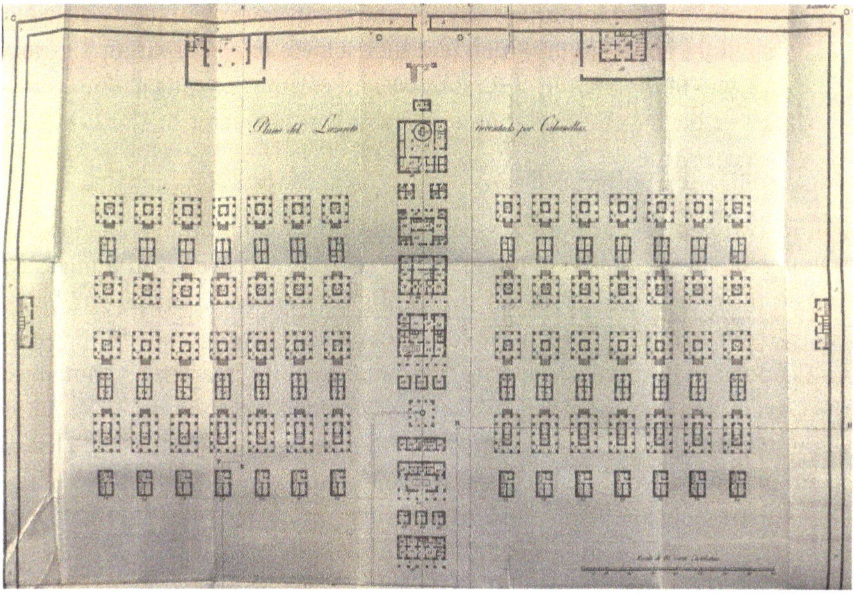

FIGURE 4.8 *Ideal Lazaretto.* Memoria. COURTESY OF BIBLIOTECA DE CATALUNYA, BARCELONA.

of the oxy-muriatic acid were also described in other cases. So fumigations may have been effective after all. As is well known, the virus that provoked yellow fever is transmitted by the bite of a tropical mosquito. The infectious mosquito crossed the Atlantic in American ships, and due to the mild and humid conditions of some Spanish towns, easily reproduced. The oxy-muriatic acid (chlorine) may have inhibited the cycle of infection when used in isolated environments, such as the closed neighborhoods of some southern Spanish towns.

Making Political Propaganda

The fumigating machine was pictured by the Government as a political weapon that could be wielded to win a battle that Spain had been forced to wage "in the middle of its sorrows and calamities". The *Memoire on the dispositions taken by the Government for introducing in Spain the fumigating method* mentioned in the former section finished with the following paragraph:

> People around the world! The annihilation of the last germ of the yellow fever is in your power. The Peruvian Bark and the mineral fumigations

> can achieve that important victory, and when you celebrate it, turn your thankful eyes to Spain, which had such a great role in assuring it to you, and who in the middle of its sorrows and calamities, enjoyed the advantage of having the firm, noble and prudent character of the Prince of Peace [ie. Godoy].[75]

Political propaganda was also disseminated through other media. In 1806, the *Semanario de agricultura y artes* published several articles on acid fumigation and on the crucial role of the Government.[76] The official *Mercurio de España* circulated the letter in which the Ministry of War recommended the use of fumigations in hospitals, lazarettos, jails, and military quarters.[77] Moreover, the Government censured opinions opposed to acid fumigations.[78] There was the notorious case of physician Juan Manuel Aréjula, director of the Health Board in Malaga, who had to cut out a whole chapter on the uselessness of the oxy-muriatic acid for disinfection in his treatise on yellow fever.[79] In December of 1803, the Government ordered the closure of the popular journal *El Correo de Madrid,* and arrested the publisher, owing to the way it depicted the Malaga epidemics.[80]

Nonetheless, some sense of the medical opposition to fumigations may be gained through the *Libro de Juntas* of the Royal Academy of Medicine in Madrid, which contains the minutes of its weekly meetings. With a membership including the most prominent doctors of the time, one of the academy's functions was to advise the Supreme Board of Health.[81] In May of 1804, the academy was consulted on whether the cases of fevers that appeared in Malaga

75 Anon., *Memoria* p. 71 (see note 58).

76 "Extracto de la Memoria que acaba de darse al publico", *Semanario de Agricultura y Artes* 20 (1806): 65-72; 89-94; 99-102; 121-128; 141-144; 159-160; 172-176.

77 Anon., *Mercurio de España* 15/8/1806: 177-80.

78 Ballester and Carrillo, "Repression" (see note 8); Belmar and Bertomeu "España fumigada" (see note 9).

79 The chapter was in Aréjula, *Breve descripción,* (see note 42). He published it when the Goverment changed: *Memoria sobre la ninguna utilidad del uso de los gases ácidos* (Esparraguera: Imprenta del Gobierno, 1821). See Juan L. Carrillo, *Juan Manuel de Aréjula (1755-1830). Estudio sobre la fiebre amarilla* (Madrid: Ministerio de Sanidad y Consumo, 1986); García Belmar and Bertomeu, "España fumigada" (see note 9).

80 AHN. Consejos, 11975: Junta de Sanidad. 26-December- 1803. All the volumes were forbidden.

81 Luis Granjel, *Historia de la Real Academia Nacional de Medicina de Madrid* (Madrid: Real Academia Nacional de Medicina, 2006).

constituted an epidemic or not.[82] The long discussions that followed spotlight the lack of agreement on the origin, treatment, and prevention of fevers that existed in eighteenth-century medical circles.[83]

In 1805, the Government encouraged the academy to respond "judiciously" to foreign works that denied the contagious character of yellow fevers. In particular, the Government specified works from "Anglo-Americans who wanted to confuse us."[84] Notwithstanding this opposition, the Academy made honorary fellows of the physicians Benjamin Rush from Philadelphia and Samuel Lathan Mitchill from New York.[85] The former defended the opinion that yellow fever was not contagious, while the latter defended a doctrine of infection incompatible with prevention by acids. Medical correspondence between physicians who worked in the infected cities also suggested disagreement on the uses of fumigations.[86]

A closer look to what lay people thought about fumigations is provided by the irreverent manuscript *Diálogo de los muertos* (*Dialogs of Dead*) written in Malaga in late 1803, when the city was still cordoned off.[87] The *Diálogo* narrates a conversation held by eight dead people – French, English, Portuguese, Italian, Catalan, Muslim, a Malaga citizen, and a "sensible-man" called Salomon. It is an exceptional document, which openly criticizes the perceived arbitrary and corrupt behavior of the authorities. The story reflects the inhabitants' fears of being unjustifiably secluded in the lazaretto, their animus against the prohibition of masses and religious parades, and their anger against bribed authorities who allowed ships to skip quarantines and facilitated the spread of contaminated merchandise. But above all, the story expressed anger with governmental measures, which the inhabitants of Malaga would later have to pay. Anger was especially directed against fumigations: "they robbed the people with the

82 Archivo de la Real Academia de Medicina de Madrid (ARAM): Junta Extraordinaria del 2 de Mayo de 1804.

83 Ward, *Desire* (see note 24). See also José Manuel López, "Dos textos epidemiologicos ineditos de Antonio Amodóvar Ruiz-bravo (1763-1823)," *Gimbernat* 2003 (39): 55-67.

84 ARAM, *Diario de Juntas*, 19 September 1805.

85 ARAM, *Diario de Juntas*, 24 October 1805.

86 José Antonio Coll, *Apuntes sobre la fiebre amarilla de Cadiz* [Manuscrito]. Biblioteca Histórica Universidad Complutense Madrid, BH Mss 853 (3). Doctor Jose Antonio Coll wrote from Cadiz on the methods for preventing the contagion, including a handkerchief of vinegar applied to the nose, quarantines, isolation of the sick, but he did not mention fumigations.

87 Juan L. Carrillo, Jesús Castellanos and María Dolores Ramos, *Enfermedad y sociedad en la Málaga de comienzos del siglo XIX: El diálogo de los muertos en la epidemia de Málaga (c.1803)* (Malaga: Universidad de Málaga, 1980), 7-11.

fumigations in the port and in the town." The French character as the stereotype of a revolutionary, proclaimed: "When the rights of the citizen and the free man are so greatly attacked, […] it is an heroic act to break the chains of so shameful and vile a slavery."[88]

These opinions forcefully expressed in the *Diálogo* suggest that people did not in fact universally perceive fumigations as either useful or liberating. Some saw them instead as a burden. Ultimately it was the people of the villages who had to pay for the cost of fumigations, mainly in the form of council taxes, so that fumigations were perceived as a new form of corruption among the authorities, as an excuse to generate wealth. Governmental propaganda was clearly not universally persuasive, and may even have had a negative effect on people's opinions.

Conclusion

The fumigating machine embodied the power of the 'new chemistry', both in its materiality – newly-formulated, manufactured gases, the thick glass of the bottle, the pneumatic techniques used for sealing its cover and controlling gas emission – and in its conception, which was grounded on the oxidation properties of acids, a feature of Lavoisier's chemistry. The machine also embodied the belief that the agents of contagion were chemically sensitive entities, the miasmas. Although their precise nature was unknown, they could be combated chemically. The choice of the oxy-muriatic acid above other acids was construed as the product of intelligent chemical design and careful experiments. Moreover, Guyton stressed that the experiments were done by a trained chemist who distinguished between "destroying an odour" and "masking it", emphasizing the distance between the muriatic oxygenated acid and other domestic methods of fumigating.

The machine was thus presented as a scientific device, whose authority as such was initially supported by its external appearance (expensive woods, convincing technology, precision of manufacture and use) and the fact that it was sold by famous instrument makers in Paris. It was advertized as a *reliable* means of fumigation. The user did not need to bother about the quantities of ingredients and time of fumigation because the machine provided a standard way to proceed. It was a ready-to-use device, and so was pictured as giving lay people operational independence from apothecaries and other knowledgeable people. However, it simultaneously contributed to increasing the gap

88 Carrillo, Castellanos and Ramos, *Enfermedad y sociedad*, p. 8 (see note 87).

between lay people and savants. As Simon Werrett argues in this volume, it was a common practice for natural philosophers to work in domestic settings, and ingeniously use everyday objects for their research. Fumigation with mineral acids could probably have been done with adapted domestic devices. But introducing a ready-to use machine eventually detached the knowledge of fumigating techniques from domestic users, so that the knowledge became a matter of expert production and commodity consumption, encouraged – to a degree enforced – by government. Thus the supposed independence of the newly responsible, operative citizen was mediated through an alienation brought about by expert manufacture, and through new degrees and forms of political-administrative control afforded to governmental and municipal authorities. Godoy and the Spanish Bourbon court effectively transformed the machine into a political tool.

Scholars have accounted for the changes that occurred in the eighteenth-century in health policies as a general strategy of power. Using the Foucauldian concept of biopower, they have documented the new practices that linked the care of the individual and the social body to the processes of state formation. However, as Claudia Stein has pointed out, it "was not a straightforward linear process replacing sovereign power [by biopower], but rather, a matter of struggle and contestation within the eighteenth-century absolutist state."[89] It was certainly the case in Spain that degrees of intra-elite contestation and popular resistance to the new regime of disinfection and public health were visibly present, despite the propagandist efforts of government, despite the figuring of the machine to convey values of patriotism, dedication and *oeconomy*, and despite the public writings and public demonstrative experimentation of chemists and physicians.

Finally, this essay has highlighted the importance of material culture for explaining the embedding of knowledge in society. The fumigating machine probably did more for spreading the new chemistry of acids and gases than any textbook. The machine afforded a particular understanding of how knowledge should be produced, a particular view of contagion and sickness, of the connections between body and environment, chemistry and life, and of sickness and social responsibility. Because it possessed these affordances, it also helped to forge a new relationship between the power of the state and the citizen.

89 Claudia Stein, "The Birth of Biopower in Eighteenth-Century Germany," *Medical History* 55 (2011): 331-37, on 335.

Acknowledgments

My heartfelt thanks to Lissa Roberts, Simon Werrett, and John Christie for being such perceptive and careful editors. I express my gratitude to Antoni Malet, who helped me to shape my argument with his witty comments on early versions. Thanks also to Oliver Hochadel, Jon Arrizabalaga, and Nina Lerman for useful discussions. My wonderful colleagues in the "Situating Chemistry" group and in my home institution MPIWG-Berlin provided inspiration and joy, while the Fritz Thyssen Stiftung provided funds for a useful workshop in 2013.

CHAPTER 5

Arsenic in France. The Cultures of Poison During the First Half of the Nineteenth Century

José Ramón Bertomeu Sánchez

This essay reviews the movement of poisons across different popular, medical and legal cultures during the 1830s and 1840s in France. Many French people at that time felt that they were living in a "wave of poisoning crimes", mostly performed by using arsenic, which was regarded as the "king of poisons" during the nineteenth century. Poisons such as arsenic were common materials employed in everyday life for different purposes in agriculture, industry and medicine. They were also frequent protagonists in popular literature, folk tales, theater plays, and other forms of popular culture. At the same time, many poisons were both objects and tools of inquiry in medicine and science. Their composition and deleterious properties had attracted the attention of doctors and natural philosophers since ancient times. With the development of animal experimentation, poisons were increasingly employed as tools for research, whose dramatic physiological effects were employed for investigating the vital functions. From a legal point of view, poisons were criminal tools for performing silent murders, which were very difficult to prove in court. The testimony of regular witnesses was useless due to the secret nature of poisoning crimes, so judges frequently requested the advice of experts in medicine and chemical analysis. Prompted by unexpected situations and puzzling questions, nineteenth-century toxicological research developed along with criminal investigations during poisoning trials.[1] The toxic effects of arsenic largely depended on the nature of the compounds, the ingested quantity, the nature of the victim and the dosage (from acute to long-term poisoning). Consequently, arsenic presented a great variety in the character, combination, and severity of symptoms, including also perplexing and misleading exceptions. For this reason, a nineteenth-century professor of legal jurisprudence dubbed arsenic as the

1 These issues are discussed in J.R. Bertomeu Sánchez, "Animal Experiments, Vital Forces and Courtrooms: Mateu Orfila, François Magendie and the study of poisons in nineteenth-century France," *Annals of Science* 69 (2012): 1-26.

 | DOI 10.1163/9789004325562_007

"very Proteus of poisons", that is, "capable of producing almost every species of poisonous action."[2]

Apart from its criminal uses, arsenic was employed in many other activities in nineteenth-century France: wallpaper pigment, embalming, agriculture, rat poison, veterinary treatments, medical drugs, and so on. Arsenic was among the regular commodities that could be easily found in a nineteenth-century rural house, commonly bought in pharmaceutical shops. And yet, arsenic never enjoyed the material "self-evidence of a slap in the face", which Lorraine Daston attributes to quotidian objects.[3] Its physical properties (white color and mild taste) were ambiguous and misleading, transforming arsenic into an elusive product, which could be confused with many other quotidian materials: flour, carbonates, salts, and so on. Poisoners largely relied on these properties and terrible accidents and false accusations of poisoning were frequent.

Arsenic was also elusive from the point of view of its detection. As in the case of many other early modern materials reviewed by Emma Spary and Ursula Klein, the existence of arsenic "was never contested, though the ways of its identification as well as its meaning and values were subject to debate."[4] Which tests were the most reliable ones and who was their right interpreter (chemists, doctors, apothecaries) were matters of contention. In short, nineteenth-century arsenic was at once a quotidian material, scientific object, criminal tool and legal concern. Its associated meanings and values were contingent and varied considerably among forensic experts, lawyers, judges or poisoners. However, the historically-located and locally-embedded ontological nature of arsenic was plastic enough to be adapted to the varied needs and expectations of different protagonists. In this sense, arsenic resembles other "boundary objects" studied by historians of science: it could inhabit different

2 "Summary of the lecture delivered by Dr. Donkin, Professor of Medical Jurisprudence at the University of Durham," *Pharmaceutical Journal* 3 (December 14, 1872): 472, quoted by James C. Whorton, *The Arsenic Century: How Victorian Britain was poisoned at home, work, and play* (Oxford: Oxford University Press, 2010), 15; For a popular account on the general history of arsenic see John Parascandola, *King of Poisons. A history of arsenic* (Washington: Potomac Books, 2012).

3 Lorraine Daston, ed., *Biographies of Scientific Objects* (Chicago: University of Chicago Press, 2000), 2; See also Lorraine Daston, ed., *Things That Talk: Object lessons from art and science* (New York: Zone Books, 2008).

4 Ursula Klein, Emma Spary, eds., *Materials and Expertise in Early Modern Europe: Between market and laboratory* (Chicago: University of Chicago Press, 2010), 7-10, on 9.

social and cultural contexts, encouraging exchanges and interactions among legal, scientific and popular cultures.[5]

This essay explores some of these interactions, circulations and unequal exchanges by following the traces of arsenic in nineteenth-century France. The first section reviews the different uses of arsenic in everyday life, from agriculture to medicine. I remark on its elusive nature regarding color, taste and smell. I discuss also the first attempts of regulating the circulation of dangerous substances in France. The next section deals with the entanglement between the ambiguous identity of arsenic and the different methods employed for detecting mineral poisons during the 1830s. These methods provided different visual and material forms of proof, which were employed for several purposes in laboratories, academies and courtrooms. Attention is paid to the new high-sensitivity chemical techniques such as the Marsh test. I also highlight the persistence of old methods, which were employed in different contexts, sometimes for different purposes. The next section deals with expert controversies emerging from poisoning trials and their circulation in newspapers, literature and other forms of popular culture. In the last section, I claim that these movements were multidirectional and involved new problems and challenges for toxicologists in courts.

Ubiquitous: The Many Uses of Arsenic

Arsenic compounds were employed for a varied range of goals during the nineteenth century. Around 286,000 kg of different arsenic compounds (oxide and

---

5 On boundary objects, see the famous essay by Susan Leigh Star and James Griesemer, "Institutional Ecology, Translations, and Boundary Objects: Amateurs and professionals in Berkeley's Museum of Vertebrate Zoology, 1907-1939," *Social Studies of Science* 19 (1988): 387-420; On issues related to scientific objects, everyday materials and commodities see Ursula Klein, Wolfgang Lefèvre, *Materials in Eighteenth-Century Science. A historical ontology* (Cambridge, MA: MIT Press, 2007) and Ursula Klein and Carsten Reinhardt, eds., *Objects of Chemical Inquiry* (Sagamore Beach, MA: Science History Publications, 2014); For a recent review of the literature see Simon Werrett, "Matter and Facts: Material culture and the history of science," Alison Wylie and Robert Chapman, eds. *Material Evidence: Learning from archaeological evidence* (London: Routledge, 2014), 339-352. I am grateful to Simon Werrett for this text; On the even more unstable identity of another toxic substance, see Astrid Schrader, "Responding to Pfiesteria Piscicida (The Fish Killer): Phantomatic ontologies, indeterminacy, and the responsibility in toxic microbiology," *Social Studies of Science* 40 (2010): 275-306; For different typologies of scientific objects see John Law and Vicky Singleton, "Object Lessons," *Organization* 12 (3) (2005): 331-355.

sulfur minerals) were consumed annually in France around 1840 (most of these products were imported).[6] Around forty percent of this quantity was "white arsenic" (or just "arsenic"), the popular term for what contemporary chemists called "arsenious oxide." The names of its sulfides ("orpiment" and "realgar") had ancient origins but these expressions were still popular in the nineteenth and even twentieth century. Other names (such as "Scheele's green") were more recent but also very popular (in contrast with the scientific names such as "arsenite of copper") due to its broad use as a pigment.[7] Apart from being used as a pigment, arsenic was employed as a component of popular nineteenth-century drugs. The most famous was the Fowler liquor, an alkaline solution of white arsenic, which was introduced by Thomas Fowler during the 1780s and soon became popular in many European countries, notably after being included in the London Pharmacopeia under the name "Liquor Arsenicalis" in early nineteenth century. It was employed for a broad range of health problems and remained in pharmacopeias until the early twentieth century, when new preparations (the most famous being "Salvarsan") expanded the therapeutic applications of arsenic.[8]

Arsenic compounds were also broadly employed in other activities, for instance, in veterinary pharmacy, taxidermy or funeral embalming. For centuries, this latter practice had been reserved for royalty, but it gained popularity during the 1830s among French bourgeois families. The new imaginary of death emerged along with the discovery of cheaper chemical methods such as those developed by Jean-Nicolas Gannal (1791-1852), a French military apothecary and entrepreneur who became famous for this work. Many of the new methods were based on arsenical solutions introduced by arterial injection. The new embalming technology was so popular and lucrative that many patent litigations took

6 Jules Barse and Adolphe Chevallier, *Manuel pratique de l'appareil de Marsh* (Paris: Labé, 1843), 8-9; Frédéric Chauvaud, *Les experts du crime. La médecine légale en France au XIXè siècle* (Paris: Aubier, 2000), 198-199, for more information on poisons in nineteenth-century France. On Britain see Katherine Watson, *Poisoned Lives: English poisoners and their victims* (London: Hambledon, 2004) and Whorton, *The Arsenic century* (see note 2).

7 See the diversity of names in a popular chemistry textbook: Thomas Brande, *Chemistry* (Philadelphia: Blanchard, 1863), 439-446.

8 More details in Parascandola, *King of Poisons*, pp. 146-151 (see note 2); Fowler's solution was included in French pharmacopeias in early nineteenth-century. See Félix-Séverin Ratier and Etienne-Ossian Henry, *Pharmacopée française ou Code des médicaments* (Paris: Ballière, 1827), 403-404; *Codex Pharmacopée Française* (Paris: Bechet, 1857), 117; A limited group of arsenical products were employed in veterinary pharmacy. See Philippe Lébas, *Pharmacie vétérinaire, chimique, théorique et pratique* (Paris: Lelong, 1836), 49.

place during the 1830s and 1840s in France.[9] This practice reinforced the popular idea that arsenic possessed the astonishing power of preserving the decay of bodies poisoned with it. Many observations of this kind had been reported in medical literature and even animal experiments were performed with poisoned dogs. After being buried for several months, their "flesh and alimentary canal were found red and fresh, as if pickled."[10] In this light, suspicions of poisoning where raised when a corpse was found un-decomposed after being inhumed several months.

The uses of arsenic in the French rural world were also varied during the nineteenth century. A mixture of white arsenic and alum was commonly employed for stepping vegetal grains ("chaulage").[11] The practice lasted throughout the nineteenth century in spite of frequent accidental poisonings produced by this method and subsequent attempts by the government to banish it. As the chemist Jean-Baptiste Boussingault (1801-1887) acknowledged in 1856, the treatment of grains with arsenic provided two important benefits to farmers: the preservation of the grains and its effects as a pesticide. Even if non-toxic products (such as sodium sulfate chalk or common salt) could easily replace arsenic regarding the first goal, these non-poisoning products could not deter the action of rats and other animals on grains, as many farmers who ever dared to abandon arsenical compounds had dramatically experienced.[12] This situation explains why arsenic pesticides (such as the popular "Paris green") lasted until the twentieth century in agriculture, in spite of official regulations and frequent accidents. One of the most dramatic cases took place as late as in 1887 in Hyères, when the contamination of vines with arsenic produced eleven deaths and poisoned more than four hundred people.[13]

---

9 On this issue, see Pascale Trompette, Mélanie Lemonnier, "Funeral Embalming: The transformation of a medical innovation," *Science Studies* 22 (2009): 9-30, on 9-14.

10 Robert Christison, "Observations on the Duration of Cholera, the Taste of Arsenic, and its Power of Preserving the Decay of the Bodies of Those Poisoned with it," *Edinburgh Medical and Surgical Journal* 28 (1827): 94-110, on 102-104, quote on 104.

11 Alphonse Chevallier, "Sur la coloration des poisons," *Journal de chimie médicale* 12 (1836): 600-609, on 605; Chevallier explains that the mixture for "chaulage" was made of 6 parts of white arsenic and 2 of alum.

12 Jean-Baptiste Boussingault, "Sur l'opportunité de faire intervenir l'arsenic dans le chaulage des grains," *Annales de chimie* 46 (1856): 458-472 ; "En définitive, le chaulage doit avoir deux buts: l'un de préserver la récolte de la carie, l'autre de la soustraire à la voracité des animaux nuisibles", quotation on 460; Frederic W.J. McCosh, *Boussingault: Chemist and agriculturist* (Dordrecht: Kluwer, 1984), 155.

13 On France see Nathalie Jas, "Publich Health and Pesticide Regulation in France before and after Silent Spring," *History and Technology* 23 (2007): 369-388; William R. Cullen, *Is*

White arsenic was commonly employed in rural life during the 1830s as rat poison, so when a person was put on trial under suspicion of a poisoning crime, the defense frequently alleged that the defendant had indeed bought arsenic for making "mort-aux-rats." There were no strict regulations concerning its commerce and it was acquired in apothecary shops all over France. Poisons only had to be in a "locked and separated" space under the surveillance of the apothecaries, who were requested to limit access to "well-known people" who could justify the use for "their profession" or other reasons.[14] The broad range of uses of arsenic made this restriction useless in practical terms, even if it somehow denied access to indigents, prostitutes, beggars and other destitute people. They could hardly enter an apothecary shop and ask for arsenic to use as rat poison. This is one reason why the profile of poisoners was so different from other common criminals during the nineteenth century, and not only from the point of view of gender. Many poisoners were respectable people who had never been imprisoned. Some of them were relatives or close friends of the victims, so they could easily obtain poison for domestic use and administer it at home without raising suspicion.

Nineteenth-century regulations also requested apothecaries to keep track of the commerce of all poisonous substances, including arsenic. These documents reveal that some defendants in poisoning trials could obtain large quantities of white arsenic without raising major suspicions. One of the most famous of them, Marie Lafarge, could easily acquire around one hundred grams of white arsenic (the lethal human dose is sometimes less than one gram) in three different purchases made in apothecary shops during December 1839 and January 1840. She was so confident of being unsuspicious that she included the following sentence in one of her letters to the apothecary: "Don't think that I want to poison the whole region of Limousin."[15] Criminal records confirm that many other defendants could easily purchase arsenic in apothecary shops. The following year, another woman accused of poisoning three direct relatives, Marie Bernardou, bought thirty grams of arsenic in an apoth-

Arsenic an Aphrodisiac? The sociochemistry of an element (Cambridge: RSC, 2008), 61-67.

14 The regulations are printed in Adolphe Trebuchet, *Jurisprudence de la médecine, de la chirurgie et de la pharmacie en France* (Paris: Baillière, 1834), 615-617. See also Guy Devaux, "Marchands de mort-aux-rats," *Revue d'histoire de la pharmacie* 92 (2004): 509-516.

15 Archives Départementales de La Corrèze (ADC), 5U88, Testimony of Jean-François Lafosse, witness number 39. His son was also interrogated (witness number 26). The register of the apothecary Eyssartier was presented during his testimony in court. See ADC, 5U88, witness number 22. More details in José Ramón Bertomeu Sánchez, *La verdad sobre el caso Lafarge. Ciencia, justicia y ley durante el siglo XIX* (Barcelona: El Serbal, 2015).

ecary shop. Again, the information was kept in a register, which was offered to the judge during the trial.[16]

After the "wave of poisoning crimes" which took place at the end of the 1830s, a magistrate and member of the Conseil d'Etat, Louis-Marie de Lahaye, vicomte de Cormerin (1788-1868), asked for more restrictive regulations concerning the commerce of poisons. He recommended that druggists, apothecaries and grocers should no longer be allowed to sell dangerous substances. But he foresaw many difficulties in replacing arsenical compounds with non-dangerous substances having a similar range of uses.[17] The French government was also concerned with the problem and requested expert reports from learned societies. By the middle of 1840s, new and more restrictive regulations concerning commerce in poisons were adopted in France, similar to the ideas adopted in the British Arsenic Act in 1851. The French decree of 29 October 1846 regulated the sale, purchase and use of poisonous substances, and explicitly forbade the use of arsenical compounds in "the stepping of grains, the embalmment of cadavers and the destruction of insects."[18] The use of arsenic compounds was only allowed for medical or industrial purposes. All arsenical preparations had to be previously approved by learned societies or governmental panels (the Paris School of Pharmacy, Alfort School of Veterinary, Minister of Agriculture, and so on). The effect of these regulations was limited, as the forensic doctor Ambroise Tardieu angrily complained: exceptions in industry and veterinary were frequent, old practices in agriculture were difficult to change and new poisonous substances were introduced without further regulation.[19]

---

16 *Gazette des Tribunaux*, 25 Janvier 1841. Another example is the trial of Victorine Jullien accussed of parricide in Ozillac. Requested by the judge, the apothecary "consulted his register" and noted that the purchase took place exactly on December 21, 1838. See *Gazette des Tribunaux*, 5 May 1839.

17 Louis-Marie de Lahaye, *Mémoire sur l'empoisonnement par l'arsenic* (Paris: Pagnerre, 1842), 25-28.

18 Louis Tripier, *Les codes collationnés sur les éditions officielles* (Paris: Cotillon, 1852), 1344-1345.

19 More details on the regulation of poisoning substances in France during the nineteenth century are in Ambroise Tardieu, *Étude médico-légale et clinique sur lempoisonnement*, (Paris: Baillière, 1875), 150-162; On England and the "arsenic act" see Peter Bartrip, "A "Pennurth of Arsenic for Rat Poison": The arsenic act (1851) and the prevention of secret poisoning," *Medical History* 36 (1992): 53-69; On India, see David Arnold, *Toxic Histories: Poison and pollution in modern India* (Cambridge: Cambridge University Press, 2016) (I am grateful to the author for sending me a preliminary version of a chapter).

Elusive and Misleading: Fallacies of the Senses and Chemical Essays

The popularity of arsenic during the nineteenth century was not only due to its ubiquity and broad use in everyday life, combined with "the easiness with which it could be obtained in stores."[20] Its features also fatally encouraged its use by criminal hands. These included its effects inside the human body, its resemblance to other common domestic products and its mild flavor, which could be masked in soups and drinks when administered to the victim.[21] All these features, along with the ambiguities and uncertainties of detecting methods, transformed arsenic into the "king of poisons" during the first half of the nineteenth century.

The flavor of arsenic was supposed to be mild, even if, for obvious reasons, information about this point was scarce. In fact, toxicologists disagreed regarding the best way of characterizing this property. The most famous British toxicologists, such as the Edinburgh professor Robert Christison (1797-1882), performed risky self-experiments in which he placed the poison on his tongue "as far back as we thought safe" and concluded that arsenic had "hardly any taste at all."[22] In contrast, his French colleague Mateu Orfila (1787-1853), and many other authors such as Foderé or Thenard, described the flavor of arsenic as "acrid" whereas other authors described it as having a "nauseous sweetish taste."[23] Be that as it may, the flavor was easily masked by food and drink, and victims rarely detected the existence of the poison in meals before it was too late. Accidents were frequently reported in newspapers. For instance, in a banquet celebrated in Sainte-Menehould, not far away from Luxemburg, a large number of participants relished a tasty beef stew, which had mistakenly been seasoned with arsenic after being cooked in an oven. Nobody realized that the white powder was not salt or pepper but white arsenic. Only a few of the poisoned guests remembered perceiving a "disgusting taste similar to sour apples."[24]

---

20 Alphonse Devergie, *Médecine légale théorique et pratique* (Paris: Germer-Baillière, 1840), vol. 3, 414.

21 Lahaye, *Mémoire*, p. 18 (see note 17).

22 Whorton, *The Arsenic Century*, p. 62 (see note 2).

23 Robert Christison, "Observations on the Duration of Cholera," quoted on p. 96 (see note 10); On the different views concerning the flavor of arsenic see Charles Flandin, *Traité des poisons* (Paris: Bachelier, 1846-1853), vol. 1, 515.

24 Devergie, *Médecine légale*, vol. 3, p. 414 (see note 20); See also René Lecanu, Adolphe Chevallier, "Rapport sur la nécessité de colorer les substances toxiques dans le but de prévenir

Another dangerous feature of arsenic was its white color and texture, which could be confused with many ordinary products such as salt, carbonates, sugar, or flour. The resemblance caused many dreadful accidents such as the ones previously mentioned. Many accidents (sometimes involving children) were reported in cases where mixtures of arsenic were confused with sugar or salt. The difficulties of reconnaissance also applied to doctors or even would-be poisoners. For instance, a woman (who apparently wanted to kill her daughter) obtained in 1835 the poison from a colporteur, but she requested him to check the nature of the white power and the colporteur poisoned a cat in front of her. The confusion created many embarrassing situations during criminal investigations. When suspicions were raised, any white and tasteless powder could be mistaken for arsenic. During the 1830s, many authors suggested adding pigments to white arsenic as a way to single it out from similar materials, in order to reduce both poisoning accidents and crimes.[25]

Another means for detecting arsenic was the nose: metallic arsenic released a garlic-like smell when sublimated. Many other chemical products were detected by the nose in nineteenth-century laboratories, for instance, lead acetate (with an odor similar to vinegar) and prussic acid (bitter almonds). As Lissa Roberts has remarked, the uses of the senses in chemical practice was broad and diverse: nineteenth-century chemists never "stopped smelling, tasting, touching, or listening in the service of their analytical activities."[26] Even if broadly employed, the sense of smell was far from being fully appreciated by toxicologists and forensic doctors, at least when credible claims had to be presented in court. The famous French toxicologist Mateu Orfila admitted having once deceived himself by his nose when performing an analysis with Nicolas Vauquelin during a poisoning trial in which arsenic was never found by means of chemical tests. He included in his textbook this cautionary tale for disapproving what was a common practice among doctors: they reported an arsenic

les empoisonnements," *Annales d'hygiène publique et de médecine légale* 24 (1840): 264-283, on 278-279.

25 *Gazette des Tribunaux* (14-15 Septembre 1835): 1098; The other examples are reported in Adolphe Chevallier, "Sur la coloration des poisons," *Journal de chimie médicale* 12 (1836): 600-609; See also Émile Grimaud, "De la coloration de l'acide arsénieux," *Bulletin de l'Académie de Médecine* 5 (1840): 403-418.

26 Lissa Roberts, "The Death of the Sensuous Chemist: The new chemistry and the transformation of sensuous technology," *Studies in History and Philosophy of Science* 26 (1995): 503-529, quoted on 507.

poisoning case just because “they had found in the digestive tract a substance which spreads [when heated] a garlic-like odor.”[27]

In spite of these warnings, the smell test was so popular that even common people performed it at home when suspicions were raised. During the Lafarge trial, relatives and friends of the victim affirmed that they had performed the smell test during the days before the death of Charles Lafarge. They took samples of drinks and meals, placed them over burning charcoal and perceived a garlic-like aroma, which they regarded as a confirmation of their worst fears. Neither of them had any previous training in chemistry or medicine, which suggests that the smell test was very popular at this time, even outside the academic community.[28] Moreover, continued criticism in academic writings suggests that the smell test was also widely used by local doctors and apothecaries during criminal investigations all through the nineteenth century.[29]

Colorful Tests and Black Stains

Arsenic could also be detected in toxicological research by means of a large group of chemical reagents yielding characteristic colors and precipitates. As Ernst Homburg has shown, the methods of analytical chemistry experienced a substantial change between 1780 and 1840. New ideas of chemical composition were developed alongside changes in material culture, the reduction of vessels and samples, new test tubes, and further sophistication and more sensitivity in analytical methods. More recently, Catherine M. Jackson has remarked that these changes were prompted by a “glassware revolution” which took place at the beginning of the nineteenth century. The changes involved a new material environment based on glass and its properties (inertness, transparency, and malleability), which prompted a new array of laboratory practices and the spread of small-scale apparatus, which could be made by professional or

27 Mateu Orfila, *Traité des poisons*, (Paris: Crochard, 1826), vol. 1, 357; See also Mateu Orfila, “Rapport médico-légal servant de base à une accusation d'empoisonnement par l'arsenic,” *Annales d'hygiène publique et de médecine légale* 2 (1829): 405-430; For more details on smells and legal medicine in nineteenth century, see José Ramón Bertomeu-Sánchez, “Smell, Chemistry and Microscopy: Bloodstains and nineteenth-century forensic medicine,” *Annals of Science* 72 (2015): 490-516.

28 ADC, 5U88. Testimony of Anna Brun and Marie-Josephine-Aména Buffière.

29 François-Vincent Raspail, *Accusation d'empoisonnement par l'arsenic* (Paris: Gazette des Hôpitaux, 1840), 33; Mateu Orfila *et al.*, “Triple accusation d'empoisonnement: condamnation à la peine de mort,” *Annales d'hygiène publique et de médecine légale* 28 (1842): 107-192, on 110-111.

even amateur chemists with the required skills in glass tube-making.[30] During the following decades, the new tests were systematized and the reagents were "increasingly seen as parts of a single, highly versatile comprehensive methodology [...] for investigating the chemical composition of a substance."[31] During the first third of the nineteenth-century, these analytical techniques were organized in handbooks and special volumes on chemical analysis were published.[32]

Chemical tests were regularly employed for toxicological research. Crime scenes were a constant source of uncertainties and challenges, involving large amounts of unknown impurities, misleading side-reactions involving complex organic products, and requiring high standards of proof for supporting verdicts, sometimes involving the death penalty. First of all, a sample had to be taken from meals, vomits or liquids found in the stomach of the victim. If looking for mineral poisons, samples were usually treated first with acids in order to destroy organic substances, then boiled in water, and the extracts submitted to the action of particular reagents. In the case of arsenic, the most common reagent was hydrogen sulfide, which was supposed to yield a characteristic yellow precipitate when the sample contained arsenic even in very small quantities. This test posed practical problems, the most important being its slowness. Some authors reported having waited hours or even days for the yellow precipitate to be formed.[33] If they were not patient enough, experts could be led astray by the lack of yellow precipitates and might wrongly conclude that there was no arsenic in the analyzed samples. Indeed, a number of such mistakes were reported in contemporary toxicological papers.[34]

More problems arose from ambiguities in the identification of colors, which could turn into false negatives or, even worse, false positives. The transmission of information concerning colors was always complex in the black-and-white

30 Catherine M. Jackson, "The "Wonderful Properties of Glass": Liebig's Kaliapparat and the practice of chemistry in glass," *Isis* 106 (2015): 43-69; For another example of the relevance of the new glass instruments, see the essay by Serrano in this volume.

31 Ernst Homburg, "The Rise of Analytical Chemistry and its Consequences for the Development of the German Chemical Profession," *Ambix* 46 (1999): 1-32, quoted on 3.

32 An example in France is Jacques Thenard, *Traité de chimie*, (Paris: Crochard, 1813-1816); This textbook became the most important reference book in France. See Antonio García Belmar and José Ramón Bertomeu-Sánchez, "Louis Jacques Thenard's Chemistry Courses at the Collège de France, 1804-1835," *Ambix* 57 (2010): 48-63.

33 Alexander Bussy, Charles P. Ollivier and Mateu Orfila, *Réponse aux écrits de M. Raspail sur l'affaire de Tulle* (Paris: Béchet, 1840), 24.

34 Mateu Orfila, "Affaire d'empoisonnement portée devant la cour royale de Maine-et-Loire," *Annales d'hygiène publique et de médecine légale* 9 (2) (1833): 410-417, on 414-415.

world of chemical literature, so toxicologists faced similar problems to other authors working in areas in which colors played a major role in the nineteenth-century.[35] The identification of characteristic tones required many hours of practical work at the laboratory and the mastering of a sophisticated chromatic language. A flavor of the copious language of colors is provided by the following examples taken from the section on arsenic in a popular nineteenth century toxicological textbook: "brownish-red colour", "yellowish-brown muddiness", "a crumbly, foliaceous mass, having a pearly lustre", "bluish-white precipitate", "a yellow colour with a faint tint of orange."[36]

Critics pointed out the difficulties of dealing with slight color nuances, whose identification largely depended on the training, experience and visual skills of experts as well as the rather capricious aptitudes of their organs of vision. François-Vincent Raspail (1794-1878), one of these critics, affirmed that "phenomena related to coloration" were very much "illusory and variable."[37] In a similar vein, Adolphe Devergie (1798-1879), a toxicologist with much experience as a forensic doctor in poisoning trials, asserted in his authoritative textbook on legal medicine, that "nothing can be less certain than the colour of a precipitate. The same colour can offer ten different nuances; four people examining the colour of a precipitate can find four different colours."[38]

Chemical tests faced many other problems related to the undisciplined nature of crime scenes. Even tiny quantities of organic substances could generate masking-effects altering the final color of solutions and precipitates. Moreover, yellow precipitates (similar to those obtained with arsenical samples) could be obtained with many non-arsenical compounds, and dangerously misled experts into false positive conclusions. The risks depended both on the employed methods and the background, skills and experience of the experts. Each test involved not only particular skills, but also different standards of

35 One of these areas was obviously spectroscopy. See Klaus Hentschel, *Mapping the Spectrum: Techniques of visual representation in research and teaching* (Oxford: Oxford University Press, 2002); and, in more general terms, his later book Klaus Hentschel, *Visual Cultures in Science and Technology* (Oxford: Oxford University Press, 2014).

36 Robert Christison, *A Treatise on Poisons* (Edinburgh: Adam Black, 1832), 258-260.

37 François-Vincent Raspail, "Sur les moyens, soit chimiques, soit microscopiques, qu'on a tout récemment proposés pour reconnaître les taches de sang en médecine légale," *Journal général de médecine, de chirurgie et de pharmacie* 102 (1828), 335-350, quoted on 335.

38 Alphonse Devergie, *Médecine légale théorique et pratique* (Paris: Baillière 1852), vol. 3, 17-18: "Rien n'est moins certain que la coloration d'un précipité; qu'un même couleur peut offrir dix nuances différentes; que quatre personnes examinant la couleur d'un précipité pourront lui trouver quatre couleurs différents."

proof and forms of evidence, which could be more or less suitable in courts, where the results had to be presented to judges, lawyers and jurors.[39]

All these issues become evident in reviewing the first years of the Marsh test for arsenic, which is commonly regarded as a landmark in the history of analytical chemistry. In fact, its introduction by no means eradicated previous methods, including other similar reduction tests. Moreover, the apparatus imagined by James Marsh in 1836 was substantially modified by chemists and toxicologists in subsequent years. It was a good example of the plasticity of the new small-scale glass apparatus. It required only a cheap, easy-to-construct vessel, but involved hours of practice and advanced laboratory skills to use. The sample was placed in a flask with zinc and sulfuric acid. If the sample contained arsenic, a thin metallic film was obtained on a porcelain vessel.[40]

There were two alleged advantages to the Marsh test: its capacity for providing "plain matters of fact" and its high sensitivity. In contrast with clinical symptoms, post-mortem examinations or color chemical tests, the Marsh test provided a material form of proof, namely the arsenical black stains obtained on a porcelain vessel, which seemed "to speak for itself" without the mediation of experts and could be dramatically presented in court as the *corpus delicti*. Toxicologists employed these dramatic effects not only in courts but also in classrooms and academies.[41] The second major advantage was its high sensitivity ("beyond any imagination" according to Justus Liebig). When skilled hands were at work, the Marsh test could detect minute amounts of arsenic, which would have remained unnoticed by earlier tests.[42]

The advent of the new test encouraged the marginalization, but never a complete abandonment, of previous methods for detecting arsenic. Like DNA-fingerprints in the 1990s, the Marsh test was employed for unveiling the fallacies and exploring the limits of previous toxicological methods (such as color tests). The introduction of the new test fueled expert controversies with the unwanted result of questioning toxicology in general and the authority of particular

39 On this issue, see Ian Burney, *Poison, Detection, and the Victorian Imagination* (Manchester: Manchester University Press, 2006).

40 For further details concerning the Marsh test see José Ramón Bertomeu-Sánchez and Agustí Nieto Galan, eds., *Chemistry, Medicine, and Crime: Mateu J.B. Orfila (1787-1853) and his Times* (Sagamore Beach, MA: Science History Publications, 2014); On the "glassware revolution" see Jackson, "Wonderful Properties of Glass" (see note 30).

41 Robert Christison, *A Treatise on Poisons*, fourth edition (Edinburgh: Black, 1845), 261; See also Ian Burney, "Languages of the Lab: Toxicological testing and medico-legal proof," *Studies in History and Philosophy of Science* 33 (2002): 289-314.

42 José Ramón Bertomeu-Sánchez, "Managing Uncertainty in the Academy and the Courtroom: Normal arsenic and nineteenth-century toxicology," *Isis* 104 (2013):197-225.

experts in courts. In contrast with DNA fingerprints, the controversy over chemical tests never reached a closure by the emergence of new technologies granted with "exceptional evidentiary status."[43] The controversies lasted for several decades and, in some cases, moved beyond the French medical community to be part of conversations in salons and other spaces of popular culture.

Expert Controversies in Courts

Expert controversies were encouraged not only by the co-existence of different chemical tests in toxicological practice.[44] Other forms of proof were also commonly employed by forensic doctors (for instance, clinical symptoms and post-mortem examinations) and sometimes this medical information was at odds with the conclusions of chemical tests. The different background, experience and location of experts also fueled many controversies. In general terms, local doctors tended to rely on clinical symptoms and autopsies, in part because they had privileged access to these data. Local apothecaries were usually requested to perform chemical tests, while experts from Paris participated in just a few of the trials, notably when different or inconclusive results were obtained by the local experts. Apart from disciplinary barriers, the heterogeneous group of nineteenth-century experts on crime was affected by huge inequalities concerning laboratory resources and academic power.

According to the French Criminal Code, the examining magistrate ("juge d'instruction") had to be assisted by one or two physicians when a violent death was suspected. Under oath, the experts produced a written report answering the questions of the magistrate concerning the circumstances and nature of the crime. At the local level, even *officiers de santé* (the lowest category of French doctors) could participate as experts in a trial. In most cases, many of the local experts were the victim's doctors and sometimes they participated in courts as both experts and as regular witnesses. When reports were inconclusive or different points of view were expressed, judges might request

43 The previous quotation was taken from the study on DNA fingerprints in Michael Lynch et al., *Truth Machine: The contentious history of DNA fingerprinting* (Chicago: University of Chicago Press, 2008), 340.

44 The co-existence of different tests was common in the analysis of many nineteenth-century products. See for instance the tests on the composition and quality of milk (including organoleptic properties, tube tests and scientific instruments) in Peter Atkins, *Liquid Materialities: A history of milk, science and the law* (Surrey: Ashgate, 2010).

new tests from another group of experts, or might accept proposals in that sense from the defense or the prosecution. As a result, the participation of several experts and the existence of multiple reports with opposing conclusions was far from unusual in French courts.[45]

Some examples demonstrate how expert controversies arose in courts. In 1834, a young woman, Zélie Pejac, was accused of having poisoned her employer in the small city of Eauze (Southwest France, near Toulouse). The victim had experienced violent vomiting followed by sudden death after having ingested a meal prepared by the defendant. A local physician ("officier de santé") and an apothecary performed an autopsy concluding that the victim had succumbed to the effects of arsenic. Another group of experts, which included physicians and apothecaries from the neighbouring village, was asked to produce a new report. They "unanimously" concluded that "there was a complete absence of any arsenical substance." In this case, the chemical proofs (mostly negative) were contrary to the medical evidence based on symptoms and autopsy (mostly positive). The young woman was acquitted but, under the pressure of the trial, she lost her mind and went insane.[46]

Apart from the magistrates, defense lawyers could also contact additional experts when they thought that the reports were "incomplete, biased, or contrary to the principles of the art." Without the constraints of official reports, these "consultations" easily turned into long research papers, sometimes published in medical journals.[47] Consultations also opened the window to the participation of experts without credentials, including those who were on the fringes of the academic world or even radical critics such as Raspail. He was rarely requested as an expert by judges during poisoning trials, in part because he never received the title of medical doctor or pharmacist. Another reason was his political activism. Raspail spent years in prison for his opposition to the French monarchy. Lacking academic degrees, experimental skills and laboratory resources, Raspail relied on skeptical arguments about toxicological methods and the limits of scientific proofs in criminal justice to make his arguments.

45 For more information about expert reports and the French legal system, see Frédéric Chauvaud, *Experts et expertise judiciaire: France, XIXe et XXe siècles* (Rennes: PUR, 2003), 192-198; The participation of multiple experts was also common in forensic psychiatry. See Laurence Guignard, *Juger la folie. La folie criminelle devant les assises au XIXe siècle* (Paris: PUF, 2010), 244, which mentions a trial in which ten different experts participated. See also 233 and 241-242; For more details see Bertomeu Sánchez, *La verdad sobre el caso Lafarge*, chapter 4 (see note 15).

46 *Gazette des Tribunaux* (11-12 August 1834).

47 Mateu Orfila, *Leçons de médecine légal* (Paris: Béchet jeune, 1823), vol. 1, 36-7.

One of the most repeated critical arguments against the new high-sensitivity methods (such as the Marsh test) concerned ironically the small quantities of arsenic detected in the analysis. In this "homeopathic legal chemistry" it was hard to avoid all possible sources of minute impurities found in reagents and vessels, graveyard soils, and so on.[48] The most challenging of these impurities was the so-called "normal arsenic", that is, the tiny amount of arsenic which was supposed to exist in normal – that is, non-poisoned- human organs. Playing with the ubiquitous nature of arsenic in nineteenth-century France, Raspail employed his creative imagination to suggest as many potential sources of arsenic contamination as possible. Arsenic might be passed onto buried corpses by natural forces which chemical experiments at the laboratory were unable to detect. Or perhaps unknown phenomena taking place during the process of putrefaction might spread the insidious normal arsenic from the bones to other parts of the corpse, making it a dangerous source of false positives. False positives were much more likely when using high sensitivity tests than when old methods were employed.[49]

Raspail employed more general epistemological questions concerning the differences between legal and scientific evidence. He highlighted tensions between the open-ended character of scientific research and the necessary closure and irreparable consequences of legal decisions, particularly in cases in which the life of the defendant was at stake. In the midst of a famous poisoning trial, Raspail affirmed: "Gentlemen, you must doubt the omnipotence of legal chemistry because it refutes itself every six months."[50] Even granting that all known sources of error had been considered, Raspail wondered who could positively affirm that subsequent studies would not discover new fallacies and problems in toxicological methods. After a death sentence, Raspail argued, who could restore the guillotined head of the defendant when the chemical error was finally acknowledged? In stressing these points, Raspail's skepticism regarding scientific evidence was in tune with growing concerns about judicial errors in French legal writings during the nineteenth century.[51]

It was not the technicalities of chemical tests, but wide-ranging concerns raised by critics such as Raspail which helped to move the controversy from courts and academies to the public arena. Raspail was willing to transport the

48 The expression was employed in an anonymous essay published in *American Journal of the Medical Sciences* 2 (October 1841), 403-417, quoted on 414.

49 See Bertomeu-Sánchez, "Managing Uncertainty"(see note 42).

50 *Gazette des Hôpitaux* (31 December, 1839): 609.

51 On the growing concern about judicial mistakes, see Chauvaud, "Experts et expertises", pp. 230-240 (see note 45).

debate to these new scenarios, whereas other experts were clearly against this circulation, maybe realizing that it was dangerous for their authority. In the midst of a fierce debate in court concerning the value of tests for arsenic, Raspail was challenged to present his conclusions before the members of the Paris Academy of Medicine, where his claims could be "judged by competent men." Raspail soon answered with a letter to the journals, announcing that he was willing to accept the challenge, but not in front of the members of the Academy of Medicine. "The judge," according to Raspail, should be "all the people, the public."[52]

From Courts to Academies and Salons

Although just a small percentage in the total number of murders in nineteenth-century France, poisoning crimes attracted a great deal of public attention. Courtrooms were crowded with large and varied audiences who avidly followed the hearings and the rather dramatic expositions of prosecutors and lawyers. The participation of local physicians or apothecaries aroused further interest, and even more when the experts were famous doctors or toxicologists from Paris. Public attention was also captured by the mysterious circumstances of the crime, the uncertainties regarding verdicts, and not the least, the embarrassing details disclosed by criminal investigations of the victims' or defendants' private lives. Newspapers often reported on these trials, sometimes including documents such as the act of accusation, the plea of the defense, or summaries of the trial offered by journalists. Judicial journals (such as the *Gazette des Tribunaux* or *Le Droit*) published verbatim transcriptions of the oral hearings and extracts from the expert reports. When poisoning trials became particularly popular, books and leaflets were published with further details about the protagonists.

The most famous nineteenth-century French poisoning trial took place during September 1840 in Tulle (La Corrèze). At the beginning of this year, Charles Lafarge, the owner of a bankrupted forge, had died after a short illness and his wife Marie was accused of poisoning him. A group of local physicians and pharmacists were consulted. The autopsy offered some evidence of poisoning, but chemical tests were inconclusive due to a common laboratory accident. New analyses were performed by another group of experts from the capital of the department (Limoges), but they could not retrieve any arsenic from the corpse, even when they used the new Marsh test. The judge requested new

52 *Gazette des Tribunaux* (6,7,8 and 10 June 1840).

tests performed by a joint group of first and second experts but they were also unable to find traces of arsenic. After a long discussion, the prosecution managed to obtain a fourth, definitive test by a group of experts lead by the most famous French toxicologist, Mateu Orfila (1787-1853). He was the dean of the Paris Medical Faculty, a member of many French advisory committees regarding medicine and education, editor of influential journals, and, last but not least, a well-reputed singer who organized popular musical soirées in his salon, which became a meeting point for musicians, physicians, lawyers, politicians and other French notables during the reign of Louis-Philippe d'Orléans.[53]

The participation of a Parisian celebrity like Orfila sparked the public's interest in the trial of Marie Lafarge even more. Moreover, Orfila introduced an unexpected and dramatic turn in the judicial developments, widely commented upon in newspapers and further publications. Contradicting previous expert reports, Orfila found very tiny quantities of arsenic in the victim's corpse, suddenly dashing the high hopes of the defense opened by previous analysis. In a desperate move, Marie Lafarge's lawyer attempted to contact Raspail, but when he arrived in Tulle the legal proceedings were over and Marie Lafarge had been indicted for murder and, later, imprisoned for life. Raspail, however, wrote a long report contradicting the conclusions of Orfila. Many other popular publications followed, in which public hearing debates were transformed into rather literary reconstructions by journalists and other commentators. Excerpts from medical reports and fragments of dialogs taken from verbatim transcriptions of oral hearings were frequently employed, so providing plausibility or amplifying the dramatic force of the narratives. Many literary genres were mobilized, from autobiographies, letters, articles in newspapers, poems, and theater plays, to different forms of academic literature, including medical papers, proceedings of learned institutions, controversial leaflets and expert reports. A review of some of the publications concerning the Lafarge affair, which appeared in an English magazine in 1841, remarked on the overlapping of fictional, medical and judicial literature dealing with poisoning trials:

> We confess to having been singularly interested in [...] the trial of Madame Lafarge for the murder of her husband. As a Romance of Real Life, it strongly exemplified the adage that Truth is stranger than Fiction; for certainly no living dramatist could have invented such a plot, or such characters, or such scenes as occurred in its progress. No extravagant

53 José Ramón Bertomeu Sánchez, "Classrooms, Salons, Academies and Courts: Mateu Orfila (1787-1853) and nineteenth-century French toxicology," *Ambix* 61 (2014): 162-186.

> German tale ever presented a wilder mixture of the revolting, the horrible, and the ludicrous. It resembled one of our own Terrific Melodramas of strong Tragic Interests.[54]

Thanks to these dramatic ingredients, the Lafarge drama soon turned into a trending topic in French salons. Many other circumstances contributed to the outstanding popularity of this affair. First, the extraordinary biography of Marie Lafarge, a well-cultivated Parisian woman who was a friend of the writer Alexandre Dumas and other French notables, and so hardly representative of the poor and unknown people who were placed on the bench of the accused. Her autobiography, published shortly after the trial, went through several editions and several translations appeared during the nineteenth century. The translations, the reports by journalists, and the publication of many polemical texts by forensic doctors and lawyers, helped to expand interest in this affair to other countries. The prosecutor could affirm in court that the trial had captured the attention of the "whole [of] Europe." Celebrated writers such as Heinrich Heine, Alexandre Dumas, Gustave Flaubert and many other authors wrote thousands of pages on the issue.[55]

The fame of the experts (such as Orfila and Raspail), who were involved in a fierce controversy, was an additional reason for the popularity of the trials. Their opposition emerged for both scientific and political reasons. Raspail was a well-known Republican activist whereas Orfila was a good representative of the group of notables supporting the new Orleanist monarchy. The public who followed the poisoning trials could not help mixing these political battles with the debates concerning the reliability of chemical tests for arsenic. French public opinion was divided into two groups: those supporting the innocence of Marie Lafarge and those accepting the guilty verdict. In general terms, most of the members of the first group were also Republican or, at least, critics of the Orleanist monarchy (like Raspail), while the second group mostly included people who accepted the political order (like Orfila). At the same time, these positions involved contrasting views concerning toxicological methods. The first group were more willing to hear the criticism of Raspail against the new high-sensitivity tests for arsenic prompted by Orfila. On 14 September 1840, almost at the same time that Orfila presented his surprizing final report in the

---

54 *The New Monthly Magazine* 3 (1841): 268; On the connections between literature and crime fiction, see Lawrence Frank, *Victorian Detective Fiction and the Nature of Evidence: The scientific investigations of Poe, Dickens, and Doyle* (London: Palgrave, 2003) and Burney, *Poison* (see note 39).

55 Bertomeu Sánchez, *La verdad sobre el caso Lafarge* (see note 15).

court of Tulle, the duchesse de Dino noted in her diary that all conversations in the salons were on the affair Lafarge. "Here," she wrote, "as everywhere, there are quite contrasting views on this issue."[56] The situation was captured by Flaubert in one of the first passages of *Sentimental Education*. The hero of the novel, the young Frédéric Moreau, who had just graduated and was about to start his studies of law in September 1840, was invited by his mother to have dinner in her house in Nogent-sur-Seine, a village situated a hundred kilometers east of Paris:

> When he [Frédéric] entered the drawing-room, all present arose with a great racket; he was embraced; and the chairs, large and small, were drawn up in a big semi-circle around the fireplace. M. Gamblin immediately asked him what was his opinion about Madame Lafarge. This case, all the rage at the time, did not fail to lead to a violent discussion. Madame Moreau stopped it, to the regret, however, of M. Gamblin; he deemed it serviceable to the young man in his character of a future lawyer, and, nettled at what had occurred, he left the drawing-room.[57]

These violent discussions in salons encouraged the popular interest in the chemistry of arsenic. In November 1840, just a few weeks after the guilty verdict, the large amphitheater of the Paris Faculty of Medicine was crowded by a varied audience following Orfila's experiments on arsenic. During this time, a Parisian pharmacist organized a soirée at which he explained in front of some twenty people, "all the experiments regarding arsenic poisoning," reconstructing the chemical analysis performed by the experts during the Lafarge trial and supporting Orfila's views on this issue.[58]

The polemical writings of Raspail were also addressed to this popular audience. As noted above, he offered not only technical details concerning the chemistry of arsenic and its methods of detection, but also dramatic details of the story (which he regarded as proofs of the innocence of Marie Lafarge) and bitter criticism against the all-mighty academic power of his opponent, Orfila. His claims could be found in newspapers, letters and fictional works in this period. One of the most popular arguments concerned the tensions between the ubiquity of arsenic and the minute quantities of poison detected by the Marsh test. When such minute quantities were involved, it seemed that arsenic

56 Quoted by Anne Martin-Fugier, *La vie élégante ou la formation de Tout-Paris, 1815-1848* (Paris: Fayard, 1990), 170-71.

57 Gustave Flaubert, *Sentimental Education* (Kent: Wordsworth, 2003), 12-13.

58 *L'Esculape* (19 November 1840): 125-126.

could be found everywhere, making the results of chemical tests either inconclusive or dangerously misleading. The writer George Sand, who like many others passionately followed the news of the Lafarge trial, affirmed that positive results in chemical tests was far from being reliable proof of poisoning crimes. "Maybe Orfila will discover in the next six months that arsenic does exist in the liver or in the brain of all corpses."[59] Combining the ubiquity of arsenic with the power of high-sensitivity tests, it seemed that arsenic could "be found everywhere", as Gustave Flaubert remarked in his unpublished *Dictionary of received ideas,* in a brief paragraph on arsenic. Similar concerns were conveyed in the engraving by Honoré Daumier published around 1841, in which Orfila says: "I am so sure of my facts that now I am going to poison my intimate friend [...] and I will find arsenic in his spectacle lenses." If self-assured experts could detect arsenic everywhere, what was the probatory value of the Marsh test in courts?[60]

The German poet Heinrich Heine (1797-1856), who was exiled in Paris in 1840, offers another example of the varied reactions caused by the trial of Marie Lafarge. Like many other left-wind activists, Heine employed the controversy in order to question the bourgeois order implemented by the Orleanist monarchy in France, in which Orfila occupied such a prominent position. Heine disqualified Orfila as "a flatterer of the powerful people and detractor of the oppressed ones", "as false in his talk as in his singing" (ironically referring to Orfila's fame as singer in salons). According to Heine, the poison was not in Charles Lafarge's remains but in "Orfila's heart." However, he was hardly convinced of the innocence of Marie Lafarge. Indeed, he thought that she had committed a desperate act of "legitimate defence" against a rude and cruel husband who had condemned her to many "moral torments and mortal deprivations." Heine aimed to transform this affair into a starting point for revisiting the situation of women in France.[61]

59 George Sand's Letter to Eugène Delacroix, 22 September 1840, quoted by Chantal Sobieniak, *Rebondissements dans l'affaire Lafarge* (Paris: Lucien Souny, 2010), 219-220.

60 Chemical Heritage Foundation, Philadelphia, Fine Art Collection (FA 2000.001.142); Whorton, *The Arsenic Century*, pp. 96-97 (see note 2), quotes a popular English novel by Humphry Sandwith, *Minsterborough: A tale of English life* (London, 1876), in which the same theme is discussed by an elderly physician "These new-fashioned chemists [...] will find arsenic [...] in your walking-stick; they will indeed. I'll lay my life, sir, that they would extract arsenic from my hat."

61 Heinrich Heine, *Lutèce: lettres sur la vie politique, artistique et sociale de la France* (Paris: Levy, 1866), 123-126.

FIGURE 5.1 *"L'amitié d'un Grand Chimiste n'est pas un Bienfait des Dieux." Hand-colored lithograph caricature by Honoré Daumier (1841). The character representing Orfila (on the right) affirms, "I am so sure of my facts that now I am going to poison my intimate friend* […] *and I will find arsenic in his spectacle lenses."* GIFT OF FISHER SCIENTIFIC INTERNATIONAL. CHEMICAL HERITAGE FOUNDATION COLLECTIONS, PHILADELPHIA (FA 2000.001.142). *For re-use, contact reproductions@chemheritage.org.*

From Salons to the Courts

The former examples show how controversies over arsenic detection in courts were intermingled with debates regarding other political and social issues. In this way, poisoning trials created a propitious context for exchanges among popular, legal and medical cultures in nineteenth-century France. Thanks to these exchanges, a large number of people were acquainted with technical details concerning the chemistry of arsenic and the methods for detecting it. In fact, some of the controversial issues were raised in courts by lawyers, magistrates and jurors without any previous training in legal medicine. Experiments on absorption of poisons developed by the French toxicologist Mateu Orfila between 1838 and 1840 offer an example. Orfila pursued a continuing program of research on the absorption of arsenic by using the new possibilities opened by the high sensitivity of the Marsh test. He performed experiments both with poisoned dogs and with samples obtained during criminal investigations from human organs. As his British colleague Robert Christison declared in 1845, his research was "pregnant alike with interesting physiological deductions and valuable medicolegal applications."[62] Regarding physiological deductions, Orfila attempted to settle old debates concerning the action of poison, opening the window to similar research on medicinal drugs. The main "medico-legal applications" concerned criminal investigations involving long-buried corpses, in which no liquids from the stomach were available and poisons needed to be retrieved from the remains of internal organs (in which poisons could have been absorbed).[63]

Several experiments performed by Orfila convinced him that arsenic might also be found in very small quantities in the bones and organs of non-poisoned animals, what came to be known as the problem of "normal arsenic." In one of these experiments, Orfila prepared a soup with beef and vegetables. After boiling it for seven hours, he took a sample and introduced it to the Marsh apparatus, so obtaining some arsenical black stains. In a letter addressed to the Academy of Medicine in April 1839, Orfila concluded that "if new experiments confirm this result, it will be demonstrated that our everyday beef soups contain an arsenical compound."[64]

---

62 See Christison, *A Treatise on Poisons*, pp. 227-228 (see note 41).

63 Bertomeu-Sánchez, "Managing Uncertainty" (see note 42).

64 *Bulletin de l'Académie de Médecine* 3 (1839), 682: "Si de nouvelles expériences confirment ce résultat, il sera démontré que le bouillon de bœuf que nous prenons tous les jours contient une préparation arsenicale." The soup was made of "cinq livres et demie de bœuf non

Although it was a minor detail in his whole research on normal arsenic, the distressing image of arsenic in regular beef soups was widely commented upon in newspapers and soon captured the public imagination. Unsurprizingly, these frightening results were also employed by Raspail as a proof that arsenic could be found everywhere. Another critic of Orfila's methods creatively transformed the experiment into alarming news for Parisian gourmand when he wrote: "On April 2, 1839, Orfila read at the Paris Academy of Medicine a paper summarizing almost two hundred experiments aimed at demonstrating that the broth consumed in different Paris restaurants was arsenical."[65] In fact, these two hundred experiments were on normal arsenic in general, and just one of them was related to soups (which by no means were taken from any Paris restaurant but, as the article indicated, they were prepared by Orfila). However, the image of "arsenic in soups" became so popular that it was even discussed during some poisoning trials at the end of this year:

> Judge: "Have you not written that arsenic could be found even in soup?"
> Orfila: "Yes, it comes from the normal arsenic contained in the bones, but, remarkably, it is never found in the liver; and we found it [arsenic] in the victim's liver."[66]

The episode reveals the unwanted consequences of the public interest in toxicological research, poisoning trials and celebrities such as Orfila. Academic meetings were usually discussed in both the medical and popular press and papers on poisons attracted further attention during the years of the famous poisoning trials, at the end of the 1830s. In this situation, a particular and inconclusive animal experiment, when moved from laboratory to the academy and from the academy to the public arena, could be transformed from an esoteric discussion of the absorption of arsenic into a frightening image concerning soups in restaurants. These images were reintroduced in courts by skeptical experts or even by lawyers, judges or jurors who read newspapers with reports on trials or excerpts from academic meetings. This example offers further evidence of how the circulation of information concerning arsenic was multidirectional and involved creative exchanges among different legal, popular

---

désossé, avec des carottes, des panais, des navet, des poireaux, de l'oignon brûlé, un clou de girofle et du sel."

65 Devergie, *Médecine légale*, vol. 3, p. 449 (note 20). "Le 2 avril 1839, M. Orfila lut à l'Académie de Médecine un travail résultant de près de deux cent expériences pour démontrer que le bouillon pris dans les divers restaurants de Paris était arsenical."

66 *Gazette des Tribunaux* (2-3 December 1839): 106.

and academic cultures. The new high-sensitivity methods for detecting arsenic afforded additional possibilities towards new interactions, introducing new images to be interpreted and discussed.

Conclusion

Arsenic in nineteenth-century France was simultaneously an object of medical and scientific inquiry, an everyday material employed for multiple purposes, a criminal tool to be detected by toxicologists during legal investigations, and a frightening ingredient in the public imagination, which aroused popular interest in chemical tests and toxicologists. Occupying so many different worlds, nineteenth-century poisons are good examples of "precarious substances" which "have no specific place in any given order of things." They were "essentially characterized by their dynamics" and by the different ways in which their effects "were perceived and framed."[67] The chemical and medical features of arsenic were crucial insofar as they raised a broad range of opportunities and constraints in connection with the interests, anxieties, and practices of diverse protagonists including physicians, pharmacists, forensic doctors, lawyers, magistrates, journalists, writers, and the general public who crowded courtrooms and perused newspapers looking for information on poisoning trials. The previous discussion offers many examples of the complex mixture of sociomaterial features and cultural representations involved in the history of materials such as arsenic and others studied in this book. From this point of view, arsenic was during the nineteenth-century "the very Proteus of poisons", not only from the point of view of its varied poisonous effects and its elusiveness to chemical tests, but also regarding its place in both academic research and public imagination.

The changing and multiple identities of arsenic were never completely isolated thanks to the frequent transit of historical actors, objects, texts, practices and values from one social setting to another. The diverse experts involved in criminal investigations (forensic doctors, chemists, toxicologists, lawyers and magistrates) played a crucial role in these transits and hybridizations. The role of journalists was also very important insofar as they wrote broadly-read texts, which both answered to and sparked public interest in poisoning trials. Surrounded by a rather theatrical atmosphere, courtrooms were the privileged

67 Viola Balz, Heiko Stoff, Alexander v. Schwerin and Bettina Wahrig, eds., *Precarious Matters: The history of dangerous and endangered substances in the 19th and 20th centuries* (Berlin: Max Planck Institute for the History of Science, 2008), 1.

places for these unequal exchanges among the different cultures of poison during the first half of the nineteenth century. Experts were requested by law to present reports on problems whose frame and focus were far from being under their control. This puzzling situation emerged not only from the impure and undomesticated nature of crime scenes, but also from the unexpected questions asked by judges, lawyers and jurors, who could appropriate the toxicological information provided by experts in very creative ways. As the last section demonstrated, by reading newspapers and judicial journals, lay people could also become acquainted with many details concerning, for instance, chemical tests for poisons and the principal delusions concerning false positives. Expert controversies in courts such as that between Orfila and Raspail generated further attention on poisoning trials and produced a broad range of publications, which contributed to the creative circulation of many details related to the toxicology of arsenic.

These circulations and exchanges had several features, including the ubiquity of arsenic in the nineteenth-century French rural world, the lack of strict regulations concerning its commerce and its elusive and somehow misleading physical nature, and the different methods for detecting it. Many tests co-existed, all of them with particular problems, assumed fallibilities, required skills and areas of uncertainty. Data was obtained by means of medical examinations, post-mortem autopsies, animal experiments and chemical tests. When different methods were employed at the same time, contrasting conclusions might fuel expert controversies. Historians have to take into account the "shock of the old" technologies and their creative interaction with the cutting edge of research in chemical analysis, moving from novelty-focused narratives to use-centered histories of toxicological methods.[68]

At the end of the 1830s, toxicological methods employed in French courts included traditional and very popular methods (such as the smell test), which were never replaced even with the advent of the tube tests and the new high-sensitivity technologies such as the Marsh test. Each method involved uncertainties, dangerous fallacies and particular forms of proof, which had to be converted into credible claims in both academic and legal contexts. Controversies were also fueled by the participation of several group of experts in poisoning trials. "Consultations" created another window for the participation of experts without credentials, including radical critics and activists such as Raspail. They introduced new epistemological concerns about the reliability of scientific proofs in criminal trials and other general topics related to science

68 David Edgerton, *The Shock of the Old: Technology and global history since 1900* (London: Profile Books, 2006), xi-xiii.

and the law. Thanks to their participation in poisoning trials, public debates concerning famous poisoning trials, such as the Lafarge affair, turned into fierce discussions about French politics, the situation of women in France, the role of jurors in poisoning trials and other issues related to criminal law, including death penalty. These debates took place in newspapers and salons and excited even more public interest in details about the toxicology of arsenic, encouraging new publications and even experimental demonstrations in salons and amphitheaters. Many of these publications included literary reconstructions of the most dramatic aspects of the poisoning crimes, along with technical information about new chemical tests and highly polemic issues concerning French politics or the administration of justice. In that sense, poisons such as arsenic created unexpected links and creative exchanges among a heterogeneous range of actors, practices and discourses in different legal, medical and popular settings during the nineteenth-century. Engaging culture and nature in such different ways, the material affordances of arsenic were rather unstable, changeable and unpredictable.

PART 2

Chemical Governance and the Governance of Chemistry

CHAPTER 6

Relations between the State and the Chemical Industry in France, 1760-1800: The Case of Ceruse

Christine Lehman

The transformation of relations between the royal government and industry is an important feature of the history of the seventeenth and eighteenth centuries in France. Of course the state's influence on industrial development started long before the second half of the seventeenth century when it was reinforced by Colbert's reforms, which retained a significant influence during the eighteenth-century.[1] For example, the new regulation issued on 26 January 1699 appointed the *Académie des sciences*, founded in 1666, to evaluate all new industrial machines: the principles of novelty and utility were reaffirmed and remained the indispensable conditions for obtaining a royal privilege.

Created in the sixteenth century in order to break the restrictive framework of corporations and guilds, a royal privilege gave an entrepreneur the right to circumvent local regulations. It created an exclusive right to use a new process for a given period of time; in other words, a temporary monopoly within a specified region. An exclusive privilege gave its recipient the freedom to manufacture and sell a commodity in France or abroad without encountering difficulties from competition. It was often complemented with production premiums and exemption from taxes on buildings and employees, even when the latter were migrant workers. Such privileges could also be granted to industries that had developed abroad, as was the case with the chemical production of ceruse, which is analyzed in this essay.[2] In France, throughout the eighteenth century, from 1722 until the creation of French patents or *brevets* in 1791,

1 Jacques Isoré, "De l'existence des brevets d'invention en droit français avant 1791," *Revue historique de droit français et étranger* 16 (1937): 94-130, on 125; Christiane Demeulenaere-Douyère and David J. Sturdy, eds., *L'enquête du Régent 1716-1718: Sciences, techniques et politique dans la France pré-industrielle* (Turnhout: Brepols, 2008); David J. Sturdy, "L'Académie royale des sciences et l'enquête du Régent de 1716-1718," Christiane Demeulenaere-Douyère and Eric Brian, eds., Règlements usages et science dans la France de l'absolutisme (Paris: Tec et Doc, 2002), 133-146.

2 Isoré, "De l'existence des brevets," pp. 97-104 (see note 1); Jeff Horn, "Privileged Enclaves: Opportunities in eighteenth-century France," *Proceedings of the Western Society for French History* 32 (2004): 29-45.

the relations between the state and industry were managed by the *Bureau du commerce*.[3] The latter's role was to substantiate inventors' requests for privileges and financial help from the Ministers of Commerce and Foreign Affairs. These requests usually required the opinion of a scientific expert who had to rule on the novelty of the inventions, their profitability and their impact on national self-sufficiency through the raw materials used. Until 1770, the requests for an expert evaluation that were successively addressed to academician chemists Jean Hellot (1685-1766) and Pierre-Joseph Macquer (1718-84), mainly concerned dyeing, coloring materials and the associated chemicals such as *vitriol* and *lessive*. From 1770, with Macquer and later with Louis-Claude Berthollet (1748-1822), his successor at the *Bureau du commerce*, they increasingly concerned the emerging chemical industry. After the Revolution, this role was taken over by the *Comité du commerce et de l'agriculture* that was itself replaced by the *Comité consultatif des arts et manufactures* at the beginning of the nineteenth century.

This essay deals with the state's influence on innovation in chemistry; it will focus on the end of the eighteenth century and on a particular chemical product: *ceruse* and/or *blanc de plomb*. Focusing on ceruse is informative for two main reasons. First, while the production of ceruse during the nineteenth century has been the subject of many studies, its history in the eighteenth century has not yet been studied.[4] Second, this example provides an in-depth description of how the French administration functioned when evaluating requests for industrial privileges, thereby setting the stage for a more accurate comparison with the British context, which is presented in the conclusion.

At the end of the Old Regime, ceruse production was still artisanal and had only reached the proto-industrial stage. Its chemical formula was unknown in France and its fabrication was not sophisticated; it required only very basic equipment and workers were few and unskilled. Ceruse production is thus a quite specific case that cannot be compared with the heavy chemical industry.

3 Pierre Bonnassieux, *Conseil de commerce et bureau du commerce 1700-1791. Inventaire analytique des procès-verbaux* (Paris: Imprimerie nationale, 1900), "Introduction," V-XXXIV; Harold T. Parker, *An Administrative Bureau During The Old Regime: The Bureau of Commerce and its relations to French industry from May 1781 to November 1783* (London: Associated University Press, 1993).

4 Laurence Lestel, Anne-Cécile Lefort and André Guillerme, eds., *La céruse: usages et effets Xe-XXe siècles* (Paris: Centre d'histoire des techniques CNAM, 2003); Thomas Le Roux, "Risques et maladies du travail: Le Conseil de salubrité de Paris aux sources de l'ambiguïté hygiéniste au XIXe siècle," A.S. Bruno, E. Geerkens, N. Hatzfeld, C. Omnès, eds., *La santé au travail, entre savoirs et pouvoirs (XIXe-XXe siècles)* (Paris: Presses universitaires de Rennes, 2011), 45-63, on 58-61.

Indeed glassmaking, porcelain manufacture, the steelmaking industries or sulfuric acid and soda production used powerful furnaces and were organized in separate workshops dealing with the successive operations.[5] On the other hand, relations between the state, that is to say the royal administration of commerce, and entrepreneurs were quite similar to the case studied in this essay.[6] A remarkable characteristic is that the examination of requests from both industrialists and craftsmen were treated by the same persons and the same structure with the same care. Based mainly on an analysis of the files of the *Bureau du commerce*, this essay presents Macquer's and Berthollet's evaluations of the processes for producing ceruse, the necessary conditions for obtaining privileges and the state's involvement in its production, which required importing lead, the customs duties of which increased the production cost.[7]

Ceruse has been known since antiquity and was used as a cosmetic until the mid-eighteenth century when, due to its recognized toxicity, it was banned by medical doctors and abandoned by *coquettes*.[8] Ceruse, or lead calx, was white lead. It had also become an important pigment used both in the East and the West over the course of centuries.[9] In 1742, the *Dictionnaire universel de commerce* described two methods of fabrication and defined it as "lead dissolved by vinegar." The first involved chopping lead into strips, which were soaked in vinegar, removed and scraped every ten days in order to collect a kind of crust, namely white lead, formed on the strips and do this again until lead has totally disappeared. The second, which we will call the "Dutch method," was predominant in the Netherlands during the eighteenth century. It involved hanging thinly beaten and rolled sheets of lead "in a pot, at the bottom of which is excellent vinegar, which is buried in dung; after thirty days the operation is over."[10]

5 Charles Coulston Gillipsie, *Science and Polity in France: The end of the old regime* (Princeton, NJ: Princeton University Press, 1980); John Graham Smith, *The Origins and Early Development of the Heavy Chemical Industry in France* (Oxford: Clarendon Press, 1979).

6 Gillipsie, *Science and Polity*, pp. 463-78 (see note 5).

7 Liliane Hilaire-Pérez, "Invention and the State in Eighteenth-Century France," *Technology and Culture* 32 (1991): 911-31, on 913-19.

8 Catherine Lanoë, "Céruse et cosmétique sous l'ancien régime, XVIe-XVIIIe siècles," Lestel, Lefort and Guillerme, *La céruse*, pp. 25-37 (see note 4).

9 Philiberto Vernatti, "A Relation of the Making of *Ceruse*," *Philosophical Transactions* 12 (1677-1678): 935-36; C.M. Wai and K.T Liu, "The Origin of White Lead – From the East or the West," *Journal of Chemical Education* 68 (1991): 25-27.

10 Jacques Savary des Bruslons, "Blanc de plomb," *Dictionnaire universel de commerce*, Tome 1 (Genève: Cramer et Philibert, 1742); The Dutch process used horse dung instead of

In spite of its harmfulness, ceruse became indispensable to painters because it yielded an irreplaceable washable and bright white paint when mixed with oil.[11] In contrast with other countries, such as Spain where lime wash was used, or the Netherlands and England where this paint was only applied to wainscoting, doors and window frames, in France ceruse was very fashionable and was used to cover both the internal and external walls of buildings. Even Toulouse, "the pink city" built in brick, complied with this fashion by the end of the eighteenth century.[12] A sign of embellishment and sanitation, ceruse paint beautified French cities. In 1787, the city of Paris alone consumed 300 tons of ceruse per year and its consumption in the rest of the kingdom amounted to about 700 tons.[13]

Because white lead was so expensive, the "ceruse" employed in France was actually not a pure product but an equal mixture of white lead, imported from Holland or England, with domestic chalk or white limestone. Consequently the word "ceruse" was misused by French painters and manufacturers during the second half of the eighteenth century. Indeed Berthollet clearly distinguished ceruse from white lead: "Ceruse always contains ground chalk with which white lead has been precisely mixed."[14] This distinction tended to disappear in the nineteenth century when white lead manufacturing became common and its cost decreased. Its availability in France meant that French painters then returned to the original meaning of "ceruse."

Ceruse Manufactures under the Old Regime

France had too few white lead manufacturing plants to meet growing domestic demand. The few attempts to import ceruse as early as in 1708 or to manufac-

manure and beer vinegar. At the end of the century, the thirty-five Dutch plants produced 4,000 tons per year of white lead, of which approximatively 1,000 tons were exported to France. Ernst Homburg and Johan H. de Vlieger, "A Victory of Practice over Science: Failed innovations in the white lead industry (1780-1850)," *Archives internationales d'histoire des sciences* 46 (1996): 95-112, on 97-102.

11 It caused the cruel colic of *Poitou* or painter's colic; François de Paule Combalusier, *Observations et réflexions sur la colique de Poitou ou des peintres*, Part I (Paris: de Bure, 1761).

12 Valérie Nègre, "La peinture à la céruse et l'embellissement des villes du Midi, aux XVIIIe et XIXe siècles," *La céruse*, pp. 39-46 (see note 4).

13 *Mémoire pour le Sieur Antoine Baille*, Archives Nationales (AN) F[12]2424.

14 Berthollet, *Rapport sur un mémoire de Mr Valentino dans lequel il propose un nouveau procédé pour fabriquer le blanc de plomb et de céruse*, 2 February 1787, AN F[12] 1507.

ture it in 1736 were strictly regulated by the *Conseil* or *Bureau du commerce*.[15] In the mid-eighteenth century royal decrees or *lettres patentes* for setting up ceruse fabrication plants could be obtained easily. For example in 1764, following a simple request, the Baron Idlinger d'Espüller and his partner the Comte de Varoc obtained permission to manufacture and sell *blanc de céruse*, minium, cinnabar, crystallized verdigris and oil of vitriol for ten years, "forbidding grocers and others from disturbing or troubling them" at a location specified by the *lieutenant général de police*.[16] Before registering d'Espüller's *lettres patentes*, however, the Parlement of Paris requested an investigation by the *Académie des sciences*.[17] Macquer and the physician and chemist Hyacinthe-Théodore Baron were appointed as commissioners. However, Macquer's favorable opinion, based solely on the need to replace foreign imports, was not shared by the *Académie* which requested that the inventors communicate their processes. The latter categorically refused because "these were the secrets on which the security and the success of their enterprise depended" and they never set up the factory.[18] Until the patent law of 1791, addressed below, commissioners generally encountered resistance when they attempted to obtain the secrets of inventions. It should also be noted that such a request for an evaluation by the *Académie* was quite exceptional for this type of chemical industry, as, with the exception of its interest in dyeing and at the end of the century in the fabrication of soda and saltpeter, the *Académie* did not show much interest in other chemical industries.[19] Thus, at the end of the old regime, developing ceruse production was based on individual initiative. This development was done

15 Bonnassieux, *Conseil de commerce*, pp. 37(b), 39(a) and 249(b) (see note 3).

16 Holker to Trudaine de Montigny, 25 October 1764, AN F[12]2424; Decree of the royal council, 15 January 1765; Pierre-Joseph Macquer, "Rapport sur les demandes des Srs *Idlinger* baron d'*Espuler* et du *comte de Varoc*," 26 May 1766, AN F[12]2424; On the location of risky factories see Thomas Le Roux, *Le laboratoire des pollutions industrielles, Paris (1770-1830)* (Paris: Albin Michel, 2011), 25-68.

17 Archives of the *Académie des sciences*, *Plumitif and Procès-verbal* of 27 November 1765, fol. 386 v.

18 Macquer, "Rapport sur les demandes des Srs *Idlinger*," (see note 16).

19 Ernest Maindron, *Les fondations de prix à l'Académie des sciences. Les lauréats de l'Académie (1714-1880)*, (Paris: Gauthier-Villars, 1881); Christine Lehman, "L'art de la teinture à l'Académie royale des sciences au XVIIIe siècle," *Methodos* 12 (2012), <http://methodos.revues.org/2874>; Between 1776 and 1782 the *Société libre d'émulation* also encouraged inventors, but attention to chemistry remained limited to dyeing and distillation vessels, AN T160[16-22]; AN T*160[4-6]; Liliane Hilaire-Pérez, *L'invention technique au siècle des Lumières* (Paris: Albin Michel, 2000), 209-20.

by little-known artisanal entrepreneurs who are difficult to identify as their names generally appear only in administrative correspondence.

Requests for privileges were usually submitted to the *Bureau du commerce* through the Intendants, powerful royal officials who had wide-ranging responsibilities for each of the *généralites* into which France was divided. As discussed below, the circulation of memoirs, reports, notes, evaluations and correspondence reflects not only the pyramidal and hierarchical structure of the *Bureau du commerce* but also the interactions between the various actors.[20] It should be noted that Macquer's mission at the *Bureau du commerce*, where he had been appointed commissioner in 1766, was very different from the function he performed at the *Académie*. It was a permanent appointment to which the title of academician was attached, but in this instance he was commissioned by the state and, like Berthollet later, he was accountable to the state – namely to the *Contrôleur général* – for his evaluations.

Before the Revolution, France imported between three and a half and four thousand tons of ceruse per year, mainly from the Netherlands.[21] Consequently all the applicants stressed the usefulness of their invention and the economic interest of the nation. This is why Jean-Guillaume Laliaud asked the *Bureau du commerce* in 1779 for the right to set up a plant at Rennes, close to the lead mines of the region, which he argued would increase local employment of unskilled manpower and the region's dynamism.[22]

His request was examined by Macquer who had to evaluate the quality of ceruse samples sent by the Intendant of commerce Jean-François Tolozan together with those of a Dutch competitor, Guillaume-Pierre d'Espar (Willem-Pieter Despar). Although Bergman had shown that ceruse was composed of fixed air and litharge in 1774 his results were not yet known in France by 1779.[23] Consequently Macquer followed established chemical knowledge and analytical practices and confined himself to identifying the nature of the earths that

20 The composition of the *Bureau* and its operation underwent many modifications between 1722 and the end of its activity in September 1791. See Bonnassieux, *Conseil de commerce*, "Introduction" (see note 3); Parker, *An Administrative Bureau* (see note 3).

21 The figures vary depending on sources but remain in the same order of magnitude: more than 1,000 tons in 1779 according to Laliaud (AN F[12]1507); 2,409 tons in 1787 according to the *Journal des mines* 1 (1794): 92; 3,000 tons in 1788 according to d'Espar (AN F[12]1507); 4,000 tons in 1790 according to Migneron de Tocqueville (AN F[12]2424); 1,200 tons average between 1787 and 1789, according to Héricart de Thury, *Rapport du Conseil des travaux publics du département de la Seine sur la… céruse de Clichy…* 1815.

22 Deliberation of the États du Languedoc, 5 January 1781, Archives départementales de l'Hérault, C7612, fols. 352-53 ; Jean-Guillaume Laliaud, Dossier "Céruse," AN F[12]1507.

23 See note 45 for details.

composed the sample substances and measuring their proportions. He deemed both men's samples to be "made of half white lead and half very white marly earth (limestone and clay), rather weighty, well cleaned of sand and soft to the touch."[24] The general qualities of the two samples seemed to be identical to or even higher than those from Holland. With regards to their practical use, Macquer relied on the opinion of professionals (house painters, merchants and colour makers), specifically on the renowned painter-gilder Jean-Félix Watin.[25] The specimens of ceruse submitted by both applicants passed Macquer's tests: they had a good covering capacity and body. Macquer gave his favorable opinion because there were "presently no manufactures for these materials in France," demand was constantly increasing, the raw materials needed for ceruse manufacturing were available domestically and the climate was more favorable than in the Netherlands. But success also depended on fresh capital, chemical knowledge and operational know-how. Laliaud had the advantage here as he had already performed full-scale operations in a plant he had acquired at Langoyran near Bordeaux. This was not the case for d'Espar, who had to prepare white lead and ceruse in the presence of specially appointed commissioners.

Macquer's opinion convinced the government, which granted Laliaud authorisation on 15 February 1780 to establish ceruse and white lead manufactures in Normandy, Orleanais, Provence, Languedoc and Guyenne, with incentives for production in the form of tax exemptions and premiums.[26] These locations had strategic importance: Orleanais and Normandy supplied Paris, Languedoc and Provence promoted trade routes to the East and Guyenne across the Atlantic. In order to balance the advantages given to Laliaud, d'Espar requested Brittany, the Ile de France, Champagne, Nivernais, Lyonnais and Dauphiné (see Fig. 6.1).[27]

For this purpose d'Espar founded a company with two nobles from the Nantes region, Jean-Louis d'Adhémar de Montréal and Philippe-Vincent-Roger de la Mouchetière, *lieutenant-général* of the Nantes Admiralty, who provided

24 Macquer, "Projets d'établissemens de manufactures de blanc de plomb et de céruse en France," 7 September 1779, AN F12 1507.

25 Jean-Félix Watin, *Art du peintre doreur et vernisseur* (Paris: Grangé, 1772).

26 AN F12 1506; "Réponse des Intendants de province à la lettre qui leur a été adressée le 24 janvier 1788," AN F12 1507.

27 D'Espar's memoir, AN F12 2424; Tolozan's response is in a letter of 23 December 1779, AN F12 1507.

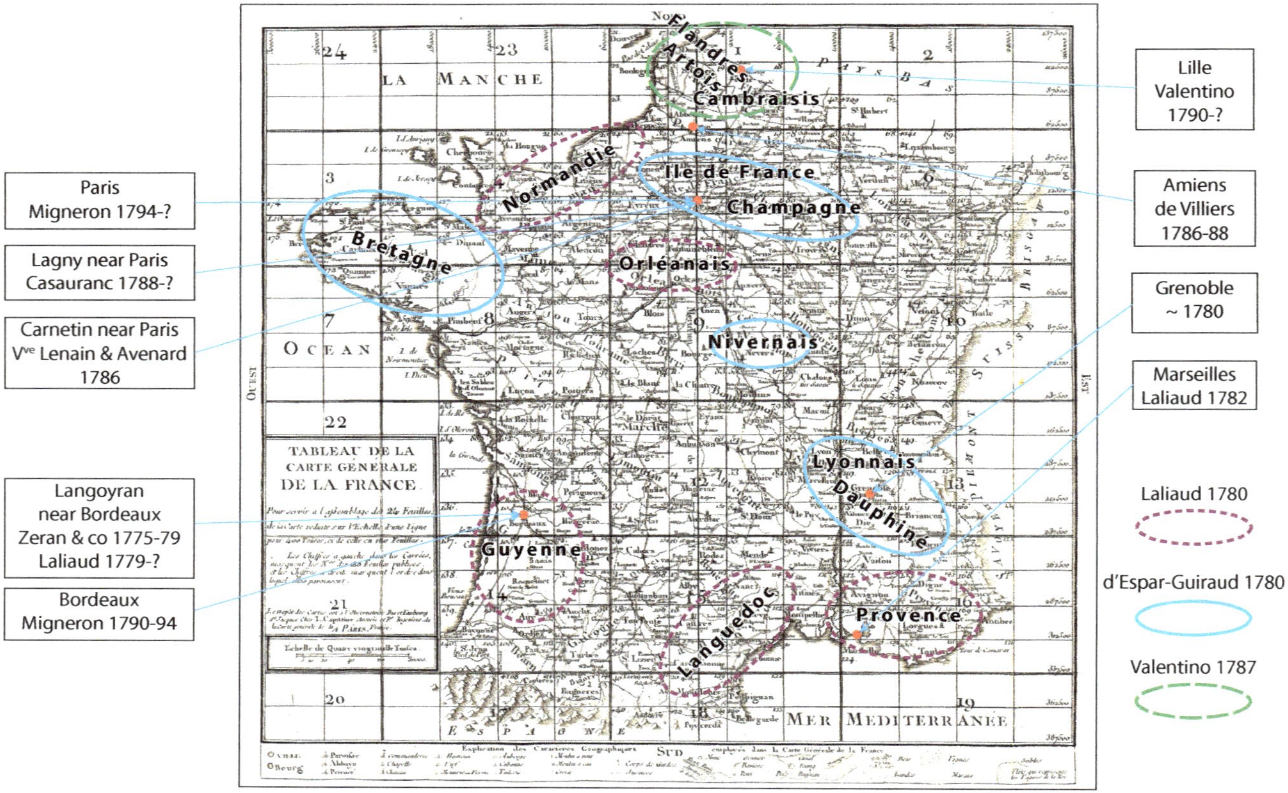

FIGURE 6.1 *Areas of white lead and ceruse privileges granted in 1780 and 1787 and locations of ceruse manufactures, 1780-1794.*

the necessary funding.[28] This company was founded under the name of *Guiraud et compagnie*. Jean Guiraud was a bourgeois of Paris, who brought his influence and notoriety, that is to say "his guarantee and his name," to support the request for a privilege.[29] On 18 July 1780, *Guiraud et compagnie* obtained the privilege to "make, distribute and retail white lead and ceruse in the requested provinces" for six years and with the same financial advantages as Laliaud's .[30] With these sites distributed around the kingdom, the state intended to match the national demand for ceruse while preventing other plants from being set up.[31] But both enterprises failed, first Laliaud's, which received 5,500 *livres* between 1781 and 1785, but produced little ceruse. The reports, issued in 1788 by the Intendants in the various provinces involved, underscored the failures: Laliaud's plant, established in Marseilles in 1782, had just changed ownership and the one in Dieppe had produced nothing yet; meanwhile d'Espar-Guiraud's manufactures had not been created. Consequently, in 1788 there was still hardly any white lead production, despite the state's will and generosity. Not only did the privileges and financial help granted to Laliaud and d'Espar and company not succeed, their exclusivity, forbidding the establishment of other plants, frustrated the industry's development.

The case of Damelon, a former cavalry officer, demonstrates this contradiction. Damelon claimed to draw his knowledge of the fabrication process from his many travels to Venice, Holland and especially Nuremberg. As Laliaud and d'Espar-Guiraud's manufactures had not been exploited after five years, on 8 august 1785 he applied to the *Contrôleur général* for the repeal of Laliaud's privilege and for the same financial aid. After gathering the necessary funds, he intended "to develop this essential business [...] that we can take from our jealous neighbours."[32] In response to this request, the requirements of the *Bureau*

28 Contract dated 9 June 1780, AN, MC/ET/XLIV/545; George V. Taylor, "Types of Capitalism in Eighteenth-Century France," *The English Historical Review* 79 (1964): 478-97, (on 495-96).

29 Contract dated 15 February 1781, MC/ET/XLIV/549; AN, MC/ET/XLIV/545; See also the contract of 4 February 1783 between a painter, Antoine Meraud who held the secret, and a priest Jean-François Girou de Montdésir, who provided funding and took care of obtaining the privilege for establishing a white lead and ceruse plant near Paris, AN, MC/ET/LXV/461.

30 AN, MC/ET/XLIV/545.

31 Such as the one from Desomer on 18 March 1780, (AN F[12]2424) or Dubreuil de la Gueronnière who on 18 July 1783 was refused permission to install a plant in the abbey of Cercanceaux near Nemours, AN F[12]1507.

32 Damelon to Merelliers?, 16 October 1785, AN F[12] 1507.

du commerce were still the same: give evidence of both knowledge of the process and support by funders.[33]

The experiments were performed before Berthollet, who had first to check whether Damelon's process was different from Laliaud's and carry out a profitability analysis. In his report, Berthollet gave a favorable opinion, although three attempts were required before getting an acceptable result from painters. On condition that Damelon improved the grinding by using new mills, Berthollet judged that the enterprise could compete with foreign competition if "prudently" managed. Damelon's process was the well-known Dutch method and therefore not original. Nonetheless Berthollet added what became a *leitmotiv*, that "it would be beneficial to take away this branch [of industry] from foreigners."[34] After Berthollet's positive assessment, Damelon still had to provide evidence of his financial support. He responded by withdrawing his request for support, arguing that:

> If it succeeds, he [Damelon] will have acted for the good of the state since he will prevent several millions per year from leaving the kingdom to purchase this substance from abroad. If it does not succeed, it will have cost the state nothing.[35]

Damelon relied instead on his influential protectors: Brissault de la Chapraie, who promised to secure the necessary funds; *Monsieur* (the King's brother), who was ready to grant a piece of land south-east of Paris, upon which to establish the manufacture; and the Comte d'Alsace, who invoked the threat of losing know-how to the benefit of England, which the French believed to be practicing freedom of trade.[36] Disappointed by the previous agreements concluded with d'Espar-Guiraud and Laliaud, who had just sold his privilege, the minister waited for a better guarantee from the funders before giving Damelon a favorable answer. But by now Damelon had competitors. In 1785 Valentino, a chemist at Lille's military hospital, submitted a request for a privilege.[37] The following year someone named Caille made a similar request for Paris and several other provinces and de Villers established a plant in Amiens without state

33 Merelliers? to Damelon, 21 January 1786, AN F^{12} 1507.

34 Berthollet, "Rapport sur une préparation de blanc de plomb et de céruse pour laquelle Mr Damelon demande un privilège," 14 July 1786, AN F^{12} 1507.

35 Damelon's memoir, undated, probably the middle of 1787, AN F^{12} 1507.

36 Comte d'Alsace to Tolozan, 5 August 1787, AN F^{12} 1507.

37 Bonnassieux, *Conseil de commerce*, p. 449(a) (see note 3).

help.[38] In 1788 d'Espar sought to recover the privilege he had been granted in 1780 and to establish a new manufacture near Bordeaux.[39] Due to the similarity of these applications, the *Bureau du commerce* decided to consider ceruse production from a more general viewpoint. In July 1788, it carried out "a new examination of the various means of introducing and maintaining the fabrication of ceruse in France" and rejected Damelon's request.[40] The commissioners realized that the profitability of ceruse fabrication was closely related to customs duties on lead. However, decreasing these duties would prejudice French mining entrepreneurs and it was decided that "lead extraction was more necessary than ceruse fabrication." Finally Damelon did not install his ceruse manufacture and withdrew his file on 2 July 1792, seven years after his initial application.

In August 1785, Liborio Philippe Valentino (1741-1803) also applied to open a ceruse manufacture. Unlike Damelon, Valentino was a chemist.[41] An Italian immigrant, he had settled in Lille in 1779 where he was apothecary at the military hospital. As the director of a large-scale manufacture of oil of vitriol and *aqua fortis* he already possessed industrial know-how.[42] In 1785 he was involved in the creation of a learned society, the *Collège des Philalèthes*, which grew out of the masonic lodge *Les amis réunis*. These ties with freemasonry and the industrial world enabled him to get financial support from three of Lille's most powerful merchants and one of the city's notables. Due to the cost of raw materials, lead and vinegar, it seemed impossible to match foreign competition. Valentino answered this challenge by proposing an innovative method for manufacturing white lead, in which vinegar was replaced with brine. Since the gas analysis of native ceruse from English mines yielded only fixed air, why not imitate nature? He therefore poured salted water onto heated lead "at a heat sufficient enough to set paper alight." After scraping the crust that formed, he exposed this lead to air for a relatively long time while drizzling it regularly with brine. The operation was repeated until all the lead was consumed.[43] (See Fig. 6.2.) Interestingly, a similar process was patented in Great Britain a few years later in 1797 by the Scottish chemist Archibald Cochrane, Earl of

38 Note of 9 August 1786, AN F12 1507; Bonnassieux, *Conseil de commerce*, p. 455(b) (see note 3).

39 Ibid., pp. 453, 454(a) and 455(b).

40 Ibid., pp. 459(b), 460(a).

41 "*Chimiste pensionné du Roy* at the Lille military hospital," AN F12 1507, dossier Valentino.

42 Valentino to the *Ministre d'état*, 1788, AN F12 1507.

43 *Nouveau procédé pour fabriquer le blanc de plomb et la céruse sans faire usage du vinaigre, par Valentino chimiste penssionné du roy à la suite de l'hôpital militaire de Lille en Flandres*, AN F12 1507.

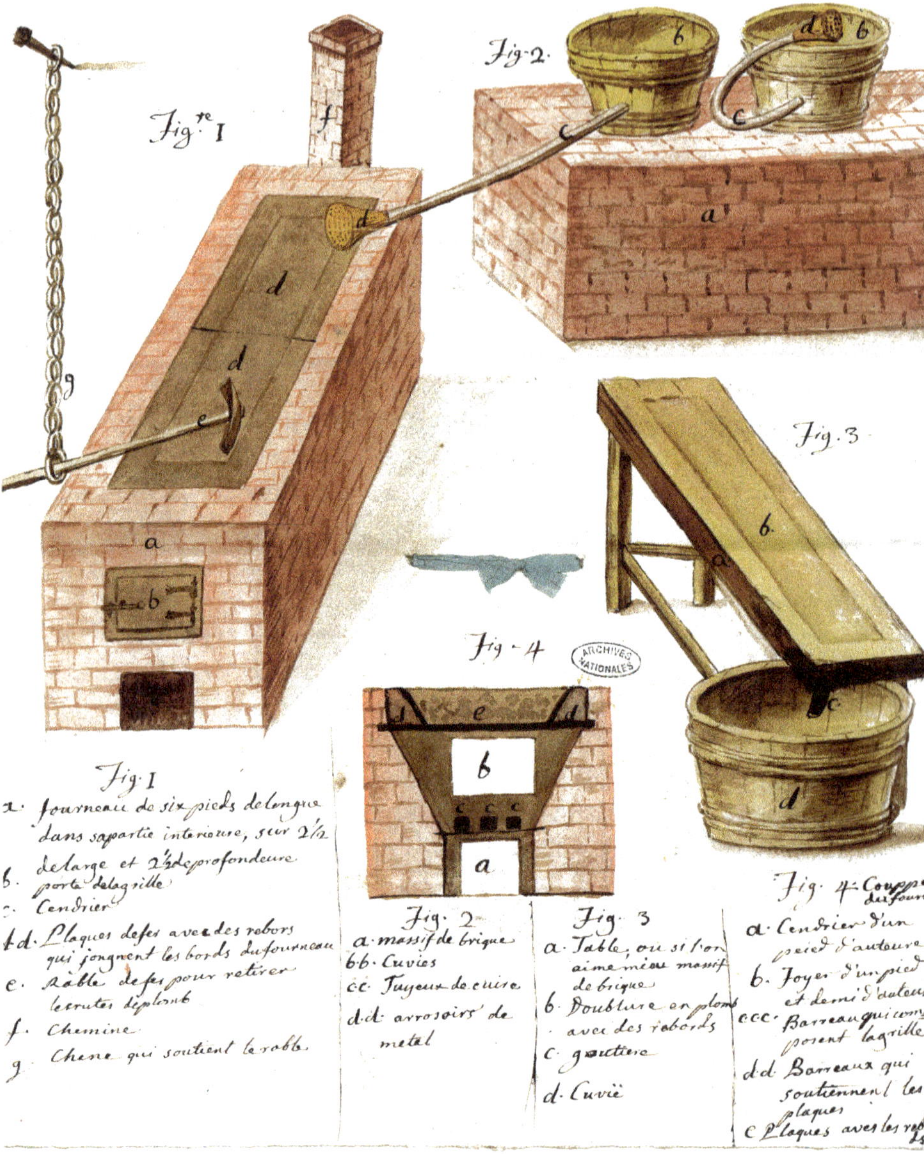

FIGURE 6.2 *Valentino, New process for manufacturing white lead and ceruse without using vinegar* (*AN F*[12] *1507*). COURTESY OF ARCHIVES NATIONALES, PIERREFITTE-SUR-SEINE.

Dundonald, by reacting muriate of potash (another chloride) instead of brine with lead calx before exposing it to a flow of carbonic acid gas or atmospheric air.[44] It was only after Torbern Bergman's analysis of white lead in 1774, which showed that it was a compound of fixed air and litharge, that these empirical methods of manufacture became understandable.[45] However this research seems to have been unknown in France, noted neither by Macquer nor Berthollet.

With the support of Charles-François-Hyacinthe d'Esmangart, *Intendant des Flandres et d'Artois*, Valentino applied for a ten year exclusive privilege and 10,000 *livres* to cover his research costs.[46] The answer was not long in coming: Valentino was required to have his process checked and "in the event that his process is recognized as new and useful by *gens de l'art* as well as by the council Sr Valentino will be treated favourably."[47]

Berthollet provided a qualified opinion, criticizing Valentino's lack of precision and adding that his experiments "did not announce a chemist enlightened by an exact theory."[48] He nonetheless advised having examinations carried out by reliable persons, and Valentino complied with this new requirement. After building a new furnace in April 1787, he again performed his process before two experts commissioned by the Intendant of Lille. The verification took place almost every day from 7am to 8pm. Seals were applied and everything was weighed and measured: the number of brushwood stacks used for fire and the quantities of heated lead, coal and marine salt. The *dephlogisticated* lead (the name often given to white lead) that was extracted during the course of operations was thrown into a barrel filled with water. On the following day the solid parts were washed, weighed and packed. The ceruse prepared by mixture with various proportions of Champagne chalk was also placed in sealed boxes to be sent to the *Bureau du commerce*. Valentino's process was deemed much more economical than the Dutch one: marine salt was less expensive than vinegar; the charcoal used in the furnace in order to melt the lead and perform the

44 Archibald Earl of Dundonald, "Methods of Making Ceruse or White-lead," 18 August 1797, *The Repertory of Arts and Manufactures* 8 (1798): 377-81.

45 Klaproth replicated this analysis and found 84% litharge (lead oxide) and 16% carbonic acid. White lead was therefore lead carbonate. Homburg and Vlieger, "A Victory of Practice," p. 103 (see note 10); Gérard Emptoz, "Un procédé de fabrication de la céruse issu de la 'chimie moderne' au début du XIXe siècle," Lestel, Lefort and Guillerme, *La céruse*, pp. 49-60 (see note 4).

46 The Intendant of Lille even proposed contributing half the financial support, d'Esmangart to Blondel on 27 December 1785, AN F[12] 1507.

47 Note by Cotte dated 22 March 1786, AN F[12] 1507.

48 Berthollet, *Rapport sur un mémoire de Mr Valentino* (see note 14).

necessary operations was less expensive than dung; lead was used in ingots as extracted from mines and did not need to be cast into thin blades and wound into spirals. It was also less time-consuming, taking twenty-four hours with a well-designed workshop versus eight to ten months for the Dutch process. Apart from the advice to operate the furnace continuously, the verdict totally favored Valentino.[49]

After complying with the orders of the government, Valentino submitted his requests to the *Bureau du commerce*. In addition to financial assistance of 20,000 *livres* that would enable him to meet his expenses, pay back his creditors and borrow again, he applied for the exclusive privilege to manufacture and sell his ceruse in the provinces of Flanders, Hainault and Cambraisis, tax exemption on drink and living costs for the workmen and especially exemption from customs duties on lead and other raw materials that would help him to compete with British ceruse. Indeed, in addition to readily available lead mines, the latter was also benefiting unfairly from premiums granted by the British government: "from five to ten per cent for some articles of all goods exported to foreign countries."[50] Moreover, as he had revealed his secret to experts who had not been sworn in, he feared being dispossessed of the secret, should it to be disclosed. Valentino therefore considered it impossible to start his enterprise without the state's help. As with Damelon, the threat of his secret being leaked to foreign countries was repeatedly invoked by d'Esmangart to pressure the government into granting the requested help.

> I know that foreign traders informed by this chemist's discovery and the advantages of his process for fabricating white lead and ceruse have made him considerable and advantageous offers in order to incite him to cross to their side and create his establish there.[51]

> When the government compels the inventor of a profitable discovery to disclose the process, it seems to tacitly commit itself to granting him compensation in proportion to the sacrifice of a secret which is his own property.[52]

---

49 Report by Merlin physician at the Lille military hospital and Boudin chemist apothecary at Lille, *Procès-verbal ordonné par Monsr Esmangart, Intendant des Flandres et d'Artois, et exécuté par le Sieur Valentino chymiste attaché à l'hopital militaire de Lille pour la fabrication de la céruse*, 25 mai 1787 (copy dated 26 September 1788), AN F^{12} 1507.

50 Valentino to d'Esmangart, 25 May 1787, AN F^{12} 1507.

51 D'Esmangart to Tolozan, 25 February 1788. See also d'Esmangart to Blondel, 27 December 1785 and to Tolozan, 27 October 1787, AN F^{12} 1507.

52 D'Esmangart to Tolozan 7 December 1787. Valentino was strongly supported by d'Esmangart who was acting for the development of his province and intervened with the *Bureau*

On 31 October 1787, the government finally responded by granting Valentino the exclusive privilege of manufacturing, selling and distributing in Flanders, Artois, Hainault and Cambraisis for five years only, while "forbidding disturbance of the owner in the operation of his manufacture." (See Fig. 6.1.) This exclusive privilege was all the more easily granted since this region did not overlap with those of Laliaud and d'Espar-Guiraud, which was not the case for Damelon. Paradoxically, Valentino refused the privilege saying that its duration was too short and would prevent him from making the installation profitable. Moreover, his request evolved over time. In 1788, he mentioned the purchase of a large piece of land on which he built his manufacture, claimed the title of Royal Manufacture and maintained his requests for the exemption of excise duty on his consumption and his workers' as well as exemption from customs duty on lead and other raw materials. It was this stubborn request for getting both the commercial privilege and the tax exemption on lead that created tension between Valentino and the government.

Let us look first at the exclusive privilege. In order to ensure equity as well as a harmonious and balanced distribution of plants of this type throughout the kingdom, the *Bureau du commerce* inquired about the presence of other ceruse plants in the north of France and the possible privileges granted to them. It was in this context that the Intendant of Amiens reported that a ceruse plant using the Dutch process had been established there by de Villers in 1786. Apart from the fact that Valentino's request seemed excessive with respect to de Villers', the deputies of the *Bureau du commerce* agreed on the position that distributing premiums should only be done in extreme situations, which was not the case for ceruse because "the art of making ceruse is not a secret, all chemists know it."[53] Although they did not jeopardise the exclusive privilege in the provinces already granted to Valentino, they declared that "in the future these kinds of favours would have no other results than holding industry back and preventing other establishments that we would like to see increase in number."[54] This shows that the commissioners had come to consider the attribution of exclusive privileges as potentially harmful to the country's industrial development. In the end, in spite of his stubbornness and supporters, four members of the *Bureau du commerce*, the Intendant of *Flandres* and the administrators of the *department du Nord*, Valentino obtained no other

du commerce on several occasions, 24 September 1788, AN F[12] 1507.

53 Draft note on a letter of 15 March 1788 and Deliberation of the *députés du commerce*, 3 June 1788, AN F[12] 1507.

54 *Avis des députés du commerce sur la demande de deux fabriquants de céruse* [*Villers et Valentino*], 9 May 1788, AN F[12] 1507.

encouragement. Indeed, from 1790, the financial situation of the central government was no longer favorable and it now became the responsibility of local administrations to support their manufactures.[55]

Furthermore the request for exemption from customs duties on foreign lead reflects a real difficulty. Valentino claimed that the problem of ceruse manufacturing came from the fact that only a part of the lead used could come from French mines, "which made lead very expensive for those who would attempt such an undertaking in France." Indeed there was an imbalance between the customs duties on lead imported from England, 4 *livres* 10 *sols* per *quintal*, and the much lower ones paid by Dutch manufacturers. This was not balanced by customs duties on foreign ceruse, 1 *livre* 2 *sols* 6 *deniers* per *quintal*.[56] Thus France could not match countries that had lead mines and granted premiums on the ceruse that they exported. During their deliberation of 3 June 1788, the commissioners recognized this difficulty. They acknowledged that the failures of ceruse plants came from the excessively high customs duties on lead and the overly low ones on ceruse, which penalized national ceruse production, and they confessed to having forgotten a basic principle of trade administration:

> This principle is that one must propose a much higher import duty on the fabricated product than on the raw material it is made from. We have lost sight of this principle when taxing lead at a rate three times higher than the one applied to ceruse & this oversight has caused all the enterprises we have established to fail.[57]

The deputies then proposed to reduce the customs duties on foreign lead and to increase the duties on ceruse. The advice of the *Bureau du commerce* reflects the commissioners' hesitations. Some of them were in favor of premiums granted to ceruse manufacturers such as exempting them from lead duties; others thought that it would probably be more advantageous to develop lead extraction in France by creating new mines in order to meet national demand.[58] The enquiry carried out by the *Inspecteurs généraux des manufactures* showed that 80,000 *quintaux* of lead and 24,000 *quintaux* of ceruse were imported in 1784.[59] One of the *Inspecteurs des mines*, Jean-Pierre-François Guillot Duhamel,

---

55 Tolozan to Valentino, 9 juillet 1790, AN F12 652.

56 *Avis des députés du commerce*, 9 May 1788 (see note 54); See the case of de Villers' establishment, draft letter by the *Bureau du commerce*, AN F12 1507.

57 *Avis des députés du commerce*, 9 May 1788 (see note 54).

58 Deliberation of 22 October 1788, AN F12 1507.

59 Antoine-Marie Héron de Villefosse, *De la richesse minérale* (Paris: Levrault, 1810), 397.

evaluated the production of national mines to be just one fourth of national consumption.[60]

Although customs duties on lead were decreased to 3 *livres* and those of ceruse increased to 3 *livres* 10 *Sols* per *quintal* on 1 June 1789, Valentino maintained his demand of a total exemption from customs duties on lead.[61] His stand was supported by all ten administrators of the *département du Nord*.[62] However, the decision of the deputies of the *Bureau du commerce* on 13 May 1791 remained the same: there was no reason to favor Valentino over other manufacturers. Indeed:

> This type of fabrication is not profitable enough to push the Treasury to make the financial sacrifice it would demand. Besides, a more important consideration would further militate against the requested exemption, that is the interest of the national mines which could no longer be operated if foreign lead, whatever its final destination, entered France without paying duties.[63]

The Minister also advised Valentino to turn to the *Comité d'agriculture et du commerce*.[64] Thus Valentino's factory, in operation from the beginning of 1790, seems to have received no help from the state. It is not known how long it remained in operation.

The increase of customs duties on ceruse imports remained a reason for refusing all new privileges for ceruse or white lead manufacture. This is why, in 1790, Migneron de Brocqueville was dismissed when he applied for an exemption from internal taxes and export duties and for a premium on the ceruse produced in his Bordeaux factory with a "Dutch *céruzier*."[65] Yet this ceruse had been favorably evaluated by the academicians of Bordeaux. It was very white and friable and its analysis showed that it was "a true lead calx, with no

60 "Avis des Inspecteurs généraux du commerce sur les demandes des Srs Valentino, de la ville de Lille; et du Sr Villers, de la ville d'Amiens," 10 July 1788, AN F[12] 1507; At the end of the eighteenth century, France produced 2,000 tons per year of raw lead, while Great Britain extracted 10,000 tons. Lynn Willies, "Derbyshire Lead Mining in the Eighteenth and Nineteenth Centuries," *Mining History* 14 (1999): 31-33.

61 *Arrêt du Conseil d'état du Roi du 23 avril 1789* (Paris: Imprimerie royale, 1789).

62 *Administrateurs du directoire du département du Nord* to Tolozan, 1 February 1791, AN F[12] 1507.

63 *Avis des députés du commerce*, 13 May 1791, AN F[12] 1507.

64 *Mémoire soumis à l'Assemblée nationale par le Sr Valentino*, 27 June 1790; Valentino to the deputies of the *Comité d'agriculture et du commerce*, 26 July 1790, AN F[12] 652.

65 Necker's answer to the letter sent by Migneron on 6 March 1790, AN, F[12] 2424.

addition of marly earth or other material, so that it could be easily reduced to lead."[66] One can see that, as early as 1790, what was called ceruse had started to become pure white lead again. In deference to the Intendant of Bordeaux who had sent the samples, Migneron's ceruse was evaluated by Berthollet but this was a mere formal exercise since the privilege had been refused from the beginning. Thus Berthollet only judged the external qualities of this ceruse that in his opinion did not look enough like the Dutch product, once again revealing the force of painters' habits. Dutch ceruse had become the standard, which created an additional handicap for all attempts at producing it nationally. In any event, the lack of competitiveness of French ceruse was irreducible. As Berthollet noted, France could not stand up to foreign competition due to "economic combinations and the modest profits accepted in several Dutch establishments," echoing Migneron's criticism, according to which increasing the customs duties on ceruse "did not balance the low cost of manpower in Holland in comparison to its high cost in France."[67]

The Revolutionary Period

From the beginning of the new Republic the problem of the supply of ceruse became crucial as France was at war with both England and the Netherlands from 1793. Paradoxically, the same Migneron who had suffered a categorical refusal by the *Bureau du commerce* now got support from the state and, instead of a financial grant, received material help to install a new factory in Paris. For example he was given 200 *quintaux* of lead thanks to the support of the *Comité d'agriculture et du commerce* and the National Convention. His process was derived from the Dutch method, but instead of earthen pots he used "lead boxes, the construction of which and their arrangement in dung were specific to him," which required a large quantity of lead. The implementation of the process was placed under the oversight of two chemists, Bertrand Pelletier and Nicolas Leblanc.[68] The country was at war and he was only granted the lead after the *Commission des armes et poudres* agreed.[69] Favors did not stop there. The *Comité des finances* assigned him a house belonging to the state, which he rented from 10 nivôse an III (30 December 1794) and in which he set up his establishment. The *Commission du commerce* supplied him wood, coal,

66 Report dated 7 February 1790, AN, F^{12} 2424.

67 Berthollet's report, 6 April 1790, AN, F^{12} 2424.

68 AN, F^{12} 2424 and AF/II/11.

69 Thury, Rapport du Conseil des travaux publics, "États de l'importation du plomb" (see note 21).

candles, oil and products of basic necessity. Finally, the Committee of Public Safety allowed him to use marble from former tomb covers in order to make the millstones for grinding the calcined lead while protecting his workers' health.[70] Thus under the auspices of the newly founded Republic, from the end of July 1794 to the end of February 1795, Migneron successfully installed a ceruse factory in Paris, which would have been practically impossible at the end of the Old Regime, given the price of lead and the absence of aid from the state.

Another entrepreneur, Simon-Léon de Casauranc de Saint Paul, was less lucky although he had been running a ceruse manufacture since 1788 at Lagny near Paris and had applied for a patent in 1792.[71] The specificity of his process was sieving limestone and washing it with water from the fountain in Lagny market before mixing it with ready-made white lead, which the patent application claimed deserved the title of *perfectionneur* (improver). His request for increasing the customs duties on ceruse was rejected by the *Comité du commerce* that wanted to ensure competition with foreign ceruse in order to maintain the quality of French ceruse. As regards his request for used lead, he was directed to the war ministry.[72] Since the lead needed for manufacturing white lead was requisitioned, its supply was left to the goodwill of the *Commission des armes et poudres*. During the revolutionary period the creation of ceruse plants was limited by a restricted supply of lead and, even with a patent like Casauranc's, getting help from the state was difficult for an entrepreneur.

The system of privileges granted by the King following the examination of a request by the *Bureau du commerce* was entirely recast in 1791 and the two laws of 7 January and 25 May reduced the power of the crown and gave increased protection to the inventor. The law of 7 January dealt with the *patente d'inventeur* and established the property rights of the inventor in his invention while that of 25 May stated that "national patents called *brevets d'invention* would be delivered by the King on a simple request and without prior examination."[73] The protection of the secret of the invention during a period of five, ten or fifteen years was a significant step forward if recalling d'Espüller's refusal to disclose his processes to Macquer in 1765 and Valentino's long hesitations, but it had to be described by a specification of the process.

70 AN, F[17] 1037; "Nouvelles manières de préparer le *Blanc de plomb* ou *Ceruse*," *Annales des Arts et manufactures* 1 (1800): 48-63, 55-58.

71 *Brevet d'invention de cinq années pour la fabrication du blanc de céruse façon de Hollande au Sr Casaurans perfectionneur*, 19 January 1792: n°1BA1942.

72 Casauranc to the representatives of the *Comité du commerce*, 19 pluviôse an III (7 February 1795), AN F[12] 2424.

73 On the change brought by the laws of 1791, see Isoré, "De l'existence des brevets," pp. 97-104 (see note 1).

In the middle of the eighteenth century, disclosing one's secret implied a risk, as during the full scale tests required for industrial implementation the inventor was not protected against possible indiscretions by workers. Moreover the granting of a royal privilege was more stringent as it required not only novelty but also examination by an expert (Macquer or, later, Berthollet) in order to evaluate its commercial profitability in view of a possible industrial implementation and, lastly, the obligation to build and operate the plants that were approved, as we have seen in the cases of Laliaud and of the association Guiraud and company founded by d'Espar. Beneficiaries of privileges also had to prove the robustness of their enterprise and of their funders. In return, the state was generous and proposed real financial support for development and production. However, after the laws of 1791, while inventors remained the owners of their secrets, they were still compelled to find private funds to found their enterprise, as the state granted no financial aid to the applicants. On the contrary, the latter had to pay a tax of 300 *livres* for a patent of five years, 800 *livres* for ten years or 1,500 *livres* for fifteen years.[74] Furthermore patents did not require strict novelty as an importer of a foreign process had the same rights as an inventor.[75] Indeed, when looking at the patents dealing with white lead and ceruse fabrication filed between 1791 and 1820, it appears that they were not always exploited and often borrowed from abroad.[76]

Conclusion

Although the ceruse industry was only emerging, the study of this particular chemical industry in the last decades of the eighteenth century reveals the conditions for obtaining a royal privilege. Without technical know-how or financial support and with no influential connections, it was difficult to get a privilege. The extensive file devoted to Valentino's case, held in the *Archives Nationales*, contains more than one hundred documents, which provide insights into the functioning of the *Bureau du commerce* and its evolution during the pre-revolutionary period until its disbanding in 1791. It can be represented in a simplified way by the following diagram. (See Graph 6.1).

74 Valérie Marchal, "Brevets, marques, dessins et modèles. Évolution des protections industrielles au XIXe siècle en France," *Documents pour l'histoire des techniques* (2009): 106-16, on 111.

75 Isoré, "De l'existence des brevets," p. 103 (see note 1).

76 Institut national de la propriété industrielle [INPI], base de données des brevets français du 19e siècle, <http://bases-brevets19e.inpi.fr/> (accessed 31 March 2016).

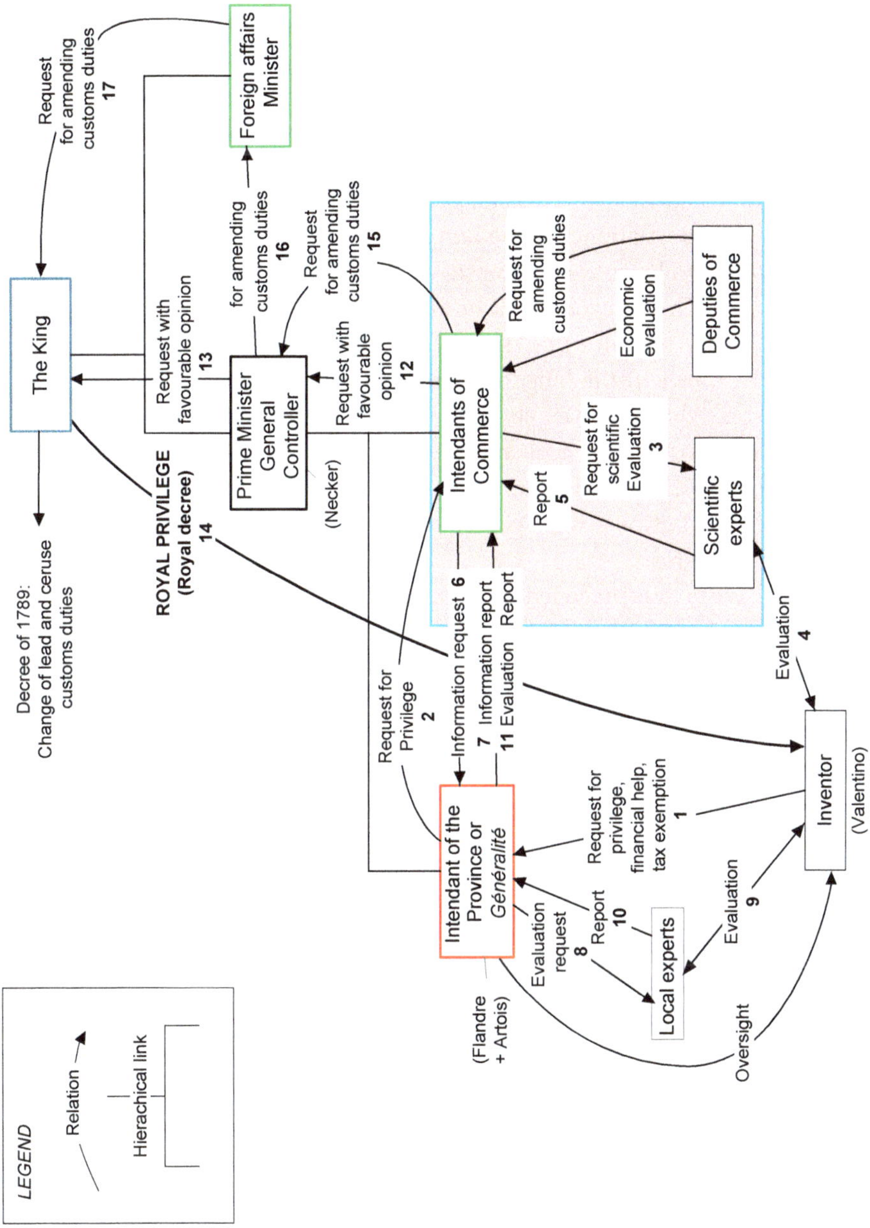

GRAPH 6.1 *Request for a privilege: decision-making route.*

This diagram shows that the Intendant in the province was only an intermediary and that the members of the *Bureau du commerce,* as well as the *Contrôleur general,* had an advisory role only. The final decision was exclusively in the king's hands and the privilege was granted by a royal decree. However the diagram also shows a real freedom in the multiple exchanges between the applicant, a private individual, such as a chemist-apothecary at a hospital in Lille, the Intendant of the city, the intendants and commissioners of commerce and the minister. This enlightened organisation contrasts with the usual image of the absolutist state. One can also note the fairness, the seriousness of the answers and the patience of the *Bureau du commerce*, which took pains to rule three times on Valentino's requests.[77] The six-year duration of Valentino's case also shows the evolving position of the government and the difficulty of reconciling the need for free trade of lead with the national income generated by the mines of the kingdom. On one hand, the applicant owned his secret and could threaten to exploit it abroad. On the other hand, the state had the power to grant the privilege and the financial aid associated with it, but it had to cope with the complexities of international trade, protective tariffs and competition between various production sectors such as lead mines and ceruse manufactures.

The exchanges that preceded granting a privilege as a compensation for a service performed by the inventor for the well-being of the nation, argue against the current idea of a French absolutist state as opposed to the British system, which was deemed to be liberal due to the fact that it was based on individual right[78]. It should be noted that eighteenth-century Britain was still not free from monopolies, the damaging effects of which were highlighted by Adam Smith.[79] Before being superseded by the 1791 patent laws, French royal privileges were very different from the monopolies to which Smith referred and should rather be compared with Britain's patent system in spite of their differences. Both were granted by the king and created a temporary monopoly. Both needed a full description of the invention but, in contrast with British patents, which required no preliminary examination, the novelty and usefulness of French inventions had to be established by means of scientific

77 On 6 March, 3 June 1788 and on 13 May 1791, just before it was disbanded on 27 December 1791, AN F^{12} 1507; Bonnassieux, *Conseil de commerce* (see note 3).

78 Christine MacLeod, *Inventing the Industrial Revolution. The English patent system (1660-1800)* (Cambridge: Cambridge University Press, 1988).

79 Monopolies were suppressed in 1624, with exception of temporary invention monopolies, which were granted for fourteen years. Adam Smith, *An Inquiry into the Nature and Causes of the Wealth of Nations* (London: Wordsworth Classics of World Literature: 2012).

evaluations.[80] Finally, British patents were not free of charge; they thus frequently entailed a heavy financial burden on provincials and were consequently often restricted to Londoners and those with wealthy patrons or local connections.[81]

In spite of the dramatic change of the French institutions during the Revolution, one can observe continuity in the governance of industrial development. The promotion of industry and the shift towards a liberal system continued to be driven by the same decision-making bodies – the *Comité d'agriculture et du Commerce* replaced the *Bureau du commerce* – and by chemists and other members of the scientific community, who believed that they could be directly useful to industry, still continuing to carry out evaluations and thus taking part in the decision-making process.

Acknowledgments

I would particularly like to thank John Perkins for his precious advice and Lissa Roberts for her precise review of this essay and for her comments and suggestions which resulted in important clarifications. I would also like to thank John Christie for his bibliographical suggestions and Ernst Homburg for the data he gave me on the ceruse industry and for the constructive exchanges we had on the various meanings of the word "ceruse".

80 Hilaire-Pérez, "Invention," pp. 93-144 (see note 7).

81 Such as official fees and accommodation in London (see note 78).

CHAPTER 7

Between Industry and the Environment: Chemical Governance in France, 1770-1830

Thomas Le Roux

Chemistry's visibility in the historiography of the first industrialization remains minimal.[1] The traditional view is that the first industrialization centered on the textile and steel industries, while energy and technological systems driven by steam engines symbolise the "revolution" of productive systems. Even the historiographies that emphasize slow, integrated changes, the intensification of know-how, and the rise of consumption, insisting on an *industrious revolution* rather than an *industrial revolution,* rarely refer to chemistry.[2] This reflects the facts that there were few large chemical companies until the mid-nineteenth century and that chemical processes are often invisible, incomprehensible and dangerous – three traits that explain why chemistry is often ignored. Yet chemistry was (and remains) at the heart of many processing operations and played a vital role in production processes and the encouragement of technical development.[3] More than physics or mechanics, the discipline of chemistry progressed rapidly in the last third of the eighteenth century, accounting for many industrial experiments and advances. While much has been written about a "chemical revolution", embodied particularly by Antoine Lavoisier, his colleagues and protagonists, chemically based activities amongst craftsmen and tradesmen are too often overlooked. Many crafts, however, used chemical substances that were indispensable for production and its "improvement". Organic and mineral acids, chlorine, ammonium chloride, various pigments and sodium hydroxide were all chemicals promoted to improve the manufacturing of consumer goods. Especially from the 1770s

1 For a more detailed presentation of this essay, see Thomas Le Roux, "Chemistry and Industrial and Environmental Governance in France, 1770-1830," *History of Science* 54 (2016): 195-222.

2 Liliane Hilaire-Pérez, *La pièce et le geste. Artisans, marchands et culture technique à Londres au XVIIIe siècle* (Paris: Albin Michel, 2013); On chemistry's importance in the first industrialization: André Guillerme, *La naissance de l'industrie à Paris. Entre sueurs et vapeurs: 1780-1830* (Seyssel: Champ Vallon, 2007).

3 Bernadette Bensaude-Vincent, Isabelle Stengers, *Histoire de la chimie* (Paris: La Découverte, 1992); Sacha Tomic, *Comment la chimie a transformé le monde. Une histoire en 7 tableaux* (Paris: Le Square, 2013).

DOI 10.1163/9789004325562_009

chemical manufacturing drastically changed industrial processes. This not only had adverse effects on workers' health, it more broadly altered European societies' relationship with their environment.[4] Revolutionary events amplified this process by freeing the productive sphere from a number of constraints, encouraging all kinds of technical improvements and giving chemists a crucial role in matters of governance.

This essay examines how chemists contributed to the technological reorganization in France at the end of the eighteenth and the beginning of the nineteenth century, how they justified using potentially harmful or polluting processes by stating that this would contribute to national prosperity, and how the idea of improvement helped legally and rhetorically to build a production regime that disqualified traditional precautionary attitudes to certain artisanal and industrial processes. This resulted in a new regime of environmental governance devoted to the advancement of chemistry and industrial production.

The Acids Revolution and Value Shift

In the 1770s in France, a silent revolution took place in the relationship between chemical production and both its environment and medicine. Alain Corbin has shown that this decade was a turning point in medical and olfactory attitudes towards certain products.[5] Broadening this line of enquiry by considering the art of governing populations, it appears that chemistry played a crucial role in social and political representations as well as in governance systems. Previously, faced with the hazards, nuisances and disadvantages involved, regulatory authorities had been wary of laboratory and artisanal chemistry. The police, who traditionally saw to matters of public health and community safety and comfort, particularly resisted the use of aggressive acids. Reflecting this distrust, several trials took place in Paris against craftsmen who made nitric acid, the only strong acid produced on an industrial scale before 1770, known then as aqua fortis. In 1768, for example, Police Superintendent Jean-Baptiste Lemaire, with the backing of the Faculty of Medicine, summoned a nitric acid distiller who operated in the city center before the police court, on a charge

4 Thomas Le Roux, *Le laboratoire des pollutions industrielles, Paris, 1770-1830* (Paris: Albin Michel, 2011).

5 Alain Corbin, *Le miasme et la jonquille. L'odorat et l'imaginaire social, XVII-XIXe siècles* (Paris: Aubier, 1982).

of endangering the public's health.[6] Under the continued influence of the miasma theory, police protocol through the end of the *ancien régime* called for keeping close tabs on manufacturers of aqua fortis and acids, viewing them as sources of ill health and pollution.[7] Workshops and factories that offended the senses and contaminated the air and water underwent strict preventive investigations known as *commodo et incommodo*.[8]

The manufacturing of chemicals, and acids in particular, was banned from cities and often carried out in small-scale and home-based production facilities, where hazards were none the less significant, as the case of nitric acid reveals. Before the growth of sulfuric acid, nitric acid was used in most industries, from tanneries to metal works[9]. A key product for industrialization and highly corrosive, it was made in Paris beyond the Porte Saint-Denis in small isolated workshops guarded by the police. Despite their product's fundamental contribution to industrialization, these spaces remained untouched by large capitalist investment and exuded a sort of toxic domesticity. In the 1770s, however, a new way of seeing chemistry was emerging. Of course, even chemists generally recognized nitric acid's corrosive nature. In 1773, describing the art of the aqua fortis distiller in *Description sur l'art du distillateur d'eau forte*, Jacques-François Demachy described its "suffocating fumes" and very dangerous manufacturing processes.[10] The accompanying illustration, however, reveals a purpose quite other than promoting care when working with dangerous substances. A worker is pictured only to show the scale of the place, which is depicted without any chemical substances; the devices shown were to be understood, not experienced, and the heat, hazards and acid-soaked atmosphere were hidden to suggest an idealized vision of technical know-how.[11] There was neither activity nor matter, just production tools, which were the pedagogical focus of this book's representation of work. (See figure 7.1)

---

6 Archives Nationales (AN), Y 9471B, report by Superintendant Lemaire, 5 August, 1768; AN, F[12] 879, *Rapport fait à la Faculté de médecine* [...] *pour examiner le laboratoire du Sieur Charlard, et juger les inconvéniens qui peuvent résulter pour les maisons voisines, de la distillation d'eau-forte*, by Bellot, de la Rivière, des Essartz, de Vallun, 1774, 16-17.

7 Nicolas Des Essarts, *Dictionnaire universel de la police*, 8 vols. (Paris: Moutard, 1786-1790), vol. 6, 1-2.

8 Le Roux, *Le laboratoire*, chapter 1 (see note 4).

9 John Graham Smith, *The Origins and Early Development of the Heavy Chemical Industry in France* (Oxford: Clarendon Press, 1979).

10 Jacques-François Demachy, *L'art du distillateur d'eaux-fortes* (Paris: impr. de Delatour 1773), 37-38.

11 Ibid., Part 2, Plate 1, Figure 2.

FIGURE 7.1 *Jacques François Demachy, "Art du distillateur d'eaux fortes etc. Laboratoire pour les eaux fortes,"* Description des arts et métiers *(Paris, 1773), Part 2, Plate 1, Figure 2.*
ILLUSTRATION COURTESY OF CONSERVATOIRE NATIONAL DES ARTS ET MÉTIERS.

Just like the *Description des arts et métiers* commissioned by the Academy of Science, which included the study of aqua fortis distillers, the *Encyclopédie*'s plates were based on facilities in Paris: their representations of work reflected a technological and universal order that wished to discipline bodies and become free from the constraints of particular locations.[12] This was a world ruled by scientists and technicians, who increasingly imposed their authority on the world of craftsmen and related physical practices. The stakes were all the higher because acids were a key industrial product, and government had begun ardently to promote acid manufacturing.

The main change came with sulfuric acid production. Despite having similar uses to nitric acid, sulfuric acid was only produced in small quantities

12 William Sewell, "Visions of Labour: Illustrations of the mechanical arts before, in and after Diderot's *Encyclopédie*," Steven Kaplan and Cynthia Koepp, eds., *Work in France. Representations, meaning, organization and practice* (Ithaca: Cornell University Press, 1986), 258-286; Georges Friedman, "L'*Encyclopédie* et le travail humain," *Annales, histoire, sciences sociales* 8 (1953): 53-68; Ken Alder, *Engineering the Revolution: Arms and enlightenment in France, 1763-1815* (Princeton: Princeton University Press, 1999).

before the late 1760s, mainly in laboratories where it was condensed in expensive and delicate glass jars during the final production stage. Sulfuric acid was absolutely essential for cotton printing, on which the government had recently lifted its ban in 1759. Simultaneously, in the United Kingdom, John Roebuck broke new ground by using lead chambers to condense sulfuric acid. The room-sized lead-lined chambers allowed sulfuric acid production on an industrial scale, which soon challenged nitric acid's preeminent position. The technology was introduced in France by the Englishman John Holker, a factory inspector employed by the French monarchy, who in 1768 set up a sulfuric acid factory next to his printed cotton factory in a suburb of Rouen.[13] Over a period of some months, the gases discharged by the chambers corroded by the strong acid caused breathing problems for neighbours and damaged surrounding vegetation.[14] According to police jurisprudence, this kind of nuisance was not tolerated near homes and, in 1772, Holker was prosecuted in France's first great industrial pollution trial. After several months of proceedings in the *Parlement de Rouen* (then called *Conseil supérieur*), the accused parties, supported by Jean-Charles Trudaine, the Commerce Director, obtained a hearing at the finance royal council. There Trudaine had to argue against Minister Henri Bertin, a former Paris Lieutenant-General from 1757 to 1759.[15] Economic interest prevailed over Bertin's arguments: in September 1774, the plaintiffs' case was dismissed and henceforth, no one was allowed to trouble or disrupt the factory's operation.[16]

The lead chamber was therefore not only a technological development: it occasioned a shift in the order of industrial and environmental governance. Firstly, it required major investment, which made any production stoppage problematic. Secondly, it was supposed to be a perfect device that replaced multiple operations by the workers with a simple system in which leaks could be better controlled. The same argument was used for both health benefits and economic profits, as any leak was treated as a loss of value.[17] Lastly, it led to a change in the representation of sulfuric acid manufacturing, presented from then on through its technology, such as by a technical drawing or a model showing only the mechanism's external envelope. Devices appeared, in the

13 Smith, *The Origins* (see note 9).

14 L.-G. de la Follie, "Réflexions sur une nouvelle méthode pour extraire en grand l'acide du soufre par l'intermédiaire du nitre, sans incommoder ses voisins," *Observations sur la physique* 4 (1774): 336.

15 AN, F[12] 879, vitriol oil factory in Rouen, letter from Bertin to Trudaine, 11 December 1773.

16 Ibid., order of the royal council, 20 September 1774.

17 de la Follie, "Réflexions," p. 336 (see note 14).

first representations of this kind of factory, like magical boxes where everything took place according to the scientific processes of physics and chemistry. Through representations of the working world and especially of artisanal and industrial chemistry, the last *ancien régime* decades witnessed the inevitable fading out of the proximity of arts and crafts. In its place arose a technical, disembodied order that would celebrate technical drawings during the nineteenth century, the seeds of which were already present in the encyclopaedic initiative and in scientific encouragement.[18]

While chemistry transformed the governance of industry and especially the government's attitude to nuisances, the root causes of this change should be sought in the government's economic policy as well as in the changes chemists were introducing to medical aetiology. The groundwork was laid by the chemist Louis-Bernard Guyton de Morveau from Dijon. In March 1773, he was contacted by the Dijon Cathedral's authorities, who could not get rid of the mephitic stench emanating from the decaying corpses in one of the building's vaults. Applying the theory on the combination of ammonia, whose presence could be deduced from the smell of decay, with an acid to produce a neutral salt, he fumigated the vault with muriatic (hydrochloric) acid and managed to neutralise the smell. In the medical community, among which the miasma theory was predominant, this removal of a smell was considered a victory over putrid infection and the experiment had a huge impact.[19] It was the first time acid fumigation was used in France as a way of controlling fermentation and its smell. The novel procedure broke with traditional conceptions about the corrosive and dangerous nature of acids. Until then, acids had never been thought of as a disinfectant; instead physicians recommended fumigation with odoriferous herbs, the spraying of vinegar or starting of a fire or a powder explosion to disperse and destroy miasmas. The fact that acid fumigation was not widely taken up, at least not immediately, is not important. The significance of these experiments and the publicity surrounding them in 1773 and 1774 was not that they immediately led to routine therapeutic use, but that they profoundly altered the perception of acids, a product that was crucial for industrial development.

18 Nicolas Pierrot, "Les images de l'industrie en France, peintures, dessins, estampes, 1760-1870," (Doctoral dissertation in History, Université Paris I Panthéon-Sorbonne, 3 vols., 2010).

19 Louis-Bernard Guyton de Morveau, "Nouveau moyen de purifier absolument, et en très-peu de temps, une masse d'air infectée," *Journal de physique* 1 (1773): 436 and 3 (1774): 73; Thomas Le Roux, "Du bienfait des acides. Guyton de Morveau et le grand basculement de l'expertise sanitaire et environnementale (1773-1809)," *Annales Historiques de la Révolution française* 383/1 (2016): 153-175; For the further history of Guyton's fumigating machine, see Elena Serrano's essay in this volume.

Because acid promotion was at the heart of a governmental scheme to encourage production in order to boost industrial development, Guyton de Morveau's experiments were a godsend and allowed medicine to progress hand in hand with economic development. In 1774, Vicq d'Azyr prescribed acid fumigation to treat epizootic diseases in the south of France, and the following year, the academicians Etienne Mignot de Montigny and Philibert Trudaine de Montigny also recommended this disinfection method in two separate notices and enquiries.[20] The chemist Antoine Parmentier, senior scientific advisor to the police lieutenant-general, observed that "acid vapours" combined with other elements in the air to "contribute to its cleanliness."[21] He further extolled the virtues of "spirit, acid and corrosive fluids, which could be released to destroy or neutralise the miasma supposed to be dispersed in the air."[22] The link between medicine and chemistry was strengthened specifically during this decade: one after the other, the physicians Claude Berthollet, Antoine Fourcroy and Jean-Antoine Chaptal stopped practising medicine to study chemistry, and all played a predominant role in the development of industrial chemistry, especially in relation to acids. Fourcroy became an expert and regular government advisor, assessing nuisances caused by chemical factories. For instance, he was commissioned by the Royal Society of Medicine in 1783 to write a report on a sulfuric acid factory in Rouen. In this report, he strongly defended the manufacturer, dismissing his opponents as prejudiced and ignorant about chemistry.[23] From then on, the *Bureau du Commerce* (Trade Office) relied on these new ideas to encourage and at times force the establishment of acid factories in close proximity to cities. To override public objections, trade officers used a combination of medical and chemical arguments, claiming that "sulphur vapours, far from being hazardous, are very healthy. They purify unhealthy air. They prevent epidemics."[24] This was a reasonable stance to take, especially as the chemists Macquer and then, from 1784, Berthollet were members of the *Bureau* and greatly contributed to spreading these ideas.

20 Etienne Mignot de Montigny, *Instruction et avis aux habitans des provinces méridionales de la France, sur la maladie putride et pestilentielle qui détruit le bétail* (Paris: Imprimerie Royale, 1775); Philibert Trudaine de Montigny, *Avis aux peuples des provinces, où la contagion sur le bétail a pénétré et à ceux des provinces voisines* (Paris: Imprimerie Royale, 1775).

21 Antoine-Augustin Parmentier, *Dissertation physique, chimique et économique, sur la nature et la salubrité des eaux de la Seine* (Paris: impr. de J.-G. Clousier, 1775), 9-10.

22 Antoine-Augustin Parmentier, *Dissertation sur la nature des eaux de la Seine* (Paris: Buisson, 1787), 104.

23 AN, F^{12} 1507, folder I-1, report by Fourcroy and Thouret, 2 November 1783.

24 AN, F^{12} 1506, folder 5, factory in Javel, department of trade documents, undated [1777-1778].

Chemistry not only contributed to the idea of improvement in arts and crafts, but also in society and the economy. Many chemical substances were used to produce semi-luxury and luxury goods, especially in the non-ferrous metal industry. This was the case, for example, for silver and gold plated products and in the new platinum industry. In a letter to Guyton de Morveau, dated November 1786, Lavoisier mentioned that he was working with the innovative gold and silversmith Jacques Daumy. To treat platinum for craftsmen and manufacturers, they refined, dissolved, precipitated and revived the metal with hydrochloric and nitric acid, ammonium chloride, borax, lead, bismuth, antimony and arsenic.[25] More generally, the integration of new chemistry with luxury goods production occurred through precision metalwork on precious metals, "the artistry of which was perfected through very delicate chemical operations and relatively challenging processes for the workers."[26] The new Paris Mint, built between 1771 and 1775, served as a laboratory, not only for making coins but also mastering the chemistry behind refining, cupellation and alloying assays to make all kinds of gold and silverwork pieces. In this field, with a similar argument as that used for sulfuric acid, the matter of goldbeating was raised before the royal council in 1773. Gold beaters in Lyons were accused by the police of using furnaces within the city, as well as treating gold with antimony and corrosive sublimate (a mercury compound), two dangerous substances. Both activities violated manufacturing and public health laws. The beaters defended themselves by appealing to the king and arguing a number of economic points: to uphold the restrictions and the "broadly prohibitive law" would condemn their industry to decline; and it was only by violating restrictions "contrary to the public interest" that their "art has improved." According to them, violations were "brutal procedures enforced by an uninformed police officer." The king was convinced and agreed to authorize the use of furnaces as well as antimony and mercury for metal refining, "to support the main factories in Lyons," by an order of the Council dated 29 April 1773. The order's preamble stated that the petitioners had "a duty to preserve their industry for the state and to perfect it."[27]

Chemistry thus contributed to transform physicians' and scientists' perception of mineral acids and other chemical substances, which until then had been feared for their corrosive effects. It demonstrated the medical usefulness of these acids, thereby helping to overcome the usual precautions and spurring their industrial use. This reversal had an effect on industrial nuisance policy in

25 Antoine Lavoisier, *Œuvres* (Paris, 1854-1868), vol. 5, 340.

26 AN, K 903, Monnaie, file 108, *Observations sur le projet d'édit de Mr de F*, undated (1770s).

27 Essarts, *Dictionnaire*, vol. 7, p. 434 (see note 7).

the medium-term. During the Revolution, the Consulate and Imperial years, it translated into fundamental reports and regulations, which tied medical expertise, chemistry and industrial development together.

Chemical Governance and the Environment (1789-1810)

In 1791 liberalism, which was already perceptible at the end of the *ancien régime*, inspired several steps to facilitate the setting up of industries whatever their nuisances. While the disruption that occurred in 1789 implicitly resulted in more freedom for industrialists, who took advantage of the dismantling of former regulatory institutions, the new legislation permanently released industry from several controlling regulations.[28] *Commodo et incommodo* investigations were stopped, and the d'Allarde Law of March 1791 abolished arts and crafts guilds and their statutes.[29] In September 1791, the *Bureau* and industry inspectorate were dismantled. In October 1791, letters patent granting exclusive privileges were abolished, which did away with preliminary investigations in use under the *ancien régime*. Consequently, industrialists were free to set up factories wherever they wanted and manufacture products using whichever processes they wished. Legislators ruled that the courts only had jurisdiction to address property damage.

However, the revolutionary period was also characterized by a strengthening of the value shift occurring in the public interest domain. Public interest was no longer concerned first and foremost with safeguarding public health, but was permanently associated with economic development. Chemists became the new official experts on assessing pollution and contributed to the policies of the successive republican governments. Thus in 1791, when the Academy of Science investigated the pollution caused by an ammonium chloride factory established in the middle of a populated neighbourhood near Valenciennes, the report's authors (chemists Louis Cadet, Fourcroy and Berthollet) conceded that pollution had disadvantages, but considered that the smoke could be tolerated in the interest of national industry and general welfare.[30]

---

28 Alain Plessis, ed., *Naissances des libertés économiques, 1791-fin XIXe siècle* (Paris: Institut d'histoire de l'industrie, 1993).

29 Philippe Minard, "Le métier sans institution: les lois d'Allarde-Le Chapelier de 1791 et leur impact au début du XIXe siècle," Steven Kaplan and Philippe Minard, eds., *La fin des corporations* (Paris: Belin, 2003), 81-95.

30 Archives de l'Académie des Sciences, *Registre des procès-verbaux de l'Académie des sciences*, fol 371-373, 28 June 1791.

Moreover, with the outbreak of war, scientists legitimized exceptional industrialization. Lazare Carnot, Fourcroy, Guyton de Morveau and Pierre-Louis Prieur took an active part in the decisions taken by the *Comité de Salut Public* (Public Safety Committee), including the decision to employ reputed chemists such as Gaspard Monge, Berthollet, Jean Darcet, Bertrand Pelletier and Chaptal in the war effort.[31] This patriotic mobilization led to the idea that national production should be boosted in the context of war and economic competition. Requisitioning and military orders caused the reconversion of factories and the adoption of foreign processes: hatters converted their workshops to make varnished helmets; the need for uniform buttons led to the setting up of workshops for copper treatment and acid gilding; and textile workshops were set up under the supervision of the agency for republican military clothing to make cloth, sheets and military dress.[32] In a more obvious way, the arms industry flourished in the capital, where the *Manufacture de Paris* was set up in the autumn of 1793 as a huge cluster of workshops including a small arms testing and improvement workshop established in April 1794 under the authority of the weapons commission, headed by Guyton de Morveau.

A few flagship products illustrate the involvement of chemists in industry. In addition to armaments manufacturing, leather, copper and pigments are worth noting. From the autumn of 1793, the revolutionary government was looking for a way to produce leather goods for the troops as fast as possible and entrusted the task to the chemists. The *Comité de salut public* instructed Berthollet "to take charge of tanning improvement," and named Armand Seguin to conduct several experiments. Fourcroy praised Seguin's "revolutionary" tanning method, which involved replacing previously used weak organic acids with a concentrated solution of sulfuric acid, in his report to the Convention, noting that the new process sped up manufacturing considerably.[33] The new method was employed at the state-financed tannery established in late 1794 on the Sèvres Island in the Paris suburbs, with used acid discharged into the Seine.

---

31 Patrice Bret, *L'Etat, l'armée, la science. L'invention de la recherche publique en France, 1763-1830* (Rennes: Presses Universitaires de Rennes, 2002); Charles C. Gillispie, *Science and Polity in France. The revolutionnary and Napoleonic years* (Princeton, NJ: Princeton University Press, 2004); Nicole Dhombres and Jean Dhombres, *Naissance d'un nouveau pouvoir: sciences et savants en France, 1793-1824* (Paris: Payot, 1989).

32 Jean-François Belhoste and Denis Woronoff, "Ateliers et manufactures: une réévaluation nécessaire," Françoise Monnier, ed., *A Paris sous la Révolution, nouvelles approches de la ville* (Paris: Publications de la Sorbonne, 2008), 79-91.

33 Antoine-François Fourcroy, *Rapport, au nom du Comité de Salut Public, sur les arts qui ont servi à la défense de la République, et sur le nouveau procédé de tannage découvert par le citoyen Armand Seguin* (Paris: impr. Nationale, 1795).

A similar mindset was applied to copper production. From 1791, the government requested gold and silversmith Daumy to melt and refine bronze bells to make coins, and then cannons, in a new factory on the Île de la Cité using chemical processes requiring large amounts of nitric, sulfuric and muriatic acid.[34] Here too, toxic remains were discharged into the river.

The example of minium, a lead-based pigment used to make porcelain, shows how chemists used pollution charges to promote industrial innovation. Neighbours alleged that the lead oxide discharged from a minium factory in the Parisian neighbourhood of Bercy in June 1793 polluted the area. Simultaneous to Bercy's council banning the factory, the government entrusted an expert report to the chemists Pelletier and Petit, who argued that the problem could be reduced by improving manufacturing processes. Guyton led a second inspection, with the understanding that minium production was "valuable for the Republic" and "useful for arts workshops." Fourcroy, then a National Convention member, advocated that the owner should be "protected in his factory given that minium could no longer be procured in Britain or Holland."[35] Confirming that there was a public health issue, Guyton's report resulted in an order to demolish the factory, but the owner was encouraged to improve his manufacturing processes with the help of well-placed chemists and physicians, who also lobbied successfully for generous government compensation to rebuild the factory.[36] This case exemplifies what became a pattern of technical improvement under the guise of chemical scientific expertise, initially only seen with sulfuric acid factories, now fixed by Guyton de Morveau and Fourcroy.[37] From the Napoleonic regime, members of the *Conseil de salubrité* would take it upon themselves to make this the core of environmental regulation. Two important considerations emerged from chemists' involvement: public interest was equated with economic development and technical solutions were proffered as the best way to reduce nuisances from craft production. It thereby became possible to divest the traditional police of its prerogative powers and to bypass the judicial reasoning of the *ancien régime*.

After peace returned in 1795, France's economic expansion was driven by its chemical industry. In Paris alone, dozens of factories were working inside the city walls and suburbs. Growth was especially embodied in four flagship plants

34 Bibliothèque Historique de la Ville de Paris, Ms 929, Manufacture Daumy.

35 *Procès-verbaux du comité d'instruction publique de la Convention nationale* (Paris: Impr. Nationale, 1891-1907), vol. 2, 792; *Archives parlementaires* (Paris: CNRS, 1980-), vol. 79, 153-154.

36 AN, F^{12} 1509, Comité agriculture, folder Ollivier, 1794.

37 On this case, Le Roux, *Le laboratoire*, pp. 204-212 (see note 4).

that were largely established between 1795-99 by chemists who were (or became) academicians. The factory owned by Chaptal in Ternes was a particular focus of public attention and began raising protests while it was being built. However, having become Interior Minister, Chaptal rejected all complaints year after year and generally became the key agent for unifying science and administration. Before the Revolution, he was a chemical entrepreneur in Montpellier, producing especially sulfuric acid, and in 1790 was sued for pollution by local residents. Berthollet recruited him to be involved in the republican administration, and he headed the gunpowder factory of Grenelle, when it exploded, killing 550 workers.[38] Prior to becoming Interior Minister in 1800, Chaptal built his famous sulfuric acid factory in Ternes and wrote *Essai sur le perfectionnement des arts chimiques* [Essay on the Means of Perfecting Chemical Arts], both a treatise on applying the latest chemical discoveries to industry and a guide for entrepreneurial leadership.[39] As Minister, academician, chemist, entrepreneur and member of the *Conseil d'Etat*, he embodied the conjunction of scientific expertise, entrepreneurial experience and emerging administrative standards through which industrialism and liberalism became associated.[40]

Between 1802 and 1804, Chaptal worked to build a coherent framework to serve industry. He began by founding the *Conseil de salubrité* in 1802, an institution with scientific expertise – mainly chemists with a soft spot for industry – to advise the Parisian authorities. In agreement with the owners of the factories and workshops, members often denied that industrial fumes were noxious or deleterious to plaintiffs' health. In the case of chemical factories, they pointed out that the waste gases were "valuable" and that it was in the interest of the manufacturer to prevent them from escaping. Pollution, thus, was construed as the result of unintended accidents rather than daily practice.[41] Meanwhile, economic affairs were entrusted to new or reorganized institutions, such as the Mint, which became a veritable laboratory for testing the

38 Thomas Le Roux, "Accidents industriels et régulation des risques: l'explosion de la poudrerie de Grenelle en 1794," *Revue d'Histoire Moderne et Contemporaine* 58/3 (2011): 34-62; Claire Barillé, Thomas Le Roux and Marie Thébaud-Sorger, "Grenelle, 1794. Secourir, indemniser et soigner les victimes d'une catastrophe industrielle à l'heure révolutionnaire," *Le Mouvement Social* 249/4 (2014): 41-71.

39 Jean-Antoine Chaptal, *Essai sur le perfectionnement des arts chimiques en France* (Paris: Déterville, undated [1799]).

40 Jeff Horn, *The Path Not Taken. French industrialization in the age of revolution, 1750-1830* (London; Cambridge, MA: MIT Press, 2006).

41 Le Roux, *Le laboratoire*, chapters 7 to 9 (see note 4).

science/administration alliance.[42] Inside the *Société d'encouragement pour l'industrie nationale* (SEIN, Society for the Encouragement of National Industry) founded in 1801 by Chaptal, Guyton chaired the committee for chemical arts. There, he regularly saw Berthollet, Fourcroy, Nicolas Vauquelin, Parmentier, Darcet and Deyeux, under the general chairmanship of Chaptal – that is, all the academician chemists of the time, all supporters of industrial development and almost all editors of the leading journal *Annales de chimie*. During these years, the state apparatus particularly supported acid chemistry. The Directory had already recognized its considerable value, emphasizing in 1798 that acids "are like a reservoir of very powerful forces, which nature has made available to man to produce effects that would be impossible to obtain using mechanical force."[43] Under the Consulate, acid promotion increased.

However, everywhere in France as in Paris, trials against owners of chemical factories accused of pollution threatened to disrupt the steady industrial production. After Chaptal was replaced by Jean-Baptiste de Champagny as the Interior Minister in August 1804, the authorities contemplated a national response to this recurring issue. In November 1804, the new Minister asked the French institute "about factories exhaling an obnoxious smell and the risk that they posed for public health"; the institute entrusted the report to Guyton de Morveau and Chaptal.[44] A second report followed in 1809.[45] Together they provided the basis for the law of 1810 on polluting industries.[46] The 1804 report argued against the necessary validity of complaints by claiming a distinction between industries with processes based on organic putrefaction, which released "smells that were disturbing or toxic fumes," and those with processes

42 Patrice Bret, "Des essais de la Monnaie à la recherche et à la certification des métaux: un laboratoire modèle au service de la guerre et de l'industrie (1775-1830)," *Annales Historiques de la Révolution Française* 320/2 (2000): 137-148.

43 AN, F^{12} 2234, information provided to the Conseil des Cinq-Cents by the Directory, 31 January, 1798.

44 "Rapport [...] sur la question de savoir si les manufactures qui exhalent une odeur désagréable peuvent être nuisibles à la santé," 17 December 1804, *Procès-verbaux des séances de l'Académie des sciences* (Hendaye: Impr. de l'Observatoire d'Abbadia, 1910-1922), vol. 3, 165-168.

45 "Rapport sur les manufactures de produits chimiques qui peuvent être dangereuses," 30 October 1809, *Procès-verbaux* vol. 4, pp. 268-273 (see note 44); Subsequent citations not referenced in the notes are from these reports.

46 For details, Geneviève Massard-Guilbaud, *Histoire de la pollution industrielle, France 1789-1914* (Paris: Editions de l'EHESS, 2010), 34-45; Le Roux, *Le laboratoire,* pp. 255-261 and 274-283 (see note 4); Jean-Baptiste Fressoz, *L'apocalypse joyeuse. Une histoire du risque technologique* (Paris: Le Seuil, 2012), 150-165.

based on fire, which emitted vapours or gases that were uncomfortable to breathe, but usually only inconvenient. In particular, factories that were well run, might release an obnoxious, but certainly not a harmful smell. As a matter of fact, they wrote, the smell released by sulfuric acid factories "was not dangerous in the least for the workers who breathed the smell daily, and no neighbours' complaint could be deemed well founded." As for nitric and hydrochloric acid factories, their characteristic smell could not affect human breathing; the men "who work there every day were not at all inconvenienced and it would be very wrong of the neighbours to complain."[47] Another hydrochloric acid expert, the industrialist Robert O'Reilly, contradicted their assertions. In his *Essai sur le blanchiment* [Essay on Bleaching], O'Reilly reported witnessing "in a very large plant near Paris, the cruel suffering endured by [the] wretched [workers] because of the suffocating fumes. I saw them writhe on the floor in pain; often these first effects of oxy-muriatic acid can cause even serious illnesses."[48] In fact, Chaptal knew that occupational health was at stake in the workplace. In 1798, in his *Essay*, he argued that "the various tasks in a workshop are not all equally easy or pleasant; and since young men are too often minded to refuse difficult or repulsive tasks, a coercive force is needed to compel them to carry out these tasks and this force can only be found in the ties that bind them to the workshop and keep them at the disposal of their superiors."[49] Politics and productivity won the day in his view.

The 1804 report's fundamental stance was that the central government needed to protect France's chemical industries. Obstructions "would be at once unfair, persecutory, harmful to the advancement of the arts and would not address the harm caused by the operation." Chaptal and Guyton thereby turned the Minister's question on its head, moving away from a public health issue to a concern of political economy by defining an entirely new program. "[P]rosperity of the crafts absolutely requires that boundaries are set to put an end to arbitrary decisions by magistrates by drawing a circle around industrialists, inside which they will be able to ply their trade freely and securely."[50] The 1809 report followed similar reasoning, but the context had changed. On the one hand, since its foundation in 1802 and its specialization in industrial affairs in 1806 (with chemists Deyeux and Cadet de Gassincourt as its authorized experts), the Paris *Conseil de salubrité* had acquired an undeniable legitimacy.

47 Quotations from "Rapport" (1804) (see note 44).

48 Robert O'Reilly, *Essai sur le blanchiment* (Paris: Bureau des Annales des arts et manufactures, 1801), 99.

49 Chaptal, *Essai*, pp. 9-10 (see note 39).

50 Quotations from "Rapport" (1804) (see note 44).

On the other hand, for several months, an ongoing problem had been caused by sodium hydroxide factories, in which sea salt was broken down by sulfuric acid using the Leblanc process, discharging large quantities of muriatic acid. Several soda plants, managed by distinguished chemists who would become members of the *Conseil de salubrité* or were very close to them, were built, in the Parisian suburbs after 1800. The irreversible damage caused by acid vapours and the utter destruction of crops and orchards around these factories was obvious. Faced with a fresh spate of pollution cases in 1809, the Minister was forced to commission a second report from the institute. The new committee membership had a similar "industrialist" flavor: alongside Chaptal and Guyton de Morveau, the entrepreneurs Fourcroy and Vauquelin also owned a sizeable chemical factory in the center of Paris, while the chemist Deyeux made no bones about his industry bias in the *Conseil de salubrité.*

The Minister urged its authors to strike a balance between the interests of industrialists and those of neighbouring property owners. No longer simply cast as victims, industrialists were required to choose factory locations carefully. The report's conclusion thus called for a consensus, proposing to group industries into three classes according to their degree of nuisance. The chemists suggested introducing specific administrative enquiries for the purpose of authorizing factories in each group, to pre-empt most pollution problems. However, the spirit of Guyton's 1793 report on minium was not forgotten; on the contrary, the report promoted technical improvement for the chemical industry as a means of moving from one class to another to lighten constraints and government control. These conclusions were included in the law of October 1810 on insalubrious industries.

Pollution and Governance through Chemistry

The decree of 1810 aimed to establish a regulatory framework by separating industries into three categories depending on their level of noxiousness. A great deal was at stake in how a factory was categorized; being moved from the first meant that the factory was no longer considered noxious and could avoid the *Conseil d'Etat*'s long and strict authorisation procedure. The decree was therefore supposed to promote innovation and the *perfection* (a word very often used) of processes.[51] *Conseil de salubrité* members, who completely sup-

51 Thomas Le Roux, "La chimie, support du développement de l'industrie perfectionnée sous la Révolution et l'Empire," Natacha Coquery, ed., *Les progrès de l'industrie perfectionnée* (Toulouse: Presses Universitaires du Midi, 2017), 26-35.

ported their founder's industrialism, were soon convinced by the principle of process improvement as a way of avoiding production restrictions. From 1811, the *Conseil de salubrité* linked industrial improvement and public health.

In the years after the decree was first implemented, *Conseil de salubrité* members expressly encouraged the building of chemical plants in Paris, as shown by numerous reports supporting the four flagship factories mentioned above. These factories belonged to the first class according to the decree, but had been set up prior to it. Their assessment by the *Conseil de salubrité* was spurred by complaints from neighbours. Impressed by these magnificent factories for which substantial capital had been raised, the *Conseil* systematically ruled in their favor. Complainants were discredited as reflecting the much-admired entrepreneurs' reverse image, their complaints deemed even less reasonable because at each inspection, improvements were observed. To explain why complaints persisted, the *Conseil* blamed exuded fumes on accidents, themselves considered rare and due to worker negligence, an increasingly standard response from the nineteenth century. Hopes about further improvement rested on the wager that scientific theorizing and laboratory tests could and would be confirmed on an industrial scale. Despite multiple protests, none of the main factories were threatened with closure.

Instead, the chemical industry became a pillar for industrial governance, with chemists and other scientists given a crucial role. On one hand, they were granted authority through claims of how they stimulated further industrial innovations. On the other hand, they were asked to exercise that authority as arbiters of the governmentally sponsored drive for national prosperity and perfection of the arts while attending to matters of public health. Like Achilles' spear, chemistry was thus poised to "cure the wound it had inflicted."[52] In calling for grouping insalubrious industries together in certain areas, for example, the *Conseil de salubrité* member Parent-Duchâtelet showed the way:

> [A] special government official will be able to supervise them *effectively* and implement the conditions required to ensure public health. We stress the importance of the latter point, to show that large manufacturing centres will not become, as we might have feared, sites of infection by expelling their poisonous atmosphere far away, but will contribute to the

52 Victor de Moléon, ed., *Rapports généraux sur les travaux du Conseil de salubrité de la ville de Paris et du département de Seine. Années 1802-1839* (Paris: Bureau du Recueil industriel, 1828-1841), vol. 1, 207-208.

> advancement and sanitizing of factories, and perhaps also to the improvement of the arts.[53]

This sanitizing by chemistry was carried out in several ways, depending on the industry, through disinfection, smoke consumption or condensation. In industries using putrescible matter, "disinfection" was one of the preferred means of applying the recommended procedures. The first large-scale trials were carried out in Parisian gut factories using chlorinated products, in a decisive battle against putrid infection. In 1820, the *Société d'encouragement pour l'industrie nationale* created an award for manufacturers who could dress guts without prolonged maceration or noxious smells. The model gut factory in Clichy near Paris became a site for testing disinfection, using the new method of the pharmacist Antoine Germain Labarraque. The guts were steeped in a soda chloride bath, which removed the smell straight away. Though expensive, the method was quicker than the old one and succeeded in sanitizing the factory. In October 1822, Labarraque was awarded the prize and the *Conseil de salubrité* recommended the method to every new gut factory, assuming that it would also be adopted in older factories in a few years.[54] The "disinfecting" properties of acids were also put to use, thanks to their powers of decomposition. Darcet tested the use of sulfuric acid himself for melting tallow in the new Parisian slaughterhouses after 1818. In the 1820s, the acid was also used to purify oils in many Parisian workshops, distilleries and potato starch factories, where it immediately turned starch to syrup, and in beet sugar refineries, where it prevented decay. Darcet began to use muriatic acid in 1815 to extract gelatine from bones, and encouraged strong glue manufacturers to adopt his method.[55] With regard to smoke consumption in furnaces, he was once more at the heart of technological change to cut down the amount of industrial smoke. To reduce the incidence of industrial smoke increasingly criticized by city dwellers, especially as the use of fossil coal had begun to spread in Parisian industries, the *Conseil d'Etat* strove to recommend the construction of smokeless furnaces. Having witnessed the first lasting attempt to build a smokeless furnace at the mint in 1808, Darcet continuously encouraged the adoption of this kind

53 Alexandre Parent-Duchâtelet, *Hygiène publique* (Paris: Baillière, 1836), vol. 2, 326.

54 Moléon, *Rapports*, vol. 1, pp. 141-142, 163 and 319-322 (see note 52).

55 Jean-Pierre Darcet, Antoine Germain Labarraque, Jean-Baptiste Huzard, and Henri-François Gaultier de Claubry, "Rapport sur l'examen comparatif de la fonte des suifs à feu nu, et par l'intermédiaire de l'acide sulfurique," *Annales d'hygiène publique et de médecine légale* 24 (1840): 54-78; "Mémoire sur divers emplois de la gélatine extraite des os, par le procédé de M. d'Arcet," *Annales de l'industrie nationale et étrangère* 7 (1822): 276-285.

of smoke "burning" furnace with improved combustion and perfected the technology.

Finally, the expansion of the chemical industry in the Paris region forced manufacturers to take technical measures to preserve the surrounding areas. Condensing, absorbing, dissolving and closed-system production were all complementary methods implemented to "coerce" or retain the vapours produced by the manufacturing or use of chemicals by industry. In the 1820s, the *Conseil de salubrité*'s efforts to condense acid vapours increased. Whenever possible, closed-system production was encouraged in acid factories. Woulfe's apparatus, in which gases were forced to pass through a series of tubes and vessels filled with water or liquid absorbents, was recommended in nitric acid workshops.[56] Other condensation devices were proposed for various industries that implemented chemicals and acids in particular. This was the case for precious metal refining, for instance. Gold and silver refining, no longer restricted by a Directory government monopoly and performed subsequently with less expensive methods using sulfuric acid instead of nitric acid, was carried out in several Parisian workshops after 1815. Having observed various technical processes at the mint, Darcet set out to prove their harmlessness provided a number of steps were followed to ensure gas condensation. Therefore, industry's presence within cities hinged on the manufacturers' ability to prevent the discharge of acid gases. In 1827, Darcet himself designed a model refining workshop and its furnishings. In the refining furnace, five closed platinum vessels allowed acid gases to discharge through a lead pipe and flow into a single pipe under the workshop towards three refrigerated lead boxes, where the sulfuric acid fumes condensed. Uncondensed sulfuric vapours remaining in the gas were then removed by directing the gas into a box filled with hydrated lime, which rotated on itself when operated by a crank and a gear system. This mixed the lime and improved contact with and absorption of the sulfuric acid. Finally, a pipe discharged any remaining vapours from the box into the main workshop stack.[57]

The same reasoning was applied to recycling. In the 1820s, the chemists Charles Derosne and Anselme Payen embarked on producing depurative

56 Moléon, *Rapports*, vol. 1, p. 160 (see note 52).

57 Jean-Pierre Darcet, *Instruction relative à l'art de l'affinage* (Paris: impr. de Huzard-Courcier, 1827); Jean-Pierre Darcet, *Seconde instruction relative à l'art de l'affinage* (Paris: Bachelier, 1828); Thomas Le Roux, "Déclinaisons du 'conflit'. Autour des atteintes environnementales de l'affinage des métaux précieux, Paris, années 1820," Thomas Le Roux and Michel Letté, eds., *Débordements industriels dans la cité et leurs conflits, XVIII-XXIe siècles* (Rennes: Presses Universitaires de Rennes, 2013), 179-198.

organic compounds (bone-black and animalized carbon) from animal residue. Recovered animal waste was made into chemicals with sanitizing properties, for example to clarify and purify beet sugar, while partly addressing the problem of refuse disposal. Against charges of polluting the neighbourhood, the *Conseil de Salubrité* praised Derosne's operation for recycling wastes, boosting production and sanitizing the environment:

> This animal matter [livestock blood], which used to be wasted and often spoiled the air as it decayed, is now carefully collected to be used in numerous sugar refineries [...] and will be turned into a worthwhile export industry; the fortuitous benefit of an industry in operation, which extracts a useful product out of a worthless substance and turns an unhealthy cause into a new source of wealth.[58]

Like Derosne, Payen was involved in the chemistry of recycling animal waste, which he distilled in his Grenelle plant to make ammonium chloride.[59] By 1820, the factory had become a huge industrial complex, also manufacturing soda chloride, lime chloride, animalized carbon, sugar, and so forth. While pollution from recycling on such a large scale was frequent and at times permanent, the *Conseil de Salubrité* found a convenient answer in proposing to recycle the recycling plant's main waste, empyreumatic oil, which they offered to gas factories. These could distil the oil into lighting gas, in exchange for which the soda chloride factory could then treat the ammoniated waste that they produced.[60] Therefore, most of the time, sanitizing processes combined waste recycling and its profitable reclamation.

This insistence on promoting technical improvement explains why chemists became so fond of engraved technical drawings, which were soon adopted by the *Bulletin de la Société d'Encouragement pour l'Industrie Nationale*. From the first issue published in 1802, the *Bulletin* included copper-engraved plates as inserts, showing the emerging graphical art form that was developing around the *Conservatoire des arts et métiers*.[61] Unlike representations by artists, who had distanced themselves from production sites during the revolutionary decades, technical drawing was a political undertaking in itself. As a tool for

58 Moléon, *Rapports*, vol. 1, p. 286 (see note 52).

59 Sabine Barles, *L'invention des déchets urbains: France, 1790-1970* (Seyssel: Champ Vallon, 2005), 38-39.

60 Moléon, *Rapports*, vol. 2, pp. 15-17 (see note 52).

61 Yves Deforges, *Le graphisme technique. Son histoire et son enseignement* (Seyssel: Champ Vallon, 1981).

rationalization, it introduced a new symbolic order that established technology as superior to work places and physical movements.[62]

Chemistry was at the heart of the combination of technical devices and law: environmental governance simply conformed to the necessities of competitive industrial production. In March 1815, to explain the shift to local prefects, the government tried to clarify the new approach and the spirit that should guide their decisions on implementing the 1810 law: "Before, the existence of chemical factories was precarious in some respects [...] In reviewing authorisation applications [the local authorities] will most certainly rise above petty interests; and driven only by reasons of public interest, they will give opinions based on considerations of a higher order."[63] Sulfuric acid production improved continuously as greater numbers of lead chambers appeared; they symbolized the analogy between economic growth, political economy, chemistry and technical and environmental devices. Increasingly effective lead chambers were one of the advanced industries that could better prevent acid vapours, and was typical of scientists' discourses. According to Chaptal, in 1819, this technology had "reached perfection, as not one sulphur atom was lost in the operation as proven by the analysis carried out on the acid produced."[64] Without any loss of acid, and therefore, no loss of value for the manufacturer, virtuous profit was combined with environmental protection, Chaptal claimed.

Conclusion

Thus, linked to industrial production and scientific improvement, chemistry contributed to change environmental perceptions of the industrial world by the turn of the nineteenth century. The mistrust widely shared by local authorities, social observers and citizens regarding factory and workshop emissions was replaced by a new definition of harmfulness and harmlessness as industrialization imposed its pace, in order to adapt to the claimed imperative of

62 Ken Alder, "Innovation and Amnesia: Engineering rationality and the fate of interchangeable parts manufacturing in France," *Technology and Culture* 38 (1997): 273-311; Eda Kranakis, *Constructing a Bridge: An exploration of engineering culture, design and research in nineteenth-century France and America* (Cambridge, MA: MIT Press, 1997); Olivier Lavoisy and Dominique Vinck, "Le dessin comme objet intermédiaire de l'industrie," Pierre Delcambre, ed., *Communications organisationnelles. Objets, pratiques et dispositifs* (Rennes: Presses Universitaires de Rennes, 2000), 47-63.

63 Archives de la Préfecture de police, DB 134, instruction by the Director-General for Agriculture, Commerce, Crafts and Industry to the department prefect, 4 March 1815.

64 Jean-Antoine Chaptal, *De l'industrie française* (Paris: Renouard, 1819), vol. 2, 65.

economic growth. While this shift was perceptible from the 1770s with the first regulatory exceptions for strategic products, the 1810 decree – imagined, designed and implemented by chemists – perpetuated chemistry's role as an environmental regulator. Chemistry and its practitioners helped build an industrial world at a time when its arrival was not universally welcomed. After 1815, there was no doubt that industrial advancement had become a value shared by many actors. Through their experiments as well as their discourses and involvement in industrial applications for their discoveries, chemists participated in this expansion more than others. The authorities provided a great deal of support, especially in resolving conflicts about pollution caused by the chemical industry, by conceiving an administered regulatory framework that justified industrialism. In 1816, in a retrospective essay on industrial growth since the Revolution, Chaptal's first assistant Claude-Anthelme Costaz sang the merits of the 1810 decree: "We are not afraid to say that it has been of great benefit to owners and manufacturers [...] [who] [...] are now assured not to be bothered when carrying out their business once it has been authorized by the authorities: which is not inconsequential for the prosperity of chemical factories."[65]

65 Claude-Anthelme Costaz, *Histoire de l'administration, en France, de l'agriculture, des arts utiles, du commerce, des manufactures* (first edition 1816; Paris: Vve Bouchard-Huzard, 1843), 375-376.

CHAPTER 8

Renegotiating Debt: Chemical Governance and Money in the Early Nineteenth-Century Dutch Empire

Andreas Weber

When Johannes Goldberg, head of the Netherlands' Department of Trade and Colonies (*Departement van Koophandel en Koloniën*), left his office in The Hague's city center in early spring 1816, he was full of expectations. After months of reflecting on how to mitigate the high level of debt accrued from the colonies in Southeast Asia, he had just dispatched instructions for a complete monetary reform program to the governor general in Batavia, the administrative seat of the Dutch in the Indonesian Archipelago.[1] Since the demise of the former Dutch East India Company (VOC) in 1799, colonial debt had become a heavy burden for the Dutch treasury.[2] Goldberg hoped that the introduction of a new currency, the colonial guilder, would help resolve this problem by capitalizing on Java and the neighbouring islands' agricultural wealth.[3] However, establishing trust in the new money in maritime Southeast Asia turned out to be a challenging endeavor, not least because the actual realization of monetary policy depended on the practices of locally situated mint masters, assayers and other chemical practitioners who engaged with the production and circulation of paper notes and metal coins both at home and in the colonies. This essay demonstrates that, instead of simply serving their metropolitan masters, these historical actors and material objects had a direct hand in shaping imperial policies through their involvement in localized processes of chemical governance.

As explained in the introduction to this volume, the concept of chemical governance enables us to follow the various ways in which chemical knowl-

1 The report, which is dated 12 March 1816, and all additional material is now stored at the National Archive [=NA], The Hague, collectie Goldberg, 162.

2 On the tremendous public debt after the Napoleonic Wars, Wantje Fritschy, *De patriotten en de financiën van de Bataafse Republiek. Hollands krediet en de smalle marges voor een nieuw beleid* (*1795-1801*) ('s-Gravenhage: Stichting Hollandse Historische Reeks, 1988), 40-41.

3 The idea of using a new currency to renegotiate debt has a long history and goes at least back to China in the ninth century AD. Philip Coggan, *Paper Promised: Debt, money and the new world order* (New York: Public Affairs, 2012).

DOI 10.1163/9789004325562_010

edge, know-how and substances were harnessed – sometimes in quite mundane ways – to realize the mediating embodiment of policy-making measures. In some cases this entailed the experimental construction of material standards and instruments for the policing of commodities through tests for 'purity' and adulteration, or the establishment of excise and taxation regimes.[4] In other cases, such as that examined in this essay, it entailed the production of objects that materialized systems of value and exchange, enabling the instantiation and authorization of these systems through the combination of their mobility and durability, on one hand, and leaving them subject to local negotiation and appropriation on the other.

By approaching Goldberg's monetary reform through the lens of chemical governance, this essay develops two closely related components discussed in this volume's introduction. First, it draws attention to the activities of actors whose practices and products were mundanely responsible for giving materialized form and substance to policies. Chemistry and knowledge of materials turned out to be critical to what might appear to be a rather abstract financial policy. Putting their specialized knowledge and skills regarding inks, papers and alloys to work, mint masters and assayers translated initial policy pronouncements into the notes and coins whose circulation determined how policies actually unfolded. This meant that those who engaged in chemical practices on a more local level – away from metropolitan centers – could nonetheless exert formative influence over global policies. This chapter moves outside Europe to show that processes of chemical governance were by no means confined to the European stage. Far from simply enabling or constraining policy enforcement, this essay argues, the locally made choices of metallic composition, symbolic decoration and the like ascribed values to currency that might differ from those which were originally conceived in the metropolitan capital.[5]

This account adds a chemical element to Michel Foucault's notion that power is better seen operating in geographically complex fields of operation than via the hands of a few government ministers and administrators. Power does not reside in individuals but is dispersed, in this case through varied acts

4 Joppe van Driel and Lissa Roberts, "Circulating Salts: Chemical governance and the bifurcation of "nature" and "society," *Eighteenth-Century Studies* 49 (2016): 233-263; William Ashworth, *Customs and Excise: Trade, production and consumption in England, 1640-1845* (Oxford: Oxford University Press, 2003).

5 Referencing the work of Madeleine Akrich, Michel Callon writes succinctly, "Without the material device the operating instructions are meaningless." Michel Callon, "What Does it Mean to Say that Economics is Performative?," *CSI Working Papers Series* 5 (2006): 1-58, on 12.

of circulating and manipulating materials.[6] In place of considering the metropolitan state as the centralized motor of empire building, this essay reveals the geography of policymaking's practical field to have been polycentric and rooted in decisions about materials. At the same time, however, the decentralized character of imperial policymaking was often obscured. The outcomes of local initiatives and acts of appropriation, which shaped the realization of policy, became associated – for better or worse – with the metropolitan actors under whose policy jurisdiction they officially fell. While chemical practice could thus alter the content and consequences of centrally conceived policy and policymakers, it takes effort to recover its traces.[7]

Goldberg's responsibilities for the establishment and policing of colonial policy began in 1814 when the Dutch king asked him – he was at that time head of the Court for Commerce in The Hague (*Rechtbank van Koophandel*) – to set up and instruct a new colonial government in Batavia. Following the British occupation of Batavia between 1811 and 1815, Willem I and his advisors hoped that Goldberg's expertise and skill as an administrator would help turn the colonies into a financially rewarding annex of the Dutch kingdom.[8] Similar to the French revolutionary government's decision to introduce *assignats* – the French revolutionary currency with which Antoine-Laurent Lavoisier was involved – two decades before, the Dutch king requested Goldberg to introduce a new paper currency in the colonies. Goldberg and the Crown hoped that the colony's proto-bankers would accept this new currency in exchange for silver and gold coins that could be used to pay colonial debts at home. In order to endow Goldberg with the authority and status to master this challenge, the king not only installed him as the well-paid head of the Netherlands' Department of Trade and Colonies but also made him a member of the Dutch nobility. Within a couple of months, Goldberg became one of Willem I's most

6 Michel Foucault, *Power/Knowledge: Selected interviews and other writings, 1972-1977* (New York: Pantheon, 1980).

7 As such, the chapter might be seen as a chemical version of Michel de Certeau's celebrated account of resistance in Michel de Certeau, *The Practice of Everyday Life*, trans. Steven Rendall (Berkeley: University of California Press, 1984).

8 For the Dutch king's high expectations regarding the colony's revenue, see Thomas Stevens, *Van der Capellen's koloniale ambitie op Java. Economisch beleid in een stagnerende conjunctuur, 1816-1826* (Amsterdam: Historisch Seminarium van de Universiteit van Amsterdam, 1982), 216-219; For Goldberg's earlier involvement in organizing the monetary system at home, see Roland Uittenbogaard, *Evolution of Central Banking? De Nederlandsche Bank 1814-1852* (Dordrecht: Springer 2015), 50-55.

important advisors regarding the management of Dutch possessions in the Indonesian Archipelago.[9]

As in the history of *assignats*, Goldberg's monetary reform program points to the inherently political nature of administrative expertise. Again similar to Lavoisier, Goldberg was confronted with the following challenge: how to convince proto-bankers and plantation owners in the colonies to link their private capital to that of the Dutch state, which faced heavy colonial debts?[10] Ironically, perhaps, there was a fundamental difference between their chosen solutions. Lavoisier considered paper money issued by a privately owned bank as the best solution to mitigate debt. Goldberg, on the other hand, opted for a program that was inseparably composed of chemical and political components. In order to convince the colony's proto-bankers to accept the new currency, Goldberg saddled the colonial government with the task of supplementing the introduction of his new paper money – in total six million guilders – with specially minted silver and copper coins. As Goldberg put it, only if the colonial government were able to prove publicly and repeatedly its ability to exchange paper money for silver and copper coins would the colony's proto-bankers gradually accept the new currency as a trustworthy medium of exchange.[11] In practice, this meant that establishing trust in the new currency rested on gaining the support and cooperation of mint masters, assayers, skilled artisans and wealthy investors.

In the end, Goldberg's efforts led to little. Owing to his inability to recruit and control local chemical expertise, the entire project turned out to be a costly failure. Instead of stabilizing his position as one of the main voices in Willem I's inner circle of advisors, Goldberg found himself shifted to a less influential position as the king decided to dissolve the Department of Trade and Colonies.[12] As a consequence, Goldberg continuously suffered from severe financial problems. Even after his death, his family was exposed to creditors claiming money from his bequest.[13]

---

9 W.M. Zappey, *De economische en politieke werkzaamheid van Johannes Goldberg* (Brussels: N. Samson, 1967), 99-100.

10 On Lavoisier's involvement in developing a new currency in France, see Charles Gillispie, *Science and Polity in France: The revolutionary and Napoleonic years* (Princeton: Princeton University Press, 2004), 84-86. For a more recent analysis, see Rebecca Spang, *Stuff and Money in the Time of the French Revolution* (Cambridge, MA: Harvard University Press, 2015), chapter 2.

11 NA The Hague, collectie Goldberg, 162A, letter Goldberg to General Committee (*Commissie generaal*), 's Gravenhage, 18 March 1816.

12 Zappey, *De economische en politieke werkzaamheid*, p. 170 (see note 9).

13 Ibid., p. 193.

By analysing Goldberg's failure through the lens of chemical governance, this essay differs from other studies on scientific inquiry in the context of the early nineteenth-century Dutch empire.[14] While older numismatic scholarship has laid important groundwork, others have based their narratives on issues such as the diffusion and failure of 'enlightened' science in the colonial Indonesian Archipelago (Lewis Pyenson, Andrew Goss and Pieter Boomgaard).[15] In particular Goss' attempt to separate 'science' and 'governance' analytically into two independent narratives has received scholarly criticism.[16] Boomgaard's approach is more nuanced but shares the disadvantage that it also relies on a normative notion of how science should have developed in the Indonesian Archipelago. For him, the key point is that the lack of metropolitan financial and organizational support led to a lack of 'scholarly excellence' in the early nineteenth-century Indonesian Archipelago.[17] Unlike historians such as Pyenson, Goss and Boomgaard, this essay does not divide the analysis of governance and science into 'metropolitan' and local 'colonial' trajectories; neither is it interested in making normative judgments about 'scientific quality'. By conceptualizing chemical governance as a process in which chemical practice was used to intermediate between local interests (for example of mint masters and assayers) and goals formulated in Europe, this essay uses Goldberg's experience to shed light on the particularities through which chemical practice

14 For a similar use of the concept in a European context, see Van Driel and Roberts, "Circulating Salts" (see note 4).

15 For the first see John Bucknill, *The Coins of the Dutch East Indies* (London: Spink, 1931); Cornelis Scholten, *De munten van de Nederlandsche gebiedsdelen oversee, 1601-1948* (Amsterdam: J. Schulman, 1951). For the latter see Lewis Pyenson, *Empire of Reason. Exact sciences in Indonesia 1840-1940* (Leiden: Brill, 1989); Andrew Goss, *The Floracrats. State-sponsored science and the failure of enlightenment in Indonesia* (Madison: University of Wisconsin Press, 2011); Peter Boomgaard, ed., *Empire and Science in the Making. Dutch colonial scholarship in comparative global perspective, 1760-1830* (New York: Palgrave Macmillan, 2013).

16 See, for instance, Robert E. Elson's review of *The Floracrats* in the *American Historical Review* 116 (2011): 1469 and Robert-Jan Wille, "The Coproduction of Station Morphology and Agricultural Management in the Tropics: Transformations in botany at the botanical garden at Buitenzorg, Java 1880-1904," Denise Phillips and Sharon Kingsland, eds., *New Perspectives on the History of Life Sciences and Agriculture* (Cham: Springer International Publishing, 2015), 253-275; For a criticism of Pyenson, see Paolo Palladino and Michael Worboys, "Science and Imperialism," *Isis* 84 (1993), 91-102.

17 Boomgaard, ed., *Empire and Science*, p. 18 (see note 15).

shaped governance in Europe and Southeast Asia in the early nineteenth century.[18]

To carry out this analysis, this essay focuses on two episodes of chemical governance. By zooming in on the production of the new money in the Netherlands, the first episode shows how Goldberg struggled to acquire local support for the introduction of a new currency. While the Haarlem printers Joh. Enschedé en Zonen considered Goldberg's governance scheme an ideal opportunity to capitalize on their accumulated chemical skills, the mint masters and assayers in Utrecht read it as a threat to their status and authority. In particular Goldberg's attempt to circumvent the Utrecht mint's monopoly by relying on a consortium of private manufacturers stirred tensions between the government and mint masters there.

While the mint in Utrecht openly refused to collaborate with Goldberg, the mint master and assayer in the Dutch East Indies chose a more subtle reaction, which provides the focus of this essay's second episode. Here, Goldberg's dream of reducing public debt in the Netherlands by introducing a new currency system in the colonies was transformed by the colonial mint master and assayer, who used chemical practices to appropriate his project for the pursuit of local interests. Taken together, the two episodes show that conferring authority over and value to money required a heavily managed and globally distributed process that was often built on quite mundane decisions about chemical practice.

Co-operation and Resistance in the Netherlands

The production of Goldberg's money first required chemical expertise in the Netherlands. While the national mint in Utrecht was tasked with producing silver and copper coins, the paper money was produced by the Haarlem company Joh. Enschedé en Zonen. Since 1795, the company regularly supplied the government with a large amount of promissory notes and paper bills.[19] It was

18 This definition resonates with what governance scholars have conceptualized as 'de facto' governance. See Arie Rip, "De Facto Governance of Nanotechnologies," Morag Goodwin, Bert-Jaap Koops, and Ronald Leenes, eds., *Dimensions of Technology Regulation* (Nijmegen: Wolf Legal Publishers, 2010), 285-308.

19 N.L.M. Arkestijn, "Met de bajonet op de keel. Inkopen met ongedekt papiergeld," in E.H.P. Cordfunke and H. Sarfatij, eds., *Van Solidus tot euro: Geld in Nederland in economisch-historisch en politiek perspectief* (Hilversum: Verloren, 2004), 139-146. For a broader history of the production of bank notes, see Colin Narbeth, Robin Hendy, and Christopher Stocker, *Historische bankbiljetten en aandelen* (Baarn: Moussault's Uitgeverij, 1979), 29-44.

therefore relatively easy for Goldberg to activate the printer's support. The production of Goldberg's new currency in Haarlem emerged from negotiations involving chemical expertise and deep knowledge of different raw materials. In order to provide printers such as Joh. Enschedé en Zonen with suitable paper, paper manufacturers in Boxtel and the Zaan area had teamed up with a chemist in Amsterdam to develop and work with liquid bleaching agents that had been described by the French chemist Claude Louis Berthollet a few years before.[20]

The production of durable and whitish paper, which was compatible with different inks and letter types, was chemically challenging for paper makers and printers. While one-guilder notes were printed in black, higher denominations were printed in red and blue, requiring consideration of the substances used to make the colours. Some of the notes were marked with a printed stamp that had a reddish-brown colour (Fig. 8.1).[21]

Moreover, since some of the bills had to be signed by government officials in the colonies, the paper had also to be coated with a special layer of gelatin to keep the ink from bleeding. Solutions to these chemical problems demanded the establishment of highly disciplined spaces and personnel. In order to guard chemical and other expertise, the company erected a 'secret' print shop (*Geheimdrukkerij*) in 1795. Access to this branch of the company was only allowed for selected staff members who had sworn an oath to remain silent about the chemical composition of value-bearing papers.[22]

The production of the new silver and copper coins turned out to involve even more challenges. Chemical practitioners might unexpectedly cease to support policy demands. Initially, the Utrecht mint master confirmed his willingness to assist with Goldberg's plan and produced at least one sample of the new silver guilder (Fig. 8.2). But subsequently Utrecht became unwilling to cooperate. Following the dissolution of district mints in 1806, the mint in Utrecht was the only site in the Netherlands where minting took place. When Goldberg requested Gideon Langerak Du Marchie Servaas, mint master in

20 NHA Haarlem, collection Van Gelder, inv. 1108: Proeven om bleekwater te vervaardigen (1813). Domestic paper money was apparently also printed on special paper. B.W. de Vries, *De Nederlandse papiernijverheid in de negentiende eeuw* ('s-Gravenhage: Martinus Nijhoff, 1957), 271.

21 For more details on the production of the bills, see Theo van Elmpt, *Netherlands East Indies. Paper currency, 1815-1827* (Uithoorn: Elran Express, 2009), 1-44; A.M. van de Waal, "De oudste bankbiljetten. Eerste relatie van de Nederlandsche Bank met Joh. Enschedé en Zonen," *Ontwikkelings- en ontspanningsvereniging "De Nederlandsche Bank"* 8 (1953), 4-14.

22 Frans Willem Lantink et al., *Voor stad en staat*, vol. 1: *Plattegrond* (Amsterdam: Joh. Enschedé, 2003), 42.

FIGURE 8.1 *Sample of paper money with the value of one colonial guilder. The musical notation in the margin was based on a special letter type which the German type cutter J.M. Fleischmann had developed exclusively for Johan Enschedé in Haarlem in the 1770s.* (*NA The Hague, collectie Goldberg, inv. 162*). IMAGE PUBLISHED UNDER CC-BY BY THE NATIONAL ARCHIVE, THE HAGUE.

FIGURE 8.2
Sample of new 'colonial' guilder produced at the mint in Utrecht in 1814 or 1815. IMAGE USED BY THE KIND PERMISSION OF DE NEDERLANDSCHE BANK.

Utrecht from 1797, to produce the necessary silver alloy, the latter rejected the order by referring to the mint's limited capacity.[23]

While historians have simply reiterated the mint's claim of limited capacity as a valid explanation, it seems more likely that the mint master's rejection mirrors growing friction between the government and the mint.[24]

At the time when Goldberg began searching for producers of the new silver coins, tensions between the mint in Utrecht and the government in The Hague were rising. In particular, the government's decision to request that a consortium of private entrepreneurs take on the production and minting of copper coins irritated Yman Dirk Christiaan Suermondt, who had succeeded Du Marchie Servaas as mint master in 1815.[25] The consortium included the Amsterdam silversmith Hendrik de Heus, who had previously produced 175 tons of copper coins for the Southeast Asian colonies in 1802, and copper millers from the Veluwe.[26] In the course of 1815 they produced thirty-five tons of copper coins for a price that undercut the Utrecht mint.[27] De Heus could offer low prices since he had successfully equipped his workshop with steam driven machinery for the flattening of coins. Circumventing the mint's monopoly was not a new phenomenon, for Suermondt and his colleagues in Utrecht had previously complained about the government's decision to hire private entrepreneurs instead of supporting their monopoly.[28]

The mint's growing resistance against Goldberg's plan to produce a new currency created other governance challenges. While paper money with the value of one guilder in the metropole was officially backed with silver coins containing 9.614 gram of fine silver, Goldberg struggled to determine the colonial

23 For a good contextualization, see Marcel van der Beek, "'s Rijks Munt en de aanmuntingen door De Heus," *De Beeldenaar* 2 (1995): 357-366.

24 For the first see, for instance, Copius Hoitsema and F. Feith, *De Utrechtsche munt uit haar verleden en heden* (Utrecht: Oosthoek, 1912).

25 Pieter Hendrik van der Kemp, *De teruggave der Oost-Indische koloniën 1814-1816* ('s-Gravenhage: M. Nijhoff, 1910), 234; Pieter Hendrik van der Kemp, "Episodes uit de geschiedenis der aanmuntingen ten behoeve van Oost-Indië, 1802-1807," *Bijdragen tot de taal-, land- en volkenkunde van Nederlandsch-Indië* 70:2 (1915): 227-440, 310.

26 J. MacLean, "Koperindustrie in Nederland, 1750-1850," *Economisch en social-historisch jaarboek* 43 (1971): 39-63, 42.

27 Van der Kemp, "Episodes uit de geschiedenis," pp. 331-338 (see note 25). On Van Heus and his company, see H.A. Diederiks, "Hendrik de Heus. Een Amsterdamse ondernemer in het begin van de 19e eeuw," *Amstelodamum* 56 (1969): 58-65, and MacLean, "Koperindustrie in Nederland," pp. 39-63 (see note 26). For a fascinating micro-history of one of the involved copper mills in the Veluwe, see Henri Slijkhuis, *De kopermolen in Zuuk* (Vorchten: De Bekenstichting, 2015).

28 Van der Kemp, "Episodes uit de geschiedenis," pp. 288-300 (see note 25).

equivalent.[29] He wished to avoid a situation whereby the new coins would simply end up in China and other parts of Asia where the silver was used for artwork or religious purposes. Eventually Goldberg advised the Dutch crown to produce silver guilders, which contained about 20 percent less fine silver than the ones which were used in the Netherlands.[30] Owing to the aforementioned tensions with the mint in Utrecht, he chose not to base this decision on the expertise of the mint master. Neither did he turn to Willem Adriaan Arnold Poelman, inspector and assayer general in Utrecht since 1814. Poelman, who had served previous governments as a mint expert, only assayed two colonial silver coins for Goldberg.[31] Since these coins had been produced in the years before the British took over Java in 1811, the mint's assay did not help Goldberg to assess the current situation in Java. Resistance from chemical practitioners led Goldberg to rework his plans. Instead of drawing upon Poelman's experience and contacts with metal traders, assayers and inspectors of weights and measures in Batavia, Goldberg chose to base his decision on policy documents taken from the archives of the VOC and the Council for the Asian Possessions and Establishments (*Raad der Aziatische Bezittingen en Etablissementen*), which contained the opinions of former colonial government officials.[32]

In turn, Du Marchie and Poelman's reluctance to support Goldberg cannot be justified by referring to the technical complexity of the chemical examination of coins. Although assays could sometimes take several days, they involved a standard technique of precision which mint masters usually employed in

29 Jean Hendrik van Swinden, *Bedenkingen over het muntwezen* (Utrecht: Het Geldmuseum, 1997; transcription of an unpublished manuscript dated 1815), 11.

30 For the first see Willem G. Wolters, "The 'Doit Infestation in Java': Exchange rates between silver and copper coins in Netherlands India in the period, 1816-1854," in Hans Ulrich Vogel, ed., *Money in Asia (1200-1900): Small currencies in social and political contexts* (Leiden: Brill, 2015), 108-139, 126-127. Pieter Hendrik van der Kemp, "De Nederlandsche-Indische proefgulden van 1815," *Tijdschrift voor munt- en penningkunde* (1913): 21-60, 24 labels Goldberg's decision as "unnatural."

31 For a short biography of Poelman, see Albert A.J. Scheffers, "Om de kwaliteit van het geld. Het toezicht op de muntproductie in de Republiek en de voorziening van kleingeld in Holland and West-Friesland in de achttiende eeuw" (PhD diss., Leiden University, 2013), vol. 2, 443-444.

32 For the global character of material expertise see Albert A.J. Scheffers, "Johan Sebastiaan van Naamen, Muntmeester van Batavia 1764-1768 en Utrecht 1782-1797 in perspectief," *De Muntkoerier* 42 (2013): 18-19; and Simon Schaffer, "Assay Instruments and the Geography of Precision on the Guinea Trade," Marie-Noëlle Bourget, Christian Licoppe and H. Otto Sibum, eds., *Instruments, Travel and Science. Itineraries of precision from the seventeenth to the twentieth century* (London: Routledge, 2002), 20-50.

their small laboratories.[33] The result of the assay depended heavily on the availability of bone ash cupels and precise balances which were used to weigh the coin material before and after the analysis.[34] In order to separate fine silver from copper and other base metals, the assayer melted each of the coins and then put them together with a piece of lead in a cupellation furnace, designed for high temperatures. After the oxidation and evaporation of the base metals, the remaining piece of fine silver ('*koninkje*') was then soaked in nitric acid in order to remove the last remnants of other materials. By comparing the weight of the *koninkje* with the weight of the original coin, assayers were able to calculate the metal content of the assayed coins.

Owing to the lack of additional information, the Dutch crown simply followed Goldberg's advice. After Willem I decided to circulate two coins (one domestically, the other in the Indonesian Archipelago) with the same denomination and weight but with a different content of silver, another governance challenge arose.[35] Although the king had reserved sufficient funds for the new money, Goldberg now struggled to find a suitable production facility. Since his initial plan to rely on the mint in Utrecht for producing silver coins did not materialize, he opted to delegate the production to the colonial mint in Surabaya on the Northeast Coast of Java. Although the Surabaya mint had only limited experience in supplying the colonies with coins, a frustrated Goldberg could do nothing more than order the colonial government in Batavia to start the minting as quickly as possible. Goldberg now had to engage with the materials of chemical practice himself in order to carry through his policy. To facilitate the production of coins in Java, Goldberg added additional minting equipment, such as castings for the production of the dies and pencil drawings to the shipment of metal he sent to get things started.[36] Chemical practitioners added further frustration to this plan. When Goldberg tried to send the Utrecht die cutter J.P. Schouberg with the silver to Batavia, Schouberg claimed that his health would not allow him to travel to Java.[37]

---

33 For a description of the process, see Scheffers, "Om de kwaliteit," vol. 1, pp. 131-134 (see note 31); and John S. Forbes, *Hallmark. A History of the London Assay Office* (London: The Goldsmith's Company, 1999), 20-24.

34 For an in-depth study on gold and silver balances, see Michael A. Crawforth, *Weighing Coins: English folding gold balances of the 18th and 19th centuries* (London: Cape Horn Trading, 1979). Contemporary assay manuals stress the importance of precise assay balances and equal cupels. Louis N. Vauquelin, *Manuel de l'essayeur* (Paris: Chez le citoyen Bernard, 1799), 6-14 and 21-24.

35 NA Den Haag, collectie Goldberg, 162: Royal decision 8 November 1815, no. 39.

36 Van der Kemp, "De Nederlandsche-Indische proefgulden," pp. 33-34 (see note 30).

37 Ibid.

As the next section demonstrates, similar problems beset Goldberg's scheme once production was transferred abroad. Instead of simply following Goldberg's orders, the mint in Surabaya and Batavia's colonial administrators became entangled in a struggle regarding the new money's composition and management. The acts of chemical practitioners operating within a complex local political environment then proved transformative for the realization of Goldberg's distantly formulated policy.

Balancing Interests in Surabaya

When Goldberg's currency arrived in Java, the colonial government in Batavia was in an ambiguous position. On the hand, there were 'men on the spot' who wished to follow Goldberg's orders and add the crates of bills and barrels of coins to the local circulation.[38] On the other hand, Goldberg's plan met severe local resistance from the colony's elite. Material concerns over silver content were at the heart of these anxieties. While Goldberg considered the low silver content of his coins as a tool for preventing them from draining off to other parts of Asia, the local elite feared that it would devalue the local currency – the rupee – which had a much higher silver content. According to British regulations, one rupee in Java had to contain 10.896 gram of fine silver, which is almost 30 percent more than Goldberg's coins.[39] Since the introduction of new money formed a serious threat to locally accumulated capital, it is unsurprizing that there was resistance. Many local elites had profited heavily from the export of cash crops such as coffee, indigo and sugar, which they produced on their large private estates in the hinterland of Batavia.[40] With those profits

38 John S. Galbraith introduced the term 'man on the spot' in his pioneering attempt to understand the active role played by locally situated colonial administrators who sought to represent colonial rule in a context of required readiness to respond to local exigencies. See John S. Galbraith, "The 'Turbulent Frontier' as a Factor in British Expansion," *Comparative Studies in Society and History* 2 (1960): 150-168.

39 Pieter Hendrik van der Kemp, "De zilveren Java-Ropijen van de jaren 1816-1817," *Bijdragen tot de Taal-, Land- en Volkenkunde van Nederlandsch-Indië* 67 (1913): 275-366, 304; N.P. van den Berg, *De kwestie over den geldsomloop in Nederlandsch-Indië* (Batavia: H.M. van Dorp, 1863), 63-66.

40 On the importance of British, American and Chinese merchants in the region, see J.R. Fichter, *So Great a Proffit. How the East India trade transformed Anglo-American capitalism* (Cambridge, MA: Harvard University Press, 2010), chapters 3-6; Leonard Blussé, *Visible Cities. Canton, Nagasaki, and Batavia and the coming of the Americans* (Cambridge, MA: Harvard University Press, 2008), 60-64.

now threatened, conflicts and negotiations would ensue over how the colony's mint should go about assaying and producing new coins. Chemical practice thus again became the focus around which Goldberg's policy would be enacted, resisted and revised.

Local resistance and redirection came in different guises. When news about Goldberg's monetary reform program spread in Java in late 1816, Batavia's merchant elite gently reminded the colonial government that money with the name of guilders was unknown in the area. Two earlier attempts to produce and circulate 'guilders' had remained on a small scale and had been short-lived.[41] Moreover, they hinted that the introduction of a new currency would further threaten the local population's belief in the government's ability to act as a trustworthy broker of value. In particular, their fears were grounded in the precedent set by earlier British attempts to issue copper coins of inferior quality, which had destabilized the situation in Java.[42]

Since the colonial government depended on the support of these elite merchants – many of whom also held high administrative positions – colonial government representatives eventually agreed with some of their concerns and decided to opt for a compromise. In order to counter their fears of devaluation, the colonial government made two important changes to Goldberg's plan. First it decreed that the raw silver which had been shipped at Goldberg's behest should be used to produce silver rupees rather than silver guilders. Moreover, they ordered the colony's mint master to increase the rupee's content of fine silver to 9.6 grams – a value which split the difference between what Goldberg had expected them to be (7.968 grams) and the local norm (10.896 grams).[43] Since silver coins in the Netherlands also contained 9.6 gram of silver, the colonial government thus hoped that this compromise would also be well received at home.

---

41 Van der Kemp, "De zilveren Java-Ropijen," pp. 291-292 (see note 39).

42 ANRI Jakarta, Arsip Varia (K 64), Stukken betreffende het muntwezen, box 2, varia 191: Resident of Rembang to colonial government, 16 July 1825; J.P. Moquette, "De munten van Nederlandsch-Indië," *Tijdschrift voor Indische Taal-, Land- en Volkenkunde* 51 (1909): 33-56, 37; For an initial glimpse into the complexity of the trade networks in the region with a special focus on coinage, see Li Tana, "Cochinchinese Coin Casting and Circulating in Eighteenth-Century Southeast Asia," in Eric Tagliacozzo and Wen-chin Chang, eds., *Chinese Circulations. Capital, commodities, and networks in Southeast Asia* (Durham: Duke University Press, 2011), 130-148.

43 Van der Kemp, "Episodes uit de geschiedenis," p. 396 (see note 25).

At the time when Goldberg's silver arrived in Surabaya in mid-1816, the city's mint was poorly set up for producing government coinage.[44] According to a survey done during the British interregnum in 1811 or 1812, the facility consisted of a smelting house (290 × 56 feet), three sheds with screw presses and other machines imported from Europe, flattening mills and a chemical workplace for the material analysis of raw materials and coins.[45] In those years, coinage equipment circulated regularly between Europe, Asia and the Americas. Matthew Boulton's workshop in Birmingham, for instance, supplied a wide range of coinage facilities in Europe, India, South America and Russia with tools and equipment.[46] Like the mint in Utrecht, the Surabaya mint was equipped with small iron assay furnaces, glassware, pans for distillation, touchstones, and several sets of balances and weights (Fig. 8.3).[47]

By 1816 production at the Surabaya mint heavily depended on private minting requests from the colony's elite. Orders were as likely to come from military men and Javanese aristocrats as local government officials.[48] In fact, the production volume for privately minted coins often surpassed production for the government. Whereas the Surabaya mint master and assayer Johan Anthonie Zwekkert manufactured 36,762 coins for the government in 1814, the following year the mint produced around 50,000 silver rupees for private individuals.[49]

For Zwekkert, the introduction of Goldberg's currency was thus a double-edged sword. On the one hand, processing a relatively large amount of silver would be a lucrative endeavor; he was allowed to keep 5 percent of the total amount of fine silver as his salary, next to expenses for the necessary labor force and equipment.[50] On the other hand, the introduction of a new currency

---

44 On the mint's history, see Elisa Netscher and Jacobus A. van der Chijs, *De munten van Nederlandsch-Indië* (Batavia: Lange & Co., 1863), appendix 2: Geschiedenis van de munt in Soerabaja; L.M.J. Boegheim, "François Loriaux, de stichter van de Duitenmunt te Soerabaja," *De Beeldenaar* 3 (1996): 131-134. On Loriaux, see also Van der Kemp, "Episodes uit de geschiedenis," pp. 250-59 (see note 25).

45 Bucknill, *The Coins*, pp. 161-172 (see note 15) and J.P. Moquette, "De munten," p. 35 (see note 42).

46 Denis R. Cooper, *The Art and Craft of Coinmaking. A history of minting technology* (London: Spink & Son 1988), 123-130.

47 Anonymous, "Verhandelingen der munten, maten, en gewigten, van Neerlandsch Indië," *Verhandelingen van het Bataviaasch Genootschap, der kunsten en wetenschappen* 6 (1824): 284-86.

48 For the most detailed overview of minting activity during the British interregnum, see Moquette, "De munten," pp. 33-96, on 53, 62 and 81 (list of names) (see note 42).

49 Ibid., pp. 33-96, appendix G.

50 See article seven of his instructions, which are reprinted in Van der Kemp "De zilveren Java-Ropijen," appendix B (see note 39).

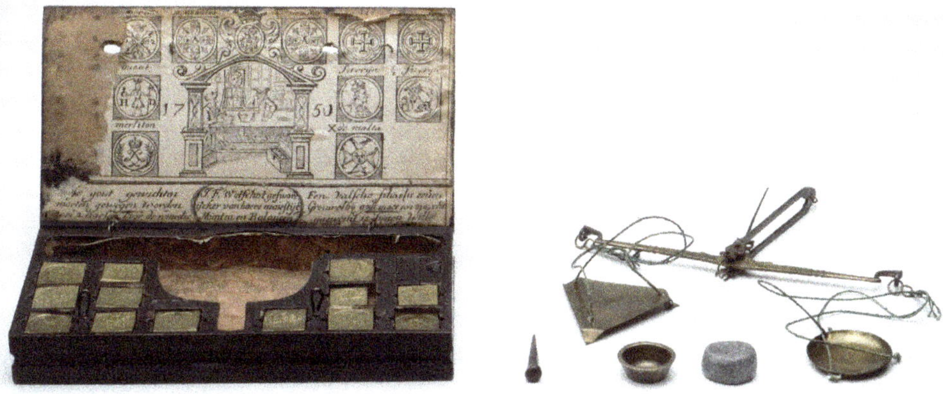

FIGURE 8.3 *Balance used for the weighing of coins, eighteenth century.* IMAGE USED BY THE KIND PERMISSION OF MUSEUM BOERHAAVE, LEIDEN.

also formed, as already explained, a threat to the status and capital of the colony's elite, who were – next to the colonial government – Zwekkert's most important customers. Zwekkert ingeniously managed to accommodate the interests of both these local clients and the distant government demands of Goldberg. He informed his superiors in Batavia that he was willing to produce silver rupees with a lower value of silver (9.614 grams) for the colonial government. But he actually used his chemical expertise to increase the silver content to the British norm (10.896 gram of fine silver), appropriating Goldberg's project for ends that protected the interests of his valued local customers.

After Zwekkert had confirmed the government's orders, he first used government funds to increase the mint's personnel.[51] While Goldberg, the government minister, had struggled to secure experts in assaying and minting in the Netherlands, Zwekkert was well-placed to find suitable skilled support, since he was linked to various networks of artisans in the relevant trades. Zwekkert assembled a group with a deep understanding of how subtle material qualities in coins would effect their success as a currency. Owing to Surabaya's function as an important trading hub in the eastern part of the Indonesian Archipelago, the city housed skilled artisans such as cupel makers

51 On Zwekkert see Frederik de Haan, "Personalia der periode van het Engelsch bestuur over Java 1811-1816," *Bijdragen tot de Taal-, Land- en Volkenkunde in Nederlandsch-Indië* 92 (1935): 477-681, on 669.

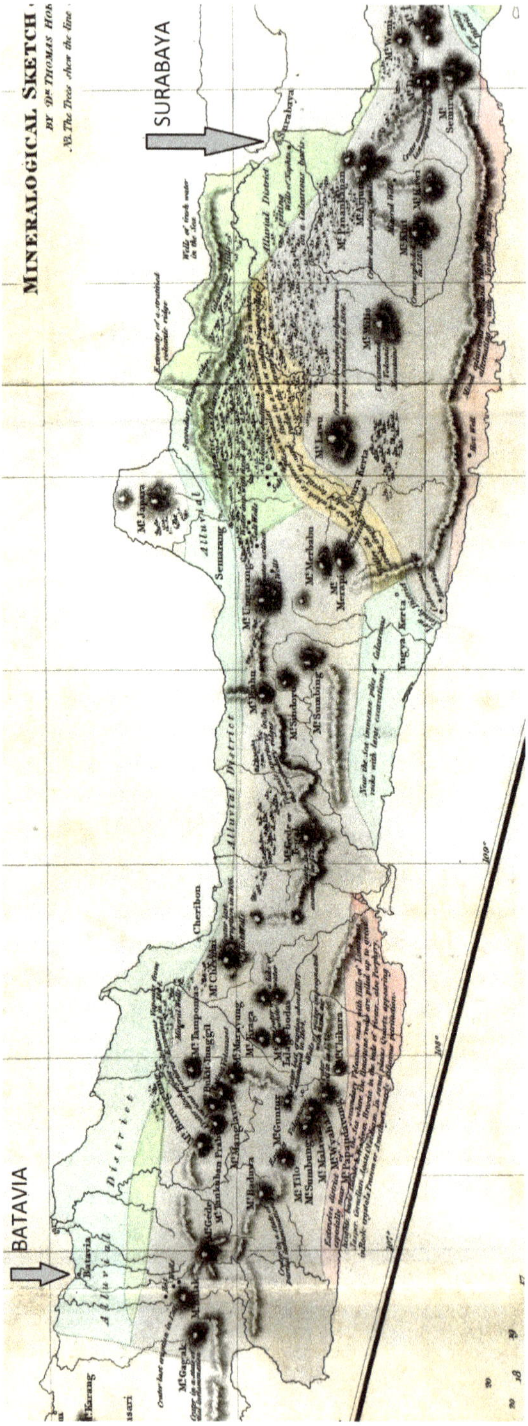

FIGURE 8.4 *Mineralogical map by Thomas Horsfield, ca. 1816. Batavia and Surabaya are marked with an arrow.*

from China and Javanese engravers.[52] Cupels (the shallow containers in which assays were carried out), together with assay balances, were crucial for the production of silver coins.[53] Before the raw materials were brought to the mint's forge, Zwekkert used them to determine the exact amount of fine silver which was necessary for the production of the new rupees. Next to local cupel makers, Zwekkert also relied on the services of the Javanese engraver Inche Maimin and four of his local helpers. Besides having a thorough material expertise, engravers such as Inche Maimin also had to be knowledgeable experts on local monetary culture. Coins such as the rupee had to display a convincing combination of linguistic and visual features in order to be accepted as a medium of exchange for local and long-distance trade. Each coin carried Arabic characters on one side and Javanese characters on the other side. Maimin used rupees produced by the British in India as a visual guide.[54] Zwekkert went to great efforts to enlist these highly skilled cupel makers and engravers, offering them permanent and well-paid positions.[55] In order to emphasize Maimin's role in relation to local money users, his initial (M) was struck on every Java rupee.

Zwekkert thus went to some lengths to enact a version of Goldberg's policy that reshaped it to serve his own ends and the interests of his private clients. It afforded him improvements to his premises and personnel, but also the opportunity to alter policy. This only became apparent to the government in mid-1817, after he had handed in detailed balance sheets. When government officials in Batavia and The Hague started to compare the amount of produced coins with the amount of silver which was sent to Surabaya, they gradually realized that Zwekkert had minted a currency that contained much more fine silver than

52 Ulbe Bosma, *The Sugar Plantation in India and Indonesia. Industrial production, 1770-2010* (Cambridge: Cambridge University Press, 2013), 107. For a fascinating insight into the Javanese metal industry, Gerret P. Rouffaer, *De voornaamste industrieën der inlandsche bevolking van Java en Madoera* ('s-Gravenhage: Martinus Nijhoff, 1904), 91-122. For the growing role of Chinese traders in the region, Kwee H. Kian, "The Expansion of Chinese Inter-Insular and Hinterland Trade in Southeast Asia, c. 1400-1850," in David Henley and Henk Schulte Nordholt, eds., *Environment, Trade and Society in Southeast Asia* (Brill: Leiden, 2015), 149-165.

53 For an in-depth description of the process, see Cooper, *The Art and Craft*, pp. 85-87 (see note 46).

54 Scholten, *De munten*, p. 84 (see note 15).

55 Van der Kemp, "De zilveren Java-Ropijen," pp. 299-301 (see note 39) and Moquette, "De munten," appendix U (see note 42). From December 1816 to June 1817, Maimin received 384 rupees. High-quality dyes were the best protection against forgery. See also G.P. Dyer and P.P. Gaspar, "Reform, the New Technology and Tower Hill, 1700-1966," in C.E. Challis, ed., *A New History of the Royal Mint* (Cambridge: Cambridge University Press, 1992), 398-606, on 409.

TABLE 8.1 *Comparing the fine metal content of silver coins involved in Goldberg's reform program. Of course, the real content of fine metal varied greatly. One ace (aas) is the equivalent of 0.048 gram.* FOR DETAILS, SEE J.H. VAN SWINDEN, BEDENKINGEN OVER HET MUNTWEZEN (UTRECHT: HET GELDMUSEUM, 1997), IV.

	Silver money in the Netherlands (1815) (guilder)	**Silver money in the colonies as projected by Goldberg in 1816 (colonial guilder)**	**Silver money produced by the British (rupee)**	**Silver money as produced by Zwekkert in Surabaya in 1816/1817.**
Amount of fine silver in aces (rounded)	200 aces	166 aces	227 aces	227 aces
Amount of fine silver in gram (rounded)	9.6 gram	7.968 gram	10.896 gram	10.896 gram
Estimation of coins minted from 1kg of fine silver (excluding loss caused by melting, etc.)	~ 104 coins	~125 coins	~ 91 coins	~ 91 coins

Owing to Zwekkert's reluctance, roughly 27 percent fewer silver coins were produced than Goldberg and the Dutch crown had projected in 1816.

either Goldberg or the government in Batavia had planned.[56] Since Zwekkert was using more silver in each coin, the total he produced with the silver sent from abroad turned out to amount to only some 75 percent of the desired quantity (Table 8.1). Producing 260,000 silver rupees in Surabaya thus increased the costs for the new money tremendously. Instead of helping to mitigate public debt at home, the production of the currency actually increased the public debt.

The reactions to these developments were harsh. The colonial government in Batavia immediately stopped the production of silver coins in Surabaya. It then intervened directly with the mint's organization. According to regulations of 1819, the mint master had to report on his activity on a weekly basis in order

56 Van der Kemp, "De zilveren Java-Ropijen," pp. 303-4 (see note 39).

to compensate for the high costs the minting of the silver rupees had created.[57] The mint master responded by claiming that the silver which Goldberg had sent to Batavia must have had a much lower silver content than metal traders in Europe had promised. But this did not alter the government's distrust of the mint.[58] Until its dissolution in 1843, the mint in Surabaya was never again asked to mint silver or gold coins for the colonial government.[59]

Material accommodations to diverse local interests over coinage in Java thus led to radical consequences for Goldberg's policy at home. Back in the Netherlands, Zwekkert's behavior and the colonial government's inability to discipline the mint's activities had fatal consequences for Goldberg's career and the treasury. Instead of maintaining his position in the king's inner circle of advisors, Goldberg was shifted to a less influential position as the king decided to dissolve the Department of Trade and Colonies. While the Dutch crown had first followed Goldberg's plan to introduce the new currency in the form of paper bills and coins with silver content far below the local norm, the king eventually opted for a less costly form of managing the colony's agricultural wealth. In order to realign the interests of proto-bankers in Batavia and at home, the Crown invited them to become shareholders in a new trading company, the so-called Netherlands Trading Society (*Nederlandsche Handel-Maatschappij*, or NHM).[60] In order to attract private capital, the NHM was awarded exclusive rights to ship cash crops and other products from and to the Indonesian Archipelago. Moreover, the king guaranteed a dividend of 4.5 percent for each share. On the day the shares were issued, wealthy individuals invested almost 70 million guilders.[61] Unlike Goldberg's attempt to reduce colonial debt by introducing a new paper currency, the NHM was an instant success because it allowed the Dutch crown to capitalize on the colony's agricultural riches. In 1840, a caricaturist had to remind the Dutch public of the opportunities merchants in Batavia once had. In order to symbolize the dominant role of the NHM, the anonymous cartoonist depicted the trading company as a sea monster controlling all ships going to and from Batavia. Seen from the

57 See the instructions of the mint master and other personnel: ANRI Jakarta, K 20 Surabaya, 1254: Ingekomen brieven bij de muntmeester te Surabaya 1818: Reglement op het beheer en de administratie van de munt in Surabaya.

58 ANRI Jakarta, Muntwezen, no. 7: Mint master to Resident of Surabaya: 15 June 1821.

59 Bucknill, *The Coins*, chapter 6 (see note 15).

60 Ton de Graaf, *Voor handel en Maatschappij. Geschiedenis van de Nederlandsche Handel-Maatschappij, 1824-1964* (Amsterdam: Boom, 2012), 39-45.

61 J. Jonker and K.E. Sluyterman, *At Home on the World Markets: Dutch international trading companies from the 16th century until the present* (The Hague: SDU Uitgevers, 2000), 160.

cartoonist's perspective, the NHM had made the accumulation of private wealth in Batavia impossible.[62]

By following Goldberg's currency to Java, the second part of this essay has shown that the colony's mint master and the chemistry of coins played an active role in the formation and execution of monetary policy. Instead of passively serving his metropolitan masters, Zwekkert appropriated Goldberg's project for his own goals, which were shaped by the local networks of which he was a part. The consequences were not only felt in the colony but also at home. In order to counter the growing cost of colonial management, the king resorted to an alternative form of monetary policy. Instead of aiming to devalue private assets in Batavia, he invited wealthy individuals to invest capital in a new trading company which received extensive trading rights to and from the Indonesian Archipelago. By reformulating the government's orders, the colony's mint master thus prevented locally accumulated private capital from being used for the tremendous public debt which had accumulated in the Netherlands.

Conclusion

In the process of governing the Dutch colonies, metropolitan ministers may have been the first to enunciate policies, but their final shape and consequences depended on the practical contributions of a variety of people and contexts. In the case of Dutch coinage in Java, the embodied formation and enactment of policy hinged on *chemical* practices that the Dutch minister Goldberg was ill-equipped to manage. Making the new coinage work to relieve government debt hinged on material considerations of the quantity of silver to be incorporated into coins, which demanded the skilled assessments and labor of expert assayers, mint masters, and engravers. So too did they depend on familiarity with more locally situated social and cultural trends, manifested in coin design and designation, that afforded local acceptance of what was

62 A digital copy of the caricature, which was titled *De Groothandelaar* (The Wholesaler) may be found here: <http://hdl.handle.net/10934/RM0001.COLLECT.521867> [accessed February 22, 2015]. Of course, this is a very simplistic picture. Research by Roger Knight and Alex Claver has shown that the Indonesian Archipelago continued to offer a fertile climate for private entrepreneurs. See Roger Knight, "Rescued from the Myths of Time: Toward a reappraisal of European merchant houses in mid-nineteenth century Java, ca. 1830-1870," *Bijdragen tot de Taal,- Land- en Volkenkunde* 170 (2014): 313-341; Alex Claver, *Dutch Commerce and Chinese Merchants in Java. Colonial relationships in trade and finance, 1800-1942* (Brill: Leiden, 2014).

otherwise seen as foreign – and therefore untrustworthy – currency. Unable to secure such skills at home, Goldberg delegated the production of new coins to colonial masters, but in so doing had to defer control to persons and institutions that were even farther from his immediate reach. These did not serve him well. Attuned to a local economy whose moral and material evaluations of currency were quite at odds with those of the Dutch government, the Surabaya mint masters reworked the new currency to suit their own ends. The result was a devastating reversal of the intended policy, costing more money than might be saved, that ultimately prompted the fall of Goldberg himself.

Approached through the lens of chemical governance, the material dimensions of colonial government become apparent. In Goldberg's case, monetary policy depended on locally situated chemical expertise. Zwekkert used his expertise to increase the silver content of the newly produced coins. Since the exact composition of the coins initially remained invisible to the colonial government, he was able for a time to secure the mint's existence and counter the threat to the wealthy colonial elite's private capital and authority, which was embodied in the introduction of a new currency.

In conclusion, this chapter has shown how modes of government and scientific practice cannot be treated separately in understanding the relations of metropole and colony in this period. Rather, an understanding of chemical governance exposes a polycentric world in which power was dispersed amongst people, skills and practices. Far from simply enabling or constraining policy enforcement, local initiatives and acts of appropriation which were informed by chemical expertise shaped the realization of policies. These chemical acts have subsequently become obscure to historians interested in colonial governance, at least in part because it was ultimately Goldberg himself who was formally recognized as responsible for the policy's failure. Nonetheless, chemical practices that were pursued at various distances from the official center of metropolitan policymaking proved critical to the formative realization of both colonial and metropolitan governance.

Acknowledgments

I would like to thank Lissa Roberts, Simon Werrett, Joppe van Driel and Esther van Gelder for their comments on earlier versions of this essay. I would also like to thank Erik van der Kam, curator of the National Numismatic Collection in Amsterdam, for providing with me with some of the visual material used in this essay, and Johan de Zoete from Museum Enschede in Haarlem for information and bibliographical references about the production of paper money by Joh. Enschedé en Zonen.

CHAPTER 9

How to Govern Chemical Courses. The Case of the Paris *École de pharmacie* During Vauquelin's Direction, 1803-1829

Sacha Tomic

The Paris School of Pharmacy (*École de pharmacie*, hereafter the École) was established by the law of 21 Germinal year 11 (April 11, 1803).[1] The new establishment succeeded the Free School (*École gratuite*) established on 3 Floreal year 4 (April 22, 1796), itself heir of the former *Collège de pharmacie* established by the reform of 1777 which separated apothecaries from grocers. This creation joined in an overall plan of education system renovation following the Revolution. Under the Consulate, pharmacists' educational fate was sealed by two chemists who held important governmental positions: the Minister of the Interior Jean Antoine Chaptal and the Council of State and General Director of Public Instruction Antoine François Fourcroy. They considered the training of the time to be too corporatist and reorganized pharmaceutical education accordingly. Henceforth, the state regulated pharmacists' training on a national scale.

The École's laboratory exemplifies a two-tiered space of chemistry and pharmacy education. Its study tackles several questions concerning governance and laboratory studies. How was the teaching of chemistry and pharmacy organized? What was the role of the laboratory's staff, including its "invisible" personnel?[2] What was the moral economy at stake in this educational context?[3] What was the budget and how was it managed? To what extent did the State intervene in the governance of the institution and its pedagogical guidance?

1 Two other schools were created, in Montpellier and Strasbourg. See Adolphe Trébuchet, *Jurisprudence de la médecine, de la chirurgie et de la pharmacie* (Paris: J.-B. Baillière, 1834). The candidates accepted into these schools became first-class pharmacists and could practice throughout France.

2 Steven Shapin, "The Invisible Technician," *American Scientist* 77 (1989): 554-63.

3 Lorraine Daston, "The Moral Economy of Science," *Osiris* 10 (1995): 2-24; French edition translated by Samuel Lézé with a presentation by Stéphane Van Damme, *L'économie morale des sciences modernes* (Paris: La Découverte, 2014).

 | DOI 10.1163/9789004325562_011

In order to tackle these questions, this essay is divided into four sections. It first examines the École's administrative and pedagogical structure and discusses the specificity of the chemistry courses organized for pharmacists. The second part insists especially on the crucial role of the lab assistant (*préparateur*) in the implementation of experimental courses. The last two sections examine in detail the École's spending management. Section three presents the École's good governance through the evolution of its accounts and gives an estimation of the staff's salaries. The final section proposes an assessment of global annual average costs for chemistry and pharmacy courses.

Through this case study, this essay provides an example of exploiting financial sources often neglected in the historiography of science. A detailed examination of annual budgets and their correlation with bills opens new perspectives in the study of the relationships between chemistry and governance.

Teaching the New Chemistry to Pharmacists

Continuity largely characterized the École's administrative and educational structure (Table 9.1). With regard to the Free School, and with the exception of Simon Morelot, a former treasurer definitively removed from the École, there was some persistence of the staff. Half of the direction (director, assistant, secretary and treasurer) was renewed, as was most of the teaching staff (full professor and his assistant) in the four chairs (chemistry, pharmacy, botany, natural and medical history or *materia medica*). The École also inherited the Free School's premises and facilities. This structural continuity explained the rapid implementation of reform. Soon after the publication of the Ministerial Order of 25 Thermidor year 11 (August 13, 1803) concerning pharmacy schools' organisation, the administration and professors were appointed by the governmental decree of 15 Vendémiaire year 12 (October 8, 1803). The administration had to meet at least once a month. Different assemblies could take place at professors' request and once a year the École's personnel met in a general assembly during the first days of Vendémiaire (September).[4] The first assembly took place on 2 Brumaire year 12 (October 25, 1803) and the École was officially settled by a decree of 3 Frimaire year 12 (November 25, 1803) – that is, eight months after law's promulgation.

4 Archives of the Bibliothèque InterUniversitaire de Santé (BIUS), register n° 25, records of the assemblies from October 1803 until May 1811.

TABLE 9.1 *The École's directors and the professors of chemistry and pharmacy.*

Direction	Chair of Chemistry	Chair of Pharmacy
Director: Vauquelin *Assistant director:* Trusson [1803-1811] Laugier [1811-1829]	*Full professor:* Bouillon-Lagrange [1803-1829]	*Full professor:* Brongniart [1803-1804] Nachet [1804-1832]
Treasurer: Chéradame [1803-1824] Robiquet [1824-1840]	*Assistant professor:* Henry *père* [1803-1826] Bussy [1826-1830]	*Assistant professor:* Bouriat [1803-1832]
Secretary: Bouillon-Lagrange [1803-1830]		

The law of 1803 did not impose a specific program but recommended chemistry teaching which was "more specially applicable to pharmaceutical science" (Title II, art. 11). How was this institutional call for a pharmaceutical chemistry put into practice? This question raises the further question of the place of pharmaceutical practice and chemical theory in the curriculum. Lavoisierian chemistry and its training model (theory before practice), which official chemists largely adopted, were not accepted by all pharmacists. Some of them claimed that pharmaceutical practice must precede theory.[5] Although the new team subscribed to this doctrine and favored "chemical philosophy" (theory), these courses were still taught after practical coursework was completed. During the deanship of Nicolas Louis Vauquelin, pupils were admitted to attend courses only after a practical internship in a dispensary (*officine*).[6] This educational sequence, based on the weight of tradition, distinguished the École from other establishments. At the Polytechnic School (*École polytechnique*) and, to a lesser extent, at the Mining School (*École des mines*) and at the

5 The pharmacist Jacques-Philibert Delunel criticized the young chemist Thenard, then examiner at Polytechnic School, for praising the teaching of theory before practice, Jonathan Simon, *Chemistry, Pharmacy and Revolution in France, 1777-1809* (Aldershot: Ashgate, 2005), 140-45; For the acceptance of Lavoisierian chemistry by official chemists, see Bernadette Bensaude-Vincent's essay in this volume.

6 Courses lasted from one to three years according to the duration of internships. Candidates with an eight year internship were allowed to take the final examination without attending courses. Assembly of Germinal 8 year 12 (March 29, 1804). Archive BIUS, register n° 25, 14b.

Faculty of Medicine (*Faculté de médecine*), pupils experimented after the lecture.[7] The École would eventually follow the same path by creating its own "Practical School" (*École pratique*) in 1831 but, though gradually shortened, internships remained mandatory during the nineteenth-century.

The teaching of "chemical philosophy" was not new in pharmacists' curriculum. The novelty is rather to be found in the teaching of the new chemistry so as to explain everyday pharmaceutical operations. Discussion of courses' experimental character and its material consequences appeared in the assembly of May 12, 1808. Both Jean-Nicolas Trusson and Jean-Pierre-René Chéradame noted that "the professors of chemistry and pharmacy [thought] that the *numerous experiments and operations performed during the courses* resulted in a rather considerable quantity of products and results [...] but they had no suitable premises to deposit and preserve them."[8]

The premises built in 1808 reflected the École's first collection of chemicals. It is important to notice that even if the chemistry and pharmacy courses were distinct, the administration considered these disciplines inseparable in their common material dimension. The École's lab was a typical chemico-pharmaceutical lab and its basic instruments and reagents were the same as those found in numerous pharmacy and chemical laboratories.

Students' notebooks give an idea of the chemistry taught at the École. The first chemistry course was given by Edmé-Jean-Baptiste Bouillon-Lagrange on 21 Germinal year 12 (April 21, 1804).[9] Bouillon-Lagrange acquired his knowledge in chemistry at the Polytechnic School where he occupied the post of lab assistant and head of chemical works (*chef des travaux chimiques*). He also taught chemistry at the *École centrale du Panthéon* and at the *Collège Henri IV*. From these experiences he wrote a successful *Manuel d'un cours de chimie*.[10] Bouillon-Lagrange was sometimes replaced by Noël-Étienne Henry *père* who was also appointed at the head of the Central Pharmacy of civilian hospitals (*Pharmacie centrale des hôpitaux civils*) in 1803.[11] Henry *père* gave similar

7 The sequence theory-practice finally became widespread and would stand out as a "natural law" by the end of the nineteenth century. Sacha Tomic, "Le cadre matériel des cours de chimie dans l'enseignement supérieur à Paris au XIXe siècle," *Histoire de l'éducation* 130 (2011): 57-83.

8 BIUS, register n° 25, 30a, my emphasis.

9 BIUS, Ms 26. Notes taken by Nicolas Denis Moutillard, BIUS, register n° 25, 28b. The manuscript contains a series of about forty lessons of approximately five pages dispensed between April and August, 1804.

10 *Manuel d'un cours de chimie* (Paris: Bernard, 1799, 1801, 1802 1809; Klostermann, 1812).

11 Antoine-François Boutron-Charlard, "Nécrologie [d'Henry *père*]," *Journal de chimie médicale de pharmacie et de toxicologie* 8 (October 1832): 703-4.

courses to the full professor's.[12] According to his colleague Louis-Antoine Planche, Henry *père* "was not a brilliant professor, but his lessons were fully attended because the example was always given next to the rule. The pupils already provided with some elementary knowledge were rewarded."[13]

What was the content of these lessons? If, as Jonathan Simon noticed, Bouillon-Lagrange asserted the dominant position of "chemical philosophy" by resuming the classification given by his master Fourcroy in the *Système des connaissances chimiques*, the professor immediately indicated to his audience that "these divisions bring nothing useful for a pharmacist, I would not give more ample details."[14] Bouillon-Lagrange adopted a mixed plan alternating general considerations and particular cases proceeding from simple to complex, following Lavoisier's pedagogical progression.[15] His lessons may be divided into three parts. In the first part he defined the main notions of chemical philosophy (analysis, attraction, aggregation, caloric, elastic fluids) and reviewed the elements (oxygen, nitrogen, carbon, phosphor, sulfur, metals). Then he considered oxides and the physical properties of liquids and water. The second and longest part was dedicated to inorganic chemistry (acids, earths, alkalis, salts, metals and mineral waters). The last and shortest part dealt with vegetable chemistry (acids, immediate materials: sugar, oil, and ether) and animal chemistry (prussic acid, bile, urine).

Bouillon-Lagrange's courses did not differ fundamentally from the structure of his book, from which he borrowed some passages. However, his courses contrasted with those clearly more exhaustive ones given by Thenard at the *Collège de France*.[16] The difference lies in the selection made by Bouillon-Lagrange to give to future pharmacists an overview of chemistry's potential to explain and improve pharmaceutical operations. The professor focused on useful compounds for pharmacy, such as arsenic, and antimony for metals, as well as camphoric acid and phosphoric ether for plant chemistry. He also addressed

---

12 BIUS, Ms 27 and Ms 33; BIUS, register n° 25, 23b; BIUS, Ms 26, 113-15.

13 Louis-Antoine Planche, "Nécrologie sur Noël-Étienne Henry," *Journal de Pharmacie* 18 (1832): 522.

14 *Système des connaissances chimiques* (Paris: Baudoin, 1800-1801); Simon, *Chemistry, Pharmacy*, pp. 140-45, 159-60 (see note 5).

15 This point marks the differences in the teaching of Bouillon-Lagrange from Berthollet at the *École normale*, see Bernadette Bensaude-Vincent's essay in this volume.

16 BIUS, Ms 22 et Ms 23, notes taken in 1809 by the pharmacist and future professor of *materia medica* Nicolas-Jean-Baptiste-Gaston Guibourt. José Ramón Bertomeu-Sánchez, Antonio García Belmar, "Les cahiers d'élèves sources pour une histoire des contenus et des pratiques de l'enseignement de la chimie," 2004, <http://rhe.ish-lyon.cnrs.fr/cours_magistral/expose_thenard/expose_thenard_complet.php> (accessed 29 September 2017).

practical subjects such as "lutes" (to assure the joint and waterproofing of devices) and capillaries. The lessons did not contain pharmaceutical recipes, but the professor handled themes of interest to pharmacists, such as vegetable and animal physiology. Directly inspired by his master Fourcroy, he taught a chemistry which did not totally break with natural history. The experimental nature of chemistry courses and the choice of subjects close to pharmacists' interests seem to have answered the government's desire for a chemistry "applicable" to pharmacy, a renewed pharmaceutical chemistry.[17]

Pharmacy teaching was entrusted to Antoine Brongniart, but he died on February 24, 1804. Louis-Isidore Nachet succeeded him.[18] Nachet had more the profile of a pharmacist than his colleague because he operated a pharmacy for fifteen years, compared to the two years of Bouillon-Lagrange. According to his biographer, Nachet was "barely assisted by his assistant Bouriat during twenty-eight years of what should have been their collaboration, Nachet carried the entire weight of teaching alone."[19] According to his former apprentice, the doctor botanist François-Victor Mérat, Nachet was a "deserving professor."[20] Mérat informs us of the practical nature of his courses:

> [Nachet was] a quite practical man, rather than a scholar, a man of the laboratory, as we say in professional terms: so the pupils looked with greediness for the details which he gave them on the particular processes in the preparation of certain medicines which were not described in books, and which pass, in a way by tradition, from laboratory to laboratory [...] He thus trained good pharmacists and good chemist-manipulators (*chimistes-manipulateurs*), who are the most useful, if not the most brilliant.[21]

Chemistry teaching and the art of chemical manipulation through pharmacy's lessons aimed more at inculcating the "new" chemistry from a practical stand-

---

17 This specialization of chemistry teaching was strongly supported by Bussy in 1850 to defend the École's chair of chemistry, archives BIUS, box n° 312, folder 312-8.

18 Patrick Bourrinet, "Nachet (Louis-Isidore) 1757-1832," *Revue d'Histoire de la Pharmacie* 93 (2005): 301.

19 Georges Dillemann, "Louis Isidore Nachet, 2e titulaire de la chaire de Pharmacie," *Produits et problèmes pharmaceutiques* 25 (1970): 1125-26.

20 François-Victor Mérat, "Notice nécrologique sur M. Nachet, professeur à l'École de pharmacie," *Journal de pharmacie* 18 (1832): 588.

21 François-Victor Mérat, "Nachet (Louis-Isidore)," Louis-Gabriel Michaud, dir., *Biographie universelle, ancienne et moderne. Supplément* (Paris, 1844), 75, 61-2.

point to an audience already well versed in practice.[22] The École also trained pharmacists who were more attracted by an effective practice of their future work, along with pupils seeking to become 'useful' chemist-manipulators. This last group would enlarge the contingent of future experts in chemical analysis and technicians needed by the growing chemical and pharmaceutical industry. Amongst the nearly 800 students trained under Vauquelin's deanship, an elite of first-class pharmacists adopted the self-proclaimed title of "pharmacist-chemist" so as to underline the profession's learned character.[23] This generation became famous for numerous discoveries in chemistry and for its members' expertise as requested by the emergent regulation of industrial society.[24]

As this description shows, assistants were not crucial in the teaching of chemistry and pharmacy. By contrast, lab assistants were strongly implicated in the implementation of courses.

The Central Position of the Lab Assistant

Prior to the revolution, chemistry courses in France were taught by a professor (a trained physician), with the assistance of a 'demonstrator' (an apothecary) who carried out the experimental demonstrations.[25] This division of labor

22 Another example illustrates this practical orientation of the École's courses. At the government's request during the assembly of April 8, 1811, Nicolas Appert was authorized to open a course "for the distribution of its process concerning the preservation of plant and animal foodstuffs." BIUS, register n° 77, document n° 39-40; register n° 25, 37b.

23 The average annual number of pupils (about forty) was stable until 1817, after which it rose to a peak of 180 pupils in 1826. By 1830, we can estimate the population of pharmacists at approximately 10,000 for France. Their number in Paris increased from c. 100 in 1800 to c. 300 in 1830.

24 Sacha Tomic, "Status and Role of French Pharmacist-Chemists in the History of Chemistry in the Early Nineteenth Century," paper presented at Leuven (Irish College) on June 1, 2013, workshop *Situating Material and Knowledge Production in the History of Chemistry: Sites and Networks of Discipline Formation and Industrial Practice, 1760-1840*. The results accumulated in a few decades by these analysts contributed to the emergence of a new speciality: organic chemistry. Sacha Tomic, *Aux origines de la chimie organique. Méthodes et pratiques des pharmaciens et des chimistes (1785-1835)* (Rennes: PUR, 2010). For the connection of the first industrial regulation law (decree of October 15, 1810) with environmental history, industrial chemistry, and pharmacists experts, see Thomas Le Roux, *Le laboratoire des pollutions industrielles. Paris, 1770-1830* (Paris: Albin Michel, 2011).

25 Christine Lehman, "Les multiples facettes des cours de chimie en France au milieu du XVIIIe siècle," *Histoire de l'éducation* 130 (2011): 31-56; John Perkins, "Chemistry Courses, the

continued following the establishment of the École out of educational necessity rather than as a statutory requirement. According to the director of the École, Gustave Planchon, it was on the wishes of Nicolas Deyeux, professor of chemistry at the Free School, that the first "official lab assistant attached to chemistry courses" was appointed in April-May, 1793.[26] The importance of lab assistants – as demonstrators were now called – was thus underlined for teaching experimental chemistry effectively. A pharmacist's diploma was not required for the post. The lab assistant was chosen from among the "rather intelligent pupils" and he was awarded fees. Four successive lab assistants served at the École under Vauquelin's deanship: Henri-Auguste Vogel, Émile Sureau, Alexandre Bussy and Louis-Dominique Guiart *fils*. Vogel and Bussy were particularly active.

After apprenticeships in different pharmacies in Hanover and Brême, Vogel came to Paris in 1802 and was appointed lab assistant on 8 Germinal year 12 (March 29, 1804). Far from being limited by a position that institutionalized the division between mind and hand, he also taught chemistry at the Lycée Napoléon before he left Paris in 1816 for the chair of chemistry at the University of Munich. During his stay at the École, Vogel published some ten articles, alone or in association with Bouillon-Lagrange or Joseph Pelletier, mainly on the chemical analysis of different substances taken from *materia medica* (saffron, scammony, juice of buckthorn, red coral, and bitter almonds, and so on). In 1804, he received a 400 francs fee and additional 100 francs with the condition that he "correctly carries out the duty of his place." Vogel did not disappoint the administrators who hired him. He entered the first graduation class of the École and received the gold medal at the École's first annual student competition in chemistry. The session was chaired by Fourcroy and took place on a symbolic day, the first year of Napoleon's reign (4th complementary day year 12 – September 21, 1804). Vogel became a pharmacist in 1808 and was soon promoted in 1809 with more duties and a pay raise. His position thus covered courses for chemistry, pharmacy, *materia medica* and, occasionally, botany.[27]

Bussy's career embodied the ascent of a pharmacist who rose through all the pedagogical ranks of the École. Appointed lab assistant in 1821, he became assistant professor in 1826, full professor in 1830 and finally managed the École for thirty years from 1844. After publishing on discoloration by animal charcoal

Parisian Chemical World and the Chemical Revolution, 1770-1790," *Ambix* 57 (2010): 27-47.

26 Gustave Planchon, *L'enseignement des sciences physico-chimiques au Jardin des apothicaires et à l'École de pharmacie de Paris* (Paris: Ernst Flammarion, 1897), 36.

27 BIUS, register n° 25, 34b. The professor of the botanical chair was also assisted by the gardener named Puyhatier dit Périgord, BIUS, register n° 25, 15a.

and the first elementary analysis of morphine in 1822, he was quickly noticed by his colleagues. In 1823, he was granted an exceptional 600 francs compensation, identical to those of professors' assistants, "for purchase of utensils and ingredients necessary for the preparation of chemistry and pharmacy courses."[28] The deliberation of August 23, 1823 signed by the secretary Bouillon-Lagrange specified the reasons for this bonus: Bussy, "having been in charge of an extraordinary work, both for the various cabinets' arrangement and chemical research, as well as the various operations, deserved to be helped."[29]

Bussy's last two years as lab assistant were productive. He supported Bouillon-Lagrange in 1825 instead of Henry *père*, a responsibility for which he received another 600 francs compensation. The following year he replaced Henry *père* who decided to dedicate all his energy to his function as chief-pharmacist of the Central Pharmacy of civilian hospitals. Also in 1825, Bussy began research with Louis-René Lecanu on the distillation of fats and acids which Michel-Eugène Chevreul had discovered the wet way, establishing the base for a renewed proximate analysis. Together with Dumas at the Athenée in the same year, Bussy introduced Félix-Polydore Boullay *fils* during chemistry courses where, according to Chevallier, he performed "practiced chemical manipulations in the École's laboratory."[30] A "skilfull and recognizable manipulator," Bussy adapted the technical part of Michael Faraday's *Chemical Manipulations* (1827).[31] He justified the necessity of such a work intended for beginners, arguing that France possessed "very few skilful workers, whereas it counts scholars in large numbers."[32]

Moving down the administrative hierarchy, we notice that other people were involved in the activities of the laboratory. In a letter of 1823, Bouillon-Lagrange specified that Bussy "was helped", which supposes that there was a kind of second lab assistant. Besides Boullay *fils*, who assisted Bussy, Planchon reported that Joseph-Bienaîmé Caventou was "lab assistant, for free, to the assistant of chemistry courses."[33] The lab assistant also had his own assistant.

---

28 AN, F/17/2326, Acounts for 1823, first article for courses' expenses.

29 BIUS, box BLII, folder 205, extract from the École register of deliberation from August 23, 1823.

30 Alphonse Chevallier, "Nécrologie de Félix Polydore Boullay (1806-1835)," *Journal de chimie médicale* 1 (1835): 390.

31 Alfred Velpeau, review of "Manipulations chimiques," *Archives générales de médecine* 6 (1828): 155-6; William B. Jensen, "Michael Faraday and the Art and Science of Chemical Manipulation," *Bulletin for the History of Chemistry* 11 (1991): 65-75.

32 Michael Faraday, *Manipulations chimiques*, trans. Raymond-Balthazar Maiseau, foreword by Alexandre Bussy (Paris: A. Sautelet, 1827), vol. 1, VIII.

33 Gustave Planchon, *L'enseignement des sciences*, p. 45 (see note 26).

The École's office boy (*garçon de bureau*), Louis Bouré, participated in several courses, his task defined in the session of 20 Floreal year 12 (May 10, 1804). Besides the duties of office and sweeping the premises, he was "also in charge of cleaning the laboratory, its vessels and utensils," becoming then a lab boy.[34] The accounts indicated that he made diverse purchases such as eggs and glue (used for "lutes"), washed the laboratory cloths (used for filtrations), and supplied coal for furnaces and heating.

Through these different activities, it appears that the laboratory was also a place of research where the École's entire staff participated in its educational activities. The courses took place only during the summer semester (1st Germinal to 1st Fructidor – March 22 to August 19), which left the winter semester (from September to March) for examinations and for the staff's own use. Bouillon-Lagrange, who had been without his own pharmacy since 1789, performed at least some of his experiments at the École, as did Vogel and Bussy. A privileged number of students also performed practical works.

To sum up, the lab assistants, assisted or not by a lab boy, did not content themselves with preparing experiments or chemicals for courses. They also performed original research and published in scientific journals and often advanced the money for and took care of the supply of equipment and chemicals.[35] They drafted "spending for the course" reports that were approved by the professor or the director.

In her seminal paper on moral economy, Loraine Daston quoted Claude Bernard and John Pond's opinions about the division of labor in science.[36] For Daston, both men considered that repetitive work (observation, data collection, calculations) must be performed by "uneducated men" and "drudges". This example of "mechanical objectivity" in which uneducated lab assistants were considered as devices contrasts with the École's pedagogical organization. The valuation of scientific work highlighted by Daston as characteristic of this period, that is to say, did not necessarily hold in an educational context, especially in pharmacy where morality was one of the qualities officially

34 BIUS, register n° 25, 15b.

35 Those technicians who became pharmacist-chemists can be considered as hybrid-experts in the sense of Ursula Klein, "Apothecary's Shops, Laboratories and Chemical Manufacture in Eighteenth-Century Germany," Lissa Roberts, Simon Schaffer, Peter Dear, eds., *The Mindful Hand. Inquiry and invention from the late Renaissance to early industrialisation* (Amsterdam: Royal Netherlands Academy of Arts and Sciences, 2007), 247-276; Ursula Klein, ed., special issue on "Artisanal-scientific Experts in Eighteenth-century France and Germany," *Ambix* 69 (2012).

36 Daston, "The Moral Economy," p. 20 (see note 3).

required by the profession.[37] Often characterized by historians as uneducated "invisible hands", lab assistants at the École were both visible and educated laboratory managers, who actively contributed to the success of experimental chemistry teaching.[38] In brief, the lab assistant was the main administrator and technician of chemical and pharmaceutical courses, who, in close collaboration with professors, contributed to the École's prestige and good governance.

The École's Good Governance

The École's accounts provide information about the moral economy and material frame of its courses.[39] Two treasurers served during Vauquelin's deanship: the former Free School's assistant director Chéradame and Pierre-Jean Robiquet.[40] The Ministerial Order of 25 Thermidor year 11 (August 13, 1803) concerning the École's regulation specified that the treasurer was appointed for three years, whereas the director had a five-year term; both were eligible for reelection.[41] Article 10 specified that "every year, in the first days of Vendémiaire [September-October], the treasurer will report the previous year's revenues and expenditures, at the École's general assembly. This account will be checked by the prefects of the department, and in Paris by the Prefect of police. It will then be submitted for the approval of the Minister of the Interior." This administrative procedure explains why certain accounts were officially validated only one

37 Pharmacists had to take an oath in front of the Prefect to validate their diploma. The École's administrators did not hesitate to dismiss pupils for "deviant behavior" as happened to a certain Gauthier in 1806, BIUS, register n° 25, 25-a-26b.

38 Another example of the visibility of laboratory workers is given by José Ramón Bertomeu-Sánchez, Antonio García Belmar, "Louis Jacques Thenard's Chemistry Courses at the *Collège de France*, 1804-1835," *Ambix* 57 (2010): 48-63. These processes of visibility/invisibility are still at stake in the evolution of contemporary science. Florence Millerand, "Les gestionnaires d'information 'invisibles' dans la production d'une base de données scientifiques," *Revue d'anthropologie des sciences* 6 (2012): 163-190.

39 It has been possible to reconstitute the various budgets thanks to the whole collection of annual accounts preserved at the Archives Nationales (series F/17/2325 and F/17/2326). Some of them have been correlated, for certain years and for certain spending, to the diverse itemized bills kept by the BIUS (box BLI, folders 1-100; box BLII, folders 101-230). Pierre Julien, "Plaidoyer pour les notes et factures anciennes de pharmacie," *Revue d'histoire de la pharmacie* 37 (1990): 81-92.

40 The assistant director André Laugier was temporary treasurer during two months and drafted the account for September and October, 1824.

41 BIUS, register n° 25, 28a, 32a.

or even two years after their writing. Chéradame presented his first account in the session of 6 Nivôse year 13 (December 27, 1804), signed by the professors and approved by the Prefect on 18 Brumaire year 14 (November 9, 1805). In a letter read at the assembly on 7 Frimaire year 14 (November 27, 1805), the Minister of the Interior Chaptal expressed through the Prefect his "satisfaction of the good order which reigns in the accounts [of 1804]."[42] With these words, the Government paid tribute to the treasurer, emphasizing his central role in the good governance of the establishment.

A typical account is divided into revenues and expenditures, themselves divided into chapters concerning a more or less precise subject. Revenues distinguished between the previous year's surplus and three items which fed the budget: pharmacists' graduation fees (fixed at 900 francs by law), by far the most important income amounting to nearly three-quarters of the revenues; visitation rights for pharmacists (six francs), druggists and herbalists (two francs); and registration fees (thirty-six francs/year). The expenditure records consisted generally of ten chapters, the most important expense items concerning renovation work.[43] Settled on the same location as the *Jardin des apothicaires*, founded in 1626 (13 rue de l'Arbalète), the École inherited buildings of the Free School and a laboratory built in 1702 and enlarged in 1760. The primitive building (c. 250 m²) contained a "big laboratory for demonstrations and public courses" (c. 90 m²).[44] Even if the overall building configuration did not change much during Vauquelin's deanship, the École underwent perpetual remodelling during this period, both inside and outside. The garden was continually in development and different spaces were created: a laboratory for receptions (1804), an amphitheater of botany and natural history (1805), a locale for chemicals (1808), a space for mineralogy (1824) and *materia medica* and mineral collections (1826).

The budgetary evolution of the École's governance reveals three phases (Graph 9.1). The average expenditures were about 36,000 francs for the period 1804-1815; they increased by 30 percent during 1816-1823 (47,000 francs) and again by 46 percent to reach about 68,000 francs at the end of Vauquelin's

42 BIUS, register n° 25, 24a.

43 Expenditure headings: 1. Examiner's rights of presence, 2. Staff fees, 3. Course expenditures, 4. Extraordinary expenditures, 5. Work expenditures, 6. Visitation fees, 7. Office fees, 8. Administrators' expenditures, 9. Gardener and lab boy expenditures, 10. Treasurer allowances.

44 BIUS, Ms 87, N.-J.-B.-G. Guibourt, "Exposé historique sur l'origine et les augmentations successives de l'immeuble affecté à l'École de Pharmacie," reproduced (without the first nine pages) in Adrien Philippe, *Histoire des apothicaires* (Paris: Direction de la Publicité Médicale, 1853), 253.

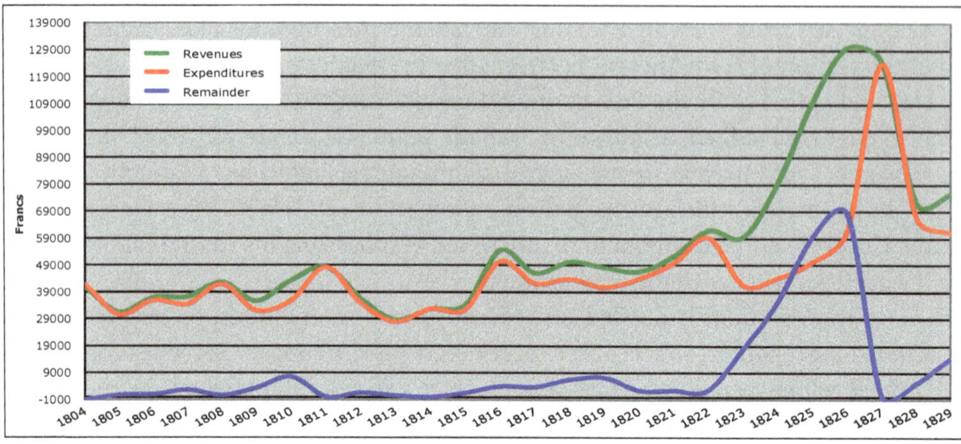

GRAPH 9.1 *Evolution of the* École's *accounts.*

deanship (1824-1829). The average revenues, following the increasing number of graduating pharmacists, became established at 37,000 francs for the first period, increased by 40 percent for the second period (53,000 francs) and jumped to 88 percent (about 100,000 francs) for the last period. For the whole period, the total average amounts were about 1.28 million francs for expenditures, and about 1.46 million francs for the revenues with a profit of about 180,000 francs. With the government's agreement, a part of the exceptional revenues accumulated in 1826 (more than 50,000 francs) was deposited in an interest-paying account.

How did this good governance apply to staff salaries? Several sources of income, divided into fixed and mobile parts, covered the staff's expenses. The fixed part did not concern the management. Professors' indemnities were fixed by the government at 1,500 francs whereas lab assistants' indemnities were decided by the École's administration and fixed at 600 francs (750 francs from 1823).[45] Lab assistants' fixed indemnity rapidly increased, going from 500 francs in 1804 to 1,200 francs from 1812. Salaries more than doubled (by 140 percent) in eight years, competing with those of a professor. The wages of the office-lab boy, who also received housing, increased more slowly. His income (without bonuses as New Year's gifts), passed from 500 francs in 1803 to 800 francs from

45 BIUS, box BLII, folder 205, extract from the École's register of deliberation from August 30, 1821. As a comparison, fixed indemnities were twice as high for professors at the Faculty of Medicine (3,000 francs), more than three times at the Polytechnic School and National Museum of Natural History (5,000 francs), and four times at the *Collège de France* (6,000 francs).

1819, with a global increase of 80 percent. As we have seen in the previous section, this appreciation of salaries underlines the fundamental role of lab assistants in the laboratory management and contributed to their visibility.

The mobile part of the École's income, which represented the directors' only revenue stream, was divided into several parts which substantially increased the total income. These sums grew with the number of candidates. The institution's life was punctuated by assemblies, examinations and other public meetings for which the staff received remuneration. The "rights of presence" (*droits de présence*) for the examinations fixed by the law represented an important part of income. For each successfully passed examination, a sixty-six franc sum was assigned to the examiners. This income was about 1,000 – 1,500 francs for a professor and the director, and about 500 – 1,500 francs for the assistant professors and the treasurer. The administrative tasks with attendance fees to the assemblies were fixed at four francs "both for the office assemblies held by the administrators, and by the committees and the assemblies of École's members," which represented about 150 francs per professor, assistant and administrator.[46] The "rights of presence" for the competitions were fixed at eight francs per session, that is twenty-four francs for three days.[47] The directors received compensation for examination expenses (thirty francs / reception) which notably increased their salary. For example, in 1826, the director, his assistant and the treasurer received about 2,000 francs. From 1826 onward (revenue peak), a secondary account appeared. It concerns the budget for the last practical examination (300 francs / candidates) initiated by Bussy.[48] The bulk of the income was used for compensation (twenty francs / sessions) and stood at about 1,500-2,000 francs for administrators, and 1,000-1,500 francs for professors and assistants.

Finally, we can estimate a 4,000 francs average salary for a full professor, 2,500 francs for an assistant professor, 1,000 francs for a lab assistant, and about 700 francs for the office/lab boys. As for the directors, whose members could also be appointed professor, the average income estimation is as follows: 3,500-4,000 francs for the treasurer, 2,500-3,500 francs for the director, and 2,000-3,000 francs for his assistant. The salaries from the lab assistant to the professor were

46 These fees were justified during the assembly of October 4, 1806 for the reason that "the École members cannot give up their own pharmacy without receiving a compensation," BIUS, register n° 25, 27b.

47 At least for 1823, BIUS, box BLI, folder 69.

48 Alfred Riche, "Notice biographique sur Bussy," *Journal de pharmacie et de chimie* 5 (1882): 303.

in a 1:4 ratio with attractive gaps so as to develop staff loyalty and encourage the personnel to climb the ladder to professorship as exemplified by Bussy.

Controlling Course Expenditures

With few exceptions (1804 and 1819), teaching expenses were grouped in annual accounts under chapter III "concerning the various purchases required for the courses of chemistry, pharmacy, natural history and other expenses relative to courses."[49] Botany expenses (gardener, maintenance of the garden, plants collections) were often scattered in other account sections. To have a closer look at the specific budget concerning chemistry and pharmacy courses (materially and financially inseparable), it is necessary to estimate the amount devoted to the cabinet of natural history and *materia medica*. Several elements enabled the minimization of this cost. First of all, the majority of reports (*mémoires* and *mémoires quittancés*), those detailing the price of every article – often simple invoices – stipulated explicitly the "supply of objects appropriate to chemistry and pharmacy."[50] The comparison of itemized bills (chemicals, vessels and utensils, instruments) with the total of courses' expenses confirms the dominating place of a budget peculiar to chemistry and pharmacy teaching.[51] At the same time, most of the expenses corresponding to *materia medica* courses offered during different periods were included under the chapter "exceptional expenses". From 1824 onward, at the instigation of Pelletier who replaced Robiquet as chair of *materia medica* in 1825, the expenses remained stable and regular (approximately 2,000 francs for minerals bought between 1824 and 1829, that is an average of 333 francs/year). *Materia medica* expenses can be estimated to approximately 5,000 francs, or nearly 200 francs/year for 1803-1829. With regard to the average course expenses (2,180 francs), this amount represents approximately 10 percent. It was then possible to deduct an annual average cost specific to chemistry and pharmacy courses of about 2,000 francs, or 4.5 percent of total expenses (Graph 9.2). One can notice that this proportion, higher at the École's beginnings (5.5 percent from 1804 till 1817), decreased from 1812 and became stable from 1818 (3.5 percent for 1818-1829). This order of magnitude corresponds to the amount mentioned by Fourcroy in

---

49 Archives Nationales series F/17/2325.

50 Archives Nationales series F/17/2326.

51 The BIUS archives provides bills for thirteen years (1806-1814, 1819, 1821, 1823 and 1824) i.e. 50% of the twenty-six years of Vauquelin's deanship.

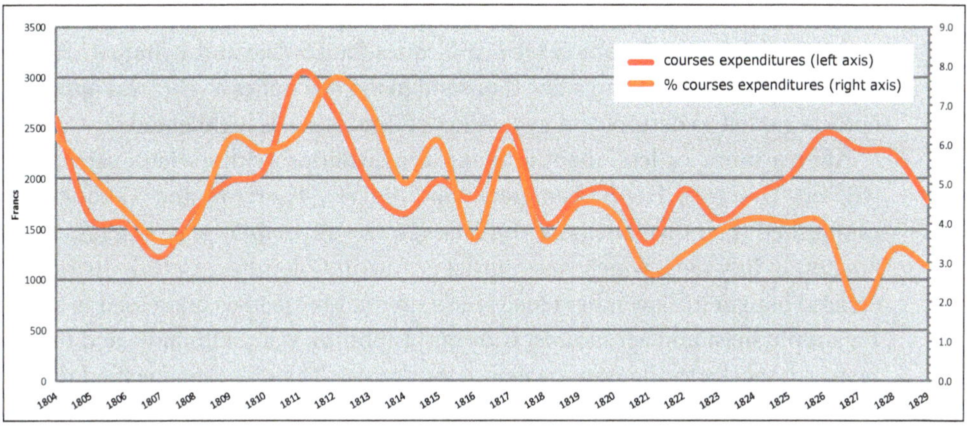

GRAPH 9.2 *Evolution of the* École's *chemical & pharmacy courses expenditures.*

the *Methodical Encyclopedia* as necessary to maintain "a daily laboratory service."[52]

From the midpoint of Vauquelin's deanship, the number of enrolled students converged toward the number of receipts, which means that practically all students followed a "normal" education of three years. By this time, the École achieved its cruising speed and administrators succeeded in mastering chemistry and pharmacy expenses, in spite of the staff's salary increases, with income from the growing number of received pharmacists. These considerations are consistent with the administration's desire to control the École's expenses. During the session of 7 Frimaire year 14 (November 27, 1805), the directors decided that the "maximum of course expenses for 1806 [will be fixed] at 2400 francs."[53] With few exceptions, this maximum was never exceeded.

The control over course-related spending was also the consequence of a faithful network of various suppliers which included some of the École's members. Pharmacist-chemists, whose name was a warranty of their products' quality, had a monopoly on the production of high value-added products. Next to a high-grade chemical industry, a market existed for teaching equipment whose development was strongly tied to pharmacists' business.[54] Not sur-

52 A.F. Fourcroy, "Opérations," *Encyclopédie méthodique – Chimie* (Paris: Panckoucke, 1808), vol. 5, 272.

53 Archives Nationales series F/17/2325.

54 Anna Simmons' discussion in this volume of the formation of a marketplace (in relation to the sale of drugs in London, in her case) can be extended to the emerging education market. This market was not specific to France as the case of Friedrich Accum suggests, but further comparative studies are needed.

prisingly, we find among the main chemicals suppliers, the director Vauquelin and the professors of *materia medica* Jacques-Paul Vallée and Robiquet. This new revenue stream (from a few thousand to 100,000 francs a year) was apparently accepted as natural and no conflict of interests was highlighted.

After a more or less chaotic financial association with various partners (Gilbert Deserres, Henry Lemercier and Hervé), Fourcroy and Vauquelin established their own business, the *Manufacture de produits et de réactifs chimiques*, in July 1804, which was housed at rue du Colombier, 23 (6th district) on land bought in November 1800.[55] In a letter of May 18, 1807 addressed to the Italian physicist and agronomist Giovanni Fabbroni, Vauquelin indicated that he was henceforth the only owner of the factory. The director specified that "Professors, Chemists, Pharmacists, Traders and Students, can be sure from now on to find in [his] stores all the reagents in a state of perfect purity, or the preparations necessary for their demonstrations, for their trade, for their studies or for their research."[56] In 1804, the factory delivered vessels and utensils valued at more than 1000 francs under the name of Deserres and Lemercier.[57] The factory also provided instruments (such as barometers, thermometers, hydrometers, instruments for mineralogy, apparatus for physics, and so on) and glassware. Vauquelin also proposed the shipping of books to clients.

In spite of its financial setbacks, the factory operated continuously thanks to the massive financial contribution of Fourcroy (150,000 francs) and Vauquelin (100,000 francs) to cope with the debts of their partner. Their associate Lemercier was mostly responsible for the bad financial situation of the factory. This commercial failure was also the consequence of the implementation of the first industrial regulation law of October 15, 1810. The factory had settled in the very heart of Paris and was the subject of periodic complaints from neighbours.[58] In 1818, Lemercier chose to move the large-scale production of acids and phosphorous substances to the suburbs and the factory was sold for 88,000 francs in August, 1822 to Jean-Baptiste Quesneville who, with his son Gustave Augustin, would make it a prosperous establishment.

This episode shows that Fourcroy and Vauquelin were, at first sight, quite bad managers. This 'bad governance' was not only due to some risky financial

55 Maurice Bouvet, "Nicolas Vauquelin, droguiste," *Revue d'histoire de la pharmacie* 46 (1958): 246-52.

56 American Philosophical Society, Mss. B F113.

57 Vauquelin sold copper, crystal, glass, porcelain and stoneware "chemical and pharmaceutical" vessels; flasks filled with emery, melting pots and capsules made of platinum, silver and china.

58 Thomas Le Roux, *Le laboratoire des pollutions*, pp. 242, 305-12, 466, 484-86 (see note 24).

choices. It was also the result of a proactive education policy. Fourcroy and Vauquelin's factory was a non-profit creation.[59] The creation of the establishment, the first one of this kind in France, was more a political act to promote a new kind of chemical education than to build a lucrative business. Moreover, different courses took place in the amphitheater located inside the plant. Rather than seek an immediate profit, the moral economy at stake behind the factory's owners was to promote a model for experimental chemical education and to support the École's good governance.

Unlike Vauquelin, two other École suppliers, Vallée and Robiquet, sold chemicals exclusively for profit. Vallée was appointed assistant professor of *materia medica* in 1803 and full professor in 1811. He was the owner of a pharmacy on rue Saint-Victor, n°98 (12th district) from 1802 to which he "annexed a small factory of chemicals" and supplied the École, even with living vipers.[60] In his obituary, Nachet insisted on "the considerable increase of this business which began with the moderate sum of 1,200 francs."[61] After his departure, Robiquet settled on the rue de la Monnaie, n°9 (1st district) and took over and supplied the majority of products from 1818. In 1826 he joined Aristide Boyveau, who possessed a factory on the rue des Fossés-Saint-Germain-l'Auxerrois, n°5 (1st district). Together with Pelletier they created a factory in 1834 on the rue des Francs-Bourgeois-Saint-Michel, n°8 (6th district).

Other suppliers of chemicals (Table 9.2) included the pharmacist Charles Baget for phosphorus and iodine, Paul Blondeau for spring waters, the apothecary-druggist Charles Bouvier for mineral acids, the director of the Faculty of Medicine's chemical laboratory Jean-Pierre Barruel for potassium, the grocer Sureau *fils* for alcohol and vinegar, and the druggist Bonnes who provided mercury in 1804 (doubtless for pneumatic tanks). Different distillers regularly supplied the École with large quantities of spirits (*eau-de-vie*) and alcohol, one of the favorite solvents used in proximate analysis. Beside Vauquelin, the École possessed another almost official supplier of vessels and utensils, the potter and master glassmaker Jean-Baptiste Acloque, who occasionally also took care of the equipment. Different potters, oven makers, boilermakers and paper-

59 Georges Kersaint, "L'usine de Fourcroy et Vauquelin," *Revue d'histoire de la pharmacie* 47 (1959): 25-30; Alain Queruel, *Vauquelin et son temps (1763-1829)* (Paris: L'Harmattan, 1994), 116-23.

60 Georges Dillemann, "Jacques Paul Vallée, 1772-1814," *Produits et problèmes pharmaceutiques* 26 (1971): 426.

61 Louis-Isidore Nachet, "Nécrologie de Vallée," *Journal de pharmacie* 6 (1814): 380-83.

TABLE 9.2 *The École's various suppliers of chemicals (main suppliers in bold).*

Suppliers	Profession	Address	Chemicals
Vauquelin	Professor of chemistry, Director of the École	Rue du Colombier, n°23	Various
Vallée	Pharmacist, Professor of *materia medica* at the École	Rue Saint-Victor, n°98	Various
Robiquet	Pharmacist-chemist, Professor of *materia medica* and treasurer at the École	Rue de la Monnaie, n°9	Various
Robiquet-Boyveau	Ibid Industrialist	Rue des Fossés-St-Germain-L'Auxerrois, n°5	Various
Baget	Pharmacist	Rue vieille du Temple, n°79	phosphorus, iodine
Blondeau	Pharmacist	Rue de Condé, n°22 & Rue de Tournon, n°17	Spring waters
Bouvier	Apothecary-druggist	Rue des Vieilles-Tuilleries, n°19	Mineral acids
Barruel	Director of the chemical laboratory at the Faculty of Medicine	Rue de l'École de Médecine	Potassium
Sureau *fils*	Grocer	Rue du Bourg-l'Abbé, n° 4	Alcohol, vinegar
Bonnes	Druggist		Mercury
Various	Distillers		Spirits

makers supplied furnaces and other accessories, as well as providing reams of Joseph's paper and other papers used in filtrations (Table 9.3).

Instruments were supplied by "patented instruments engineers", especially Dumotiez and his successor and nephew Nicolas Constant Pixii (Table 9.4).[62] They supplied generally expensive devices such as a pneumatic machine (220 francs in 1808), a device for water decomposition (120 francs in 1813), and a Volta eudiometer (40 francs in 1819). Jean-Gabriel-Augustin Chevallier said the

62 Dumotiez was one of the first to receive an "engineer's certificate in instruments of optics, physics, and mathematics" with privilege delivered in April, 1788 by the Committee of the artists of the Royal Academy of Sciences, created at the instigation of the astronomer Dominique Cassini, *Mémoires pour servir à l'histoire des sciences et à celle de l'Observatoire Royal de Paris* 1810 (Paris: Bleuet), 86-94.

TABLE 9.3 *The École's various suppliers of utensils (main suppliers in bold).*

Suppliers	Profession	Address	Utensils
Vauquelin	Professor of chemistry, Director of the École	Rue du Colombier, n°23	Various
Acloque	Glassmaker	Rue de la Barillerie, n°22	Various
Blanc	Potter and oven maker (*fournaliste*)	Rue de Larbralètre, n°12	Furnaces
Gaillard	Boilermaker		Furnaces
Petit *aîné*	Trader (*négociant*)	Rue Saint-Jacques-la-Boucherie, n°23 & 25	Test tubes
Paillard	Paper-maker		Joseph paper

TABLE 9.4 *The École's suppliers of instruments (main suppliers in bold)*

Suppliers	Profession	Address	Instruments
Dumotiez	Patended instrument engineer	Rue du Jardinet, n°2	Pneumatic machine, air thermometer, pyrometer
Pixii	Patended instrument engineer	Rue du Jardinet, n°2	Pneumatic machine, Volta's eudiometer, galvanic devices
Chevallier (Le Chevalier)	Patended instrument engineer	Quai & Tour de l'Horloge du Palais, n°1	Hydrometers, Descroizille's stills
Chemin *fils* and Chevillé *fils*	Mechanics (balanciers-mécaniciens)		Balances
Dourches	Instrument maker		Precision balance

engineer Chevallier (Le Chevalier), sold hydrometers and Descroizilles stills for wine assays (54 francs in 1823). The mechanics (*balancier-mécanicien*) Chemin *fils* and Chevillé *fils* sold and repaired balances, and Dourches provided an expensive precision balance (330 francs) in 1828.[63] Some of these instrument-makers, such as Pixii, built an international reputation and provisioned universities including Leuven, Turin, Moscow, Warsaw, and Coïmbra in Europe, as well as across the Atlantic.

63 Louis Marquet describes a Fourché-Chemin balance, "Balances et boîtes de poids de la collection de la Société d'Histoire de la pharmacie," *Revue d'histoire de la pharmacie* 34 (1987): 41-46, see 43-4.

In the end, and in order to promote an updated, high-quality chemical education, the good management of the École ensured a reconciliation between educational requirements and equipment. Quality reagents and instruments placed at the staff's disposal were provided by the best traders according to market prices and a policy of loyalty. Alongside this liberal management of material resources, the École's direction also promoted an attractive internal promotion policy based on a moral economy that provided its reputation and identity.

Conclusion

This essay ends as the École, supervised by the new director André Laugier, was about to enter a new phase. Thanks to the surplus of the previous years, the École added two new wings. Under the impulse of Bussy, a Practical School was created in 1831. This paid training addressed pupils duly selected by competition and constituted, after the Polytechnic School in its early stages, the first regular training of practical chemistry within higher education. The administration was to confirm and endow this training with a specific budget during the École's affiliation with the University by the ordinance of September 27, 1840.[64]

Despite State control, the École kept a large degree of autonomy. This was especially true for the training program but also for the École's management. Its good governance rested on financial management but also on the sharing of common values. Administrators and professors established a moral economy founded on a compensation policy for the staff, on promotion of moral and scientific "excellence" amongst students and lab assistants, and on the adaptation of chemistry courses intended for the profession. The promotion from lab assistant to professor as embodied by Bussy in this essay, was not specific to pharmacy teaching. During the nineteenth-century this was the standard promotion at the Faculty of Science, at the Faculty of Medicine and at the *Collège de France*. Yet, the École was not integrated into the University system in 1808

64 This integration was led by the physician-chemist and dean of the Faculty of Medicine, Mateu Josep Bonaventura (fr. Mathieu) Orfila. Meanwhile, the establishment underwent a "sort of metamorphosis" with the creation in 1834 of two additional chairs (toxicology and elementary physics) which would redefine the use of the premises and introduce a second semester (1st November – 1st April), Louis-René Lecanu, "Rapport sur l'état actuel de l'enseignement dans l'École de pharmacie de Paris," *Journal de Pharmacie* 22 (1836): 703-14.

as was the Faculty of Medicine. The École's scientific character was however recognized by the government in the 1840s when the administration adopted, without making it official, the name of École *supérieure de pharmacie*.[65] This recognition was tied to the weight of chemistry in the scientific rise of the establishment. With the exception of Laugier's brief deanship (1829-1832), the three other directors who followed until 1873 were professors of chemistry (Vauquelin, Bouillon-Lagrange, and Bussy).

For these men, chemistry was considered as an "accessory science" (*science accessoire*) to pharmacy, with a utilitarian value for society and a route to employment opportunities. Some of the brightest students established schools of research specializing in industrial analytical chemistry. These sites cultivated an almost exclusively industrial chemistry, developed side-by-side with the "academic chemistry" developed in 'research schools' led by eminent chemists such as Thenard, Dumas and Pelouze in France, and Thomas Thomson and Justus Liebig abroad.[66] A pharmacist-chemist like Jean-Baptiste Alphonse Chevallier, who was trained at the École before becoming a professor of pharmacy, was at the head of a laboratory where he trained numerous analysts.[67] The generalization of such spaces of practical teaching or 'schools of analytical chemistry' significantly contributed to the emergence of the future generations of technicians, expert chemists, and chemical engineers who found numerous career opportunities in the industrial society of the nineteenth and early twentieth-century. These men embodied links between practitioners and scholars and contributed to chemical and sanitary risk management by the development of food science and occupational toxicology where chemistry, politics, justice, and economy were intimately intertwined.[68]

65 The École received the title of Faculty in 1920. Georges Dillemann, "Les établissements d'enseignement pharmaceutique de 1803 à 1994," *Annales pharmaceutiques françaises* 53 (1995): 1-7.

66 Gerald L. Geison, Frederic L. Holmes, eds., "Research Schools: Historical Reappraisals," *Osiris* 8 (1993).

67 Alex Berman, "J.B.A. Chevallier, Pharmacist-Chemist: A major figure in nineteenth-century French public health," *Bulletin for the History of Medicine* 52 (1978): 200-13.

68 Soraya Boudia and Nathalie Jas, eds., special issue, "Risk and Risk Society in Historical Perspective," *History and Technology* 23 (2007).

PART 3

Revisiting the History of Production

CHAPTER 10

Teaching Chemistry in the French Revolution: Pedagogy, Materials and Politics

Bernadette Bensaude Vincent

By the end of 1794 – Year III of the revolutionary calendar, the young "one and indivisible" French republic established a Normal School (*École normale*) to "provide the French people with a system of instruction worthy of its novel destiny."[1] More concretely the purpose was to train a number of citizens quickly so that they could train the future primary and secondary teachers in all the districts of the French territory. Chemistry was an integral part of the curriculum, which included both science and humanities. The term '*école normale*' clearly conveys a project of normalization or standardization of education providing a uniform approach to all the sectors of knowledge previously covered by the *Encyclopédie*. This teaching institution was dedicated to "shaping the man and the citizen." It was meant to coproduce knowledge and citizenship according to the republican ideals of liberty, equality, and fraternity. Well-trained teachers (*instituteurs*) would be "capable of being the executives of a plan aimed at regenerating the human understanding in a republic of 25 million people, all of whom democracy rendered equal."[2]

Claude-Louis Berthollet participated in this state initiative to assume his social and political role as a teacher. As one of the supporters of Lavoisier's chemistry and a member of the group of academicians who reformed the

1 Anon., "Arrêté des représentants du peuple près les Écoles normales du 24 nivôse an III de la République française une et indivisible," *Séances des Écoles normales* (Paris: Imprimerie du Cercle social, 1800-1801). All lessons are now available in print: Jean Dhombres, ed., *L'École normale de l'An III. Leçons de mathématiques. Laplace, Lagrange, Monge* (Paris: Dunod, 1992); Daniel Nordman, ed., *L'École normale de l'An III. Leçons d'histoire, de géographie, d'économie politique. Volney, Buache de la Neuville, Mentelle, Vandermonde* (Paris: Dunod, 1994); Etienne Guyon, ed., *L'École normale de l'An III. Leçons de physique, de chimie, et d'histoire naturelle. Haüy, Berthollet, Daubenton* (Paris: éditions ENS Rue d'Ulm, 2006); Béatrice Didier, Jean Dhombres, eds., *L'École normale de l'An III. Leçons littéraires, art de la parole, morale, analyse de l'entendement* (Paris: éditions ENS Rue d'Ulm, 2009). On the ambitions of the educational projects of the French Revolution, see Mona Ozouf, *L'homme régénéré. Essais sur la Révolution française* (Paris: Gallimard, 1989).

2 Debate at the *Convention nationale* in September 1794 quoted in Jean Dhombres, "Introduction générale," Dhombre, ed., *Cours de l'école normale de l'An III*, p. 1 (see note 1).

chemical language in the 1780s, he could be expected to contribute to the national effort of normalization by teaching the new chemistry, known as the "theory of French chemists". Given the synchronism of the chemical and political revolutions, chemistry featured as an ideal discipline to prompt the transformation of the socio-political order by making up citizens capable of transforming natural substances into national resources for the nation's warfare and public welfare.

However Berthollet's chemistry lectures suggest a quite different endeavor. He did not take on the role of spokesman for a chemical community engaged in a deep transformation of this discipline. Far from aiming at stabilizing the new language of chemistry and promoting the theory outlined in Lavoisier's *Traité élémentaire de chimie* (1789), Berthollet developed personal views about chemical theories and utilitarian applications. He did nothing to spread the gospel of two synchronic revolutions.

This essay stresses the gap between the national aspiration to shape the future citizens of a new sociopolitical order and the pedagogical practice of the chemistry lecturer. While the *École normale* was meant to instill in children and their teachers the normative frame according to which they should live and work in a republican society, Berthollet's lessons rather aimed at shaping competent citizens by sharing his experience of the world as an expert in chemical theory and arts. While revolutionary education was institutionally projected as possessing a disciplinary power with which to govern both social and material constituents, in Berthollet's design it consisted in transmitting one's own knowledge and expertise in the art of governing the multitude of chemical substances.

The first section of this essay presents the goals and ambitions of the creation of the Normal School in the dual context of the political and chemical revolutions. The second section focuses on the course of chemistry and describes the site, the audience, and what we know about the teacher's performance. In the third section I argue that Berthollet's pedagogical choice expressed his personal vision of chemistry, which deeply differed from Lavoisier's chemistry. It reveals a tension between Berthollet's aspiration to a world ruled by general laws and his concern with exceptions. While he stressed the deductive power of chemical theory for enlightening chemical arts, Berthollet nevertheless focused on anomalies and peculiar circumstances. Berthollet's lessons reflected his own experience of the world as a chemist oscillating between the power of disciplining material substances conferred by chemical theory and the art of paying attention to circumstances.

Normalizing Science and Education: A Revolutionary Ambition

The National Convention was a site for legislation intended to discipline society and the economy through the organization of centralized structures and practices of administration and education. It established the Normal School a few months after the end of the Terror, which had generated strong campaigns of slander against scientists.[3] It was crucial for the young republic to create trust between the elites and the new regime following the dissolution of the Paris Royal Academy of Sciences together with all other academies in 1793 and the execution of Lavoisier together with twenty-one tax collectors of the *ancien régime*. "The republic has no need of savants" ("*la république n'a pas besoin de savants*"): although this alleged response of the revolutionary tribunal to Lavoisier's request of a delay of capital punishment proved to be a legend, it certainly expressed the general opinion and fear of the elite who had initially supported the revolutionary movement. In fact the revolutionary government badly needed the knowledge and skills of scientists for warfare. Chemists were mobilized to produce large quantities of hydrogen for balloons, to transform the church bells into canons and to manufacture saltpeter. So pressing was the lack of explosive powder that in February 1794, the French government mobilized leading chemists such as Gaspard Monge, Jean-Henri Hassenfratz, Louis-Bernard Guyton de Morveau, Antoine de Fourcroy and Berthollet to teach French citizens how to collect and process saltpeter in their basements.[4]

In addition to warfare, the young republic needed scientists for the institution of public service in general, and education in particular. The decree creating the Normal School aimed at replacing the educational system, which had been in the hands of the Church during the *ancien régime*. For this purpose it was crucial to train primary and secondary teachers. The Church educational system had already inaugurated this kind of institution in 1685 with Jean Baptiste de la Salle's *Institute of the Brothers of the Christian Schools*. The National Convention decreed: "An *école normale* will be established in Paris where citizens from all parts of the republic already instructed in useful

3 Charles Coulston Gillispie, "Science in the French Revolution," *Proceedings of the National Academy* 45 (1959): 677-84. Charles Coulston Gillispie, *The Edge of Objectivity: An essay in the history of scientific ideas* (Princeton: Princeton University Press, 1966).

4 *Programmes des cours révolutionnaires sur la fabrication des salpêtres, des poudres et des canons; faits à Paris dans l'amphithéâtre du Muséum d'histoire naturelle et dans la salle des Electeurs ... les 1, 11, 21 ventôse et 5 germinal, 2ème année de la République française ... par les citoyens Guyton, Fourcroy, Dufourny, Berthollet, Carny, Pluvinet, Monge, Hassenfratz et Perrier*, 2nd edition (Paris: Comité de Salut public, 1794).

sciences will be called in order to learn the art of teaching under the most skilled teachers in all genres."[5]

In 1791-92, Nicolas de Condorcet submitted a plan of education to the Legislative Assembly.[6] It was inspired by the Enlightenment ideal of education for citizenship and Condorcet's conviction that education is the key for the progress of civilization. Although Condorcet's *Plan* was formally rejected, his ambition to build up a complete national, free and compulsory system of secular schools to provide equal opportunity for all children, served as a guideline for establishing the normal school in 1794. The purpose was to set the standards or norms for a universal plan and methods of education.

> The unanimous concourse of wills and the convergence of efforts toward the same goal being the safest guarantee of social order, schools, which are the nurseries of the people bounded to serve the state, must equally provide a uniform instruction, based on fixed and common rules for all teaching houses.[7]

The Normal School founded in 1794 was short-lived but it initiated a process of standardization that would be implemented on its re-opening in 1808 under the First Empire. This process was to be achieved by the students themselves. Those who came to the first training course at the Normal School in Paris had been recruited in each of the eighty seven geographic districts (*départements*) created in December 1789. On their return to their districts they would create local normal schools to train dozens of primary and secondary teachers.

The Convention wanted to train primary and secondary teachers for the entire territory of the Republic.

---

5 Article 1 of the Decree of the National Convention, 9 Brumaire An III (October 30, 1794).

6 Nicolas Caritat de Condorcet, *Cinq mémoires sur l'instruction publique* (1791), new edition Charles Coutel and Catherine Kinzler eds., (Paris: Garnier Flammarion, 1994). See also Keith M. Baker and William A. Smeaton, "The Origins and Authorship of the Educational Proposals Published in 1793 by the *Bureau de Consultation des Arts et métiers* and Generally Ascribed to Lavoisier," *Annals of Science* 21 (1965): 33-46. Catherine Kinzler, *Condorcet. L'instruction publique et la naissance du citoyen* (Paris: Folio/Essais, 1984).

7 This statement about the driving forces behind the creation of the Normal School was formulated on the occasion of its re-opening in 1808. See the introduction to the edition of the lectures in 1808 quoted by Guyon, ed., *L'École normale de l'An III*, p. 3 (see note 1): "Le concours unanime et uniforme des volontés, la réunion des efforts vers un même but étant le plus sûr garant de l'ordre social, les écoles, qui sont les pépinières des hommes destinés à servir l'Etat, doivent également offrir une instruction uniforme, basée sur des règles invariables et communes à toutes les maisons d'enseignement."

> Such is the goal of the institution of the *écoles normales*. In other schools, only the various branches of human knowledge are taught; in the *écoles normales*, one will teach the most useful knowledge of each kind and insist on the method of exposition. This will essentially distinguish the *écoles normales*; this will fulfill their denomination.[8]

In keeping with the previous systematizations of natural history and chemical languages and the simultaneous on-going project of standardization of weights and measures, the normal schools standardized both the contents and the methods of teaching. The contents were encyclopedic. According to Condorcet's grand views, all citizens had to be taught how to use reason in order to overcome prejudices due to ignorance and superstition, to know their rights and how to exercise them, as well as how to manage their household and to advance the arts for national welfare.[9] Similarly in his lectures on physics, René-Just Haüy insisted that primary teachers needed a broad instruction in order to eradicate the prejudices spread by ordinary people, the "vulgar". They had to acquire a bird's-eye view of the contents "in order to distinguish the most direct and easiest route from among all those leading to the targeted goal."[10]

The Normal School recruited the leading figures of the former academies to deliver an up-to-date survey of all the knowledge embedded in Diderot and D'Alembert's *Encyclopédie*. It also clearly recommended a linkage between science and arts in education and the concern for public utility. However, the revolutionary Normal School departed from Diderot's *Encyclopédie* in adopting a disciplinary division of knowledge like the *Encyclopédie méthodique*.[11] The introductory lessons carefully delineated the contemporary – albeit practically contested – boundaries between physics, chemistry, and natural history, and tried to inculcate them in the future teachers' minds.[12] Natural sciences

8 *Arrêté du 24 nivôse an II*, quoted in Guyon, ed., *L'École normale de l'An III*, p. 33 (see note 1).

9 Nicolas Caritat de Condorcet, *Esquisse d'un tableau des progrès de l'esprit humain* (Paris: Au bureau de la bibliothèque choisie, 1795) Engl. Transl. J. Barraclough, *Sketch for a historical picture of the progress of the human mind* (London: William Clowes and Sons, 1955), 182-184.

10 René Just Haüy, 3rd lecture, in Guyon, ed., *L'École normale de l'An III*, p. 61 (see note 1).

11 Charles-Joseph Panckoucke ed., *Encyclopédie méthodique par ordre de matières par une Société de gens de Lettres, de Savans et d'Artistes* (Lille: Librairie Panckoucke, 1782-1832); Claude Blanckaert, Michel Porret, eds., *L'Encyclopédie méthodique, 1782-1832. Des Lumières au positivisme* (Genève: Librairie Droz, 2006).

12 See René-Just Haüy and Berthollet's first lectures, in Guyon, ed., *L'École normale de l'An III*, pp. 44 and 256 (see note 1).

were essentially promoted for the enlightenment of workshop practices from a utilitarian perspective.

The project further entailed standardizing the method of learning. This was first to be achieved by teaching all subjects according to the "method of analysis", the method derived from algebra recommended by Etienne Bonnot de Condillac.[13] In relation to empirical data, the standard method was inspired by a kind of vulgate of Newton's method. Hypotheses were not banned but strictly confined to predictions; theories, clearly distinguished from "systems", could play a key role because they subsume a large number of particular phenomena under one single general law.

To sum up this section, the educational program of the Normal School of Year III was part of the political agenda of the French Revolution. It was one aspect of a huge and multifaceted effort of standardization aimed at making a homogeneous republican society. This effort encompassed the constitution of a national territory divided into 89 administrative units (*départements*) and the standardization of weights and measures thanks to the institution of republican units of measurement (the metric system). The disciplinary power of these political and institutional measures had to be reinforced by education. Pedagogy and epistemology would shape the future citizens of the French Republic. The academic elites mobilized in the Normal School were expected to contribute to this national effort with their own sense of disciplinarity.

Actual Practices: Site and Courses

In January 1795, 1400 citizens selected to become primary and secondary teachers came to Paris from all over the country for an intensive four-month course. The training included mathematics, physics, natural history, history, geography, political economy, literature, and what is today known as rhetoric (*art du discours*), ethics (*morale*) and philosophy (*analyse de l'entendement*). The contrast is striking between the ambition of the program and the time allocated:

13 William Albury, "The Order of Ideas: Condillac's method of analysis as a political instrument in the French revolution," J.A. Schuster, R. Yeo, eds., *The Politics and the Rhetoric of Scientific Method. Historical studies* (Dordrecht: Reidel, 1986), 203-226. Lissa L. Roberts, "Condillac, Lavoisier, and the instrumentalization of science," *Eighteenth-Century, Essays and Interpretation* 33 (1992): 252-271. Marco Beretta, *The Enlightenment of Matter* (Sagamore Beach: Science History Publications, 1993). Bernadette Bensaude Vincent, "Lavoisier, lecteur de Condillac," *Dix-Huitième Siècle* 42 (2010): 49-65.

less than a semester from January 20 to May 15, 1795. The winter season being extremely cold, the circumstances were not ideal.

The course included twelve to sixteen sessions per subject, with an average of two to three lectures on each subject per *décade* (the ten day week of the revolutionary calendar) every day except on the fifth and tenth day between 11am and 2pm.[14] The sessions took place in a large auditorium of the former *Jardin du Roy*, renamed *Jardin des plantes*.[15] The auditorium, being too small to accommodate 1400 people, was uncomfortable and on some occasions the attendance became irregular and sparse.[16]

Well-trained stenographers took the minutes of each lecture in shorthand and transcribed them. The transcriptions were revised by the professors, then printed and circulated among the audience. For a number of subjects, in particular chemistry, the published version of the oral lectures delivered in 1795 included an additional lecture delivered in 1801 for updating the contents. More than half of the pages of the printed course were devoted to scientific subjects, although rhetoric got the most number of pages of any single discipline (eighteen percent of the whole). Among the scientific disciplines mathematics received many more than chemistry (sixteen percent versus ten percent).

The students of the first Normal School had been selected on the basis of unclear criteria. They had to be well educated in all subjects, but it seems that their political inclinations also mattered a lot. They were preferably young but half of them had teaching experience, including a good proportion of clergymen. Eighty percent of them became teachers in the *écoles centrales* (high schools) created by a decree in October 1795. For instance, Antoine Libes (1752-1832), who had taught physics at the *collège* (high school) of Toulouse, would later get the chair of experimental physics and chemistry at the *École centrale* of the Saint-Antoine suburb in Paris. So too did he go on to author a three-volume *Traité élementaire de physique* (1801) and a *Histoire philosophique des progrès de la physique* (1810).

The lecturers were selected from among the leading figures of science and humanities, the members of the former academies that had been abolished by a revolutionary decree in 1793. Most of them – Monge, Haüy, Bernardin de St

14 École normale supérieure, *Le Livre du Centenaire. Edition du bicentenaire* (Paris: Presses de l'École normale supérieure, 1994); Paul Viallaneix and R. Elmoznino eds., *L'apprentissage du savoir savant* (Paris: PUF, 1995).

15 Yves Laissus, *Le Muséum national d'histoire naturelle* (Paris: Gallimard, 1995). Stéphane Deligeorges, Alexandre Gady et Françoise Labalette, *Le jardin des plantes et le Muséum national d'histoire naturelle* (Paris: Monum, 2004).

16 Dominique Julia, "L'École normale de l'an III: bilan d'une expérience révolutionnaire," *Revue du Nord* 78 (1996): 853-886.

Pierre – had already taught in higher education institutions. In this respect, Berthollet was an exception. Trained in medicine in Torino he became the private physician of the lover of the Duc d'Orléans on his arrival in Paris in 1772 and for fifteen years he conducted chemical research in the duke's private laboratory.[17] Like many chemists of the time, he collected airs and investigated their chemical properties.[18] In 1779, he became *docteur-régent* of the Paris faculty of medicine. He entered the *Académie des sciences* as adjunct chemist in 1780 where he regularly interacted and collaborated with Lavoisier. Typical of chemists' and chemistry's fluid movement between material and knowledge production at that time, he became Director of Dyes at the Gobelins Royal Manufacture in 1784. This position implied that he had to visit the workshops to explore dyeing practices and to write a treatise on this art. It was in this context that he invented a bleaching process (involving what later came to be known as chlorine) and published the *Eléments de l'art de la teinture* in 1791.[19] Berthollet was thus not only a leading figure in academic chemistry but also an expert in chemical arts. However, unlike his colleagues at the Normal School, he had no teaching experience prior to the revolutionary course on saltpeter he co-taught and his course at the *Ecole des travaux publics*.

According to the testimonies of some students of the Normal School, Berthollet was not a brilliant teacher. Joseph Fourier, who later became a famous physicist, wrote in 1785:

> Berthollet is the greatest chemist of today, whether in France, or abroad; he is not aged (forty-seven years) and with an ordinary look. He speaks with the most extreme difficulty, hesitates and repeats ten times the same words in a sentence and he seems embarrassed in the least details of experiments.[20]

This view is confirmed by another testimony from Jean-Michel Raymond-Latour, who praised Berthollet as "the Newton of transcendental chemistry," but remembered his poor teaching performances. He remembered especially

17 Michelle Sadoun Goupil, *Le chimiste C.L. Berthollet, 1748-1822, sa vie, son oeuvre* (Paris: Vrin, 1977).

18 Claude-Louis Berthollet, *Observations sur l'air* (Paris: Firmon Didot, 1776).

19 Agusti Nieto Galàn, *Colouring Textiles. A history of natural dyestuffs in industrial Europe* (Boston: Boston Studies in the Philosophy and History of Science, 2001).

20 Joseph Fourier, letter dated from March 18, 1785, quoted by Sadoun Goupil, *Le chimiste Claude-Louis Berthollet*, p. 37 (see note 17).

that when Berthollet presented his own work on ammonia, he was so shy and embarrassed that Monge had to replace him for the rest of the lecture.[21]

All lecturers at the Normal School had to follow three specific rules to conform to the novel republican pedagogy, which broke with the traditional dogmatic style of teaching based on authority. First, lecturers should not read their lectures. Second, they were asked to focus on the basics of their subject matter, the idea being that knowledge for all would reduce social inequalities. Most importantly, entire sessions were to be dedicated to debates with the audience. The printed transcriptions of these special sessions show that the debates were not just meant for clarifying basic concepts. Some of the students who had teaching experience did not hesitate to discuss the professors' theoretical and pedagogical choices and even criticized them.

The printed version of the oral lectures, however, also demonstrates that the lecturers did not strictly follow the revolutionary guidelines. Indeed, in their introductory lectures they claimed that scientific theories were keys for the progress of human welfare and that a good instruction should proceed gradually from elementary notions to more complex ones. However these were essentially rhetorical claims. The messages delivered in some of their lectures could be summarized by the motto. "Follow what I say. Do not follow what I do."[22] For instance, while Haüy emphasized the importance of mathematics for physics he nevertheless adopted a descriptive approach to natural phenomena and insisted on the practice of experimental methods.[23] Louis Jean-Marie Daubenton who was both a professor and former director of the *Jardin du Roy* was extremely concerned with didactics, but his lectures were full of digressions and details, which in some cases obliterated their theoretical consistency.[24] More generally, the transcribed lectures suggest that the professors struggled to resolve the tension between two conflicting demands: to promote a new didactics of science and to provide a review of the most recent

21 Jean-Michel Raymond-Latour, *Souvenirs d'un oisif* (Lyon: Chez Ayne fils, Isidore Person,1836), 51-52.

22 Guyon, ed., *L'École normale de l'An III*, p. 3 (see note 1).

23 For instance Haüy's second lecture on the general properties of bodies (impenetrability, divisibility, mobility, gravity), presented the experiments conducted for determining the republican kilogram. Guyon, ed., *L'École normale de l'An III*, pp. 50-55 (see note 1); The fourteenth lecture on electricity consisted in a detailed description of Coulomb's experiments. Ibid., pp. 174-184.

24 For instance in his first lecture meant to define the object and limits of natural history, Daubenton interrupted the presentation of veterinary art with a lengthy description of the efforts of acclimatization of useful animals. Guyon, ed., *L'École normale de l'An III*, pp. 420-22 (see note 1).

developments in their field. In other words, the pedagogical practice of the teachers responded to the requirements of their respective discipline more than to the centralizing policy demands to discipline future citizens. The constraints of individual scientific disciplines and personal styles of individual lecturers clashed with the overall disciplinary framework that was supposed to create a new sociopolitical order.

Disciplinary Power or Art of Governance?

In January 1795 Lavoisier, who had been among the victims of the Terror in May 1794, could not deliver the chemistry lectures. This was unfortunate for the Normal School because his view of chemistry matched its pedagogical style. Lavoisier, as it is well known, promoted his own theory as a 'revolution' in chemistry.[25] Also well known is that contemporary historians of chemistry legitimately question whether the term 'revolution' is warranted for various reasons, including the Kuhnian view of scientific revolutions as involving paradigm shifts.[26] However neither this Kuhnian definition of the term 'revolution' nor the idea that it necessarily involved a radical break with the past prevailed at the end of the eighteenth century.[27] In a sense, as I have argued elsewhere, Lavoisier's own revolutionary ambition and the subsequent controversy that he sparked together with the reform of the chemical language contributed to stabilizing the current meaning of the phrase 'scientific revolution'.[28] In any event, the elements that he and his champions viewed as constituting the 'chemical revolution' served the project of normalization carried out by the *École normale* in three ways. First, Lavoisier's method of balancing the inputs and outputs of chemical reactions could be perceived as a kind of disciplinary

25 In 1772, Lavoisier deposited a sealed note at the Paris Academy of Sciences describing the experiments of calcination, which led him to question the phlogiston theory and mentioning that they might bring about "a revolution in physics and chemistry." Henry Guerlac, *Lavoisier- The Crucial Year: The background and origin of his first experiments on combustion, in 1772* (Ithaca, NY: Cornell University Press, 1961).

26 See for instance Ursula Klein, "A Revolution that Never Happened," *Studies in History and Philosophy of Science* 49 (2015): 80-90.

27 I. Bernard Cohen, *Revolution in Science* (Cambridge MA: Harvard University Press, 1985). Rolf Reichardt Rolf, H.J. Lüsebrink, "Révolution à la fin du 18e siècle. Pour une relecture d'un concept-clé du siècle des Lumières," *Mots* 16 (1988): 35-68. Alain Rey, *Révolution: Histoire d'un mot* (Paris: Gallimard, 1989).

28 Bernadette Bensaude Vincent, *Lavoisier. Mémoires d'une révolution* (Paris: Flammarion, 1993).

power over both materials and the minds and bodies of those who were charged with performing these balancing acts. Whether or not Lavoisier actually matched his claims of quantitative precision, he used them as a rhetorical argument to convert his fellow chemists to his views as instantiated in his spectacular experiment on water, which determined Berthollet to rally to him.[29] Second, the reform of chemical nomenclature provided a revolutionary model of standardization; the new language of chemistry submitted to the Paris Academy in 1787 by Guyton de Morveau, Lavoisier, Fourcroy and Berthollet was an artificial language, breaking with the language forged by those who had worked with chemical substances over many centuries.[30] The new language was promoted as a way to prompt a radical break with the past and in particular with the phlogiston theory, although the appropriation of this nomenclature by European chemists did not always match this ideal.[31]

Third and more importantly, the normalization of linguistic habits could be used as a prelude to the normalization of the republican educational system. Lavoisier's *Traité élementaire de chimie* promoted a new pedagogy of chemistry based on Condillac's claim that words, facts and ideas were the three ingredients of the formation of ideas in children's minds. Lavoisier presented his *Traité* as an extension of the memoir on the reform of nomenclature:

> The impossibility of separating the nomenclature of a science from the science itself is owing to this, that every branch of physical science must consist in three things: the series of facts which are the objects of the science; the ideas which represented these facts; and the words by which these ideas are expressed. Like three impressions of the same seal, the word ought to produce the idea, and the idea to be a picture of the fact. And as ideas are preserved and communicated by means of words, it necessarily follows that we cannot improve the language of any science without at the same time improving the science itself. [...] When we

29 See Bernadette Bensaude Vincent, "Between Chemistry and Politics: Lavoisier and the balance," *Eighteenth-century, Essays and Interpretation*, 33 (1992): 217-237. On the challenges to Lavoisier's precision see Jan Golinski, "'Fit Instruments': Thermometers in eighteenth-century chemistry," Frederic L. Holmes and Trevor H. Levere, eds., *Instruments and Experimentation in the History of Chemistry* (Cambridge, MA: MIT Press, 2000), 185-210.

30 Louis-Bernard Guyton de Morveau, Antoine Lavoisier, Claude-Louis Berthollet, Antoine de Fourcroy, *Méthode de nomenclature chimique* (Paris: Cuchet, 1787), new edition (Paris: Seuil, 1994).

31 Bernadette Bensaude Vincent and Ferdinando Abbri, eds., *Lavoisier in European Context. Negotiating a new language for chemistry* (Sagamore Beach: Science History Publications, 1995).

> begin the study of any science, we are in a situation respecting that science similar to children's and the course by which we have to advance, is precisely the same, which Nature follows in the formation of their ideas.[32]

According to this formulation, learning – the formation of complex ideas in children's minds – results from the association of simple sensations just as chemical compounds are formed through the association of simple substances and compound words by the addition of simple words. In Lavoisier's view, the material elements of chemistry (simple bodies) were one and the same as chemistry's pedagogical elements. From the simple to the complex, the order of learning mirrored the order of material composition in nature.[33] Thus chemistry could elegantly solve the dilemma that lecturers at the Normal School had to face: promoting a new didactics of science and at the same time providing a review of the most recent advances in their field.

Yet Berthollet did not follow this elegant pattern. On the one hand, he deliberately refused to give an elementary course. He assumed that his mission was to provide a survey of recent advances in chemistry. The future teachers in the audience were supposed to rephrase them later into a more didactic form. Indeed, Berthollet recommended proceeding from the simple to the complex in elementary courses of chemistry, but he clearly stated that he was not in the same situation as teachers addressing an audience of pupils. On the other hand, he did not follow Lavoisier's metaphysical assumption that the formation of ideas from simple to more complex ones was the mirror image of the composition of words in human language or of the formation of compound substances in nature. He firmly rejected the parallel between the order of learning and the order of nature, which was the basic pre-assumption of Lavoisier's revolutionary pedagogy. Berthollet claimed instead that natural phenomena should be approached from various standpoints in order both to facilitate the organization of ideas in the pupils' minds and to develop further investigations. To the linear analytical order (from the simple-to-the-complex), he preferred displaying a broad spectrum of perspectives on each subject. Lavoisier, for example, had refused to discuss affinities in his *Traité élémentaire*, arguing that the subject was too difficult and required too many pre-

32 Antoine Lavoisier, *Traité élémentaire de chimie* (Paris: Cuchet, 1789); *Elements of Chemistry*, translated by R. Kerr (Edinburgh : W. Creech, 1790), xiii.

33 Bernadette Bensaude Vincent, "A View of the Chemical Revolution Through Contemporary Textbooks: Chaptal, Lavoisier, Fourcroy," *British Journal for the History of Science* 23 (1990): 435-60.

conceptions to be included in an elementary course of chemistry.[34] Berthollet, contrariwise, dedicated his first lecture to the theory of affinity, based on his own claim that affinities constituted the core concept of chemistry. His justification was that their attractive action balanced by the repulsive force of caloric provided "the basic principle of all chemical phenomena." The description of the properties of simple substances and their compounds thereby entailed the application of this general principle.[35]

The structure of Berthollet's course (table 10.1) reveals both an interest in the theoretically arcane and a concern for practical aspects of the social practice of chemistry. In his first programmatic lecture Berthollet advocated the view of technological applications and improvements deduced from theoretical understanding. He intended to provide the broad deductive framework on which basis future teachers would have to develop more particular aspects according to their audience. But in none of the following lectures did he deduce practical applications from theory.

Strikingly, far from dogmatically imposing the revolutionary norms on chemistry, Berthollet demanded that his audience be able to interpret knowledge by themselves. In particular, he did not devote a single lecture to the new language of chemistry. His silence is all the more striking since he was a member of the group of chemists who had reformed the language of chemistry in 1787.[36] The Normal School was an ideal place to spread the new nomenclature and the underlying "theory of French chemists." Judging from the transcription of the first session of debates, such was the expectation of a number of the students in the audience. While Berthollet announced that the session would be devoted to a discussion of chemical attractions, Etienne-Bernard Guillemet from the district of Besançon opened the debate with this critical remark:

> Would it not have been more appropriate, or would it not be more appropriate, before proceeding further in your lectures, to talk about the words and rules on which the new chemical nomenclature is based? When one has to talk about a new theory, it is necessary to render the adopted

34 Lavoisier, *Traité élémentaire de chimie* (see note 32).

35 Berthollet, in Guyon, ed., *L'École normale de l'An III. Leçons de physique, de chimie, d'histoire naturelle*, p. 109 (see note 1).

36 Berthollet was in fact the first chemist who adopted Lavoisier's oxygen theory. In 1785 he used it to account for the formation of dephogisticated marine acid (future chlorine) in reacting manganese dioxide on marine acid (future chlorhydric acid). In 1787 he joined Guyton de Morveau, Lavoisier and Fourcroy to compose the *Méthode de nomenclature chimique*. In 1789, he was part of the editorial committee of the *Annales de chimie*, a new journal promoting the antiphlogiston theory.

TABLE 10.1 *Berthollet's Chemistry Course at the* École normale de l'An III

First Lecture: 6 *Pluviôse an III* (January 26, 1795)
- Order to be followed in an elementary course
- Indispensable preliminary: chemical attraction or affinity

Second Lecture: 7 *Pluviôse an III* (January 27, 1795)
- Anomalies in the affinities

Third Lecture: 22 *Pluviôse an III* (February 10, 1795)
- Nature and properties of heat: caloric and its action presented through their analogy with dissolutions

First Debate: *27 Pluviôse an III* (February 15, 1795)
- On nomenclature, on the laws of affinities, and on phlogiston as well as caloric

Fourth Lecture: *2 Ventôse an III* (February 20, 1795)
- The chemical action of light

Second Debate: *7 Ventôse an III* (February 25, 1795)
- On heat and light

Fifth Lecture: *12 Ventôse an III* (March 2, 1795)
- Analysis and synthesis of water

Sixth Lecture: 22 *Ventôse an III* (March mars, 1795)
- Carbonic gas

Seventh Lecture: 2 *germinal an III* (March 22,1795)
- Composition and properties of combustible bodies; their domestic applications

Eighth Lecture: *12 Germinal an III* (April 1, 1795)
- Niters and saltpeters

Ninth Lecture: 22 *Germinal an III* (April 11, 1795)
- Oxygenated muriatic acid

Tenth Lecture: 2 *Floréal an III* (April 21, 1795)
- Chemical properties of atmospheric air

Eleventh Lecture : *12 Floréal an III* (April 31,1795)
- Hydrogen combinations: ammonia, action of electricity on various substances

Twelfth Lecture: 22 *Floréal an III* (May 11, 1795)
- Acidity: influence of oxygen and its limits

Additional Lecture: *21 Prairial an IX* (June 10, 1801)
- Further thoughts about chemical action in gas, plants and animals

> phrases perfectly intelligible, in establishing, prior to using them, the rules of the recent convention, which alone had imagined them. I know that you suppose elementary notions of chemistry in the discussion; but one can have these notions and ignore the meaning of *carbonic acid*, *bombiate carbonate*, etc. I will therefore ask you whether it would not be more appropriate to say a few words about the rules of this convention. Your lecture would be more successful, and fruitful....[37]

Guillemet rightly expected rules in order to extend the process of normalization of chemistry in his own teaching. The emphasis on the term 'convention', which had a clear political connotation at the time of the National Convention, suggests a close association of the chemical revolution and the political revolution. After the end of the Terror fighting external and internal enemies no longer was a priority. The time had come to explain clearly the foundations of the new sociopolitical order. Such was Guillemet's expectation: a revolutionary teacher had to explain clearly the meaning and foundations of the new linguistic order. Berthollet ignored the criticism and answered that the new language was already embedded in the exposition of the recent advances in chemistry. It would suffice, he replied, to juxtapose the old and the novel denominations in the rest of his lectures.

Berthollet did not care to normalize chemistry. Quite the contrary, he was interested in anomalies. Following the first lecture on the theory of elective affinities and the construction of affinity tables, he devoted an entire lecture to the cases that opposed the order of affinities in the tables. He thus detailed a number of reactions where the products reacted with reagents and offered two possible explanations for such anomalies: either a variation in temperature or the proportion of substances.[38] His take-home message to the future instructors was that they should pay attention to the circumstances of reaction. He repeatedly emphasized their peculiarity more than their conformity with the general rule. This attention to the local and particular circumstances would lead him to revise the standard theory of affinity in his *Essai de statique chimique* (1803) and to oppose Joseph Proust regarding the general law of fixed proportions.[39]

---

37 Berthollet, First debate, 27 pluviôse/February 15 in Guyon, ed., *L'École normale de l'An III. Leçons de physique, de chimie, d'histoire naturelle*, p. 275 (see note 1).

38 See Pere Grapi and Merce Izquierdo, "Berthollet's Conception of Chemical Change in Context," *Ambix* 44 (1997): 113-30.

39 Kiyohisa Fujii, "The Berthollet-Proust Controversy and Dalton's Atomic Theory, 1800-1820," *British Journal for the History of Science* 19 (1986): 177-200.

More broadly Berthollet did not use the opportunity of this course to spread the "theory of French chemists". He surveyed state of the art European chemistry and downplayed the divide between so-called 'phlogistonist' and 'antiphlogistonist' chemists. Strikingly his lectures portrayed chemistry as an international and communal science, a position that opponents to the 'new chemistry' had sought to use as a foil to what they saw as its imperialistically French character. While he hardly mentioned his own work and contributions (on ammonia, for instance), he nevertheless ventured personal theoretical views in the twelfth lecture on acidity. Here he began with two examples that instantiated Lavoisier's theory of acids, followed by the presentation of oxygenated compounds without acidic properties and one acid that did not contain oxygen. Berthollet clearly expressed his own view: "This is a question on which I disagree with most chemists."[40]

Conclusion

This essay presented Berthollet's lectures at the *École normale de l'An III* as one face of chemistry in the public sphere, which has been eclipsed by the standard narratives of the chemical revolution shaped by Lavoisier. Berthollet's lectures clearly indicate that there was no uniform paradigm of chemistry even among the French antiphlogistonist school. He disagreed with Lavoisier not only about theoretical aspects (such as the nature of acids) but also about the moral economy of chemistry.[41] The values underlying Lavoisier's chemistry were simplicity, standardization, abstraction from local contexts and circumstances. By contrast, Berthollet put the emphasis on non-standard cases and local circumstances. Through his engagement in both the chemical and the French revolutions, Berthollet was trying to promote an alternative model of order – one that would take into account the complexities of the real world rather than promoting an ideal of normality.

Although Lavoisier's chemistry provided a form of disciplinary power over materials and language, which was in full agreement with the project of normalizing science and education, Berthollet promoted chemistry as an art of circumstances requiring attention and personal experience. To the discipli-

40 Berthollet, Premier débat, 27 pluviôse/February 15 in Guyon, ed., *L'École normale de l'An III. Leçons de physique, de chimie, d'histoire naturelle*, p. 350 (see note 1).

41 The moral economy of science in Lorraine Daston's sense is "a balanced system of emotional forces, with equilibrium points and constraints." Lorraine Daston, "The Moral Economy of Science," *Osiris* 10 (1995): 2-24, 2.

nary power of chemistry balancing chemical equations and ruling all chemical arts, he preferred a governance of materials and reactions based on negotiations with their peculiar behaviors in specific circumstances. It is not that Berthollet was against all systematic order. Rather for him disciplining materials by submitting them to the rule of general laws did not make sense as long as it introduced an arbitrary decision of minimizing the significance of exceptions. In his view anomalies were interesting and informative as much as the standard cases. In more general terms, Berthollet was against the revolutionary plea for normalization because it entailed a process of abstraction and idealization of the actual behavior of materials in the real world. He sought a more inclusive order that would take the particularities of ambient circumstances into account. In recognizing the mundane complexities of the material world, he invited future citizens to negotiate between the goal of disciplining knowledge, on one hand, and both nature and society on the other.

Lavoisier's analytic-synthetic ideal of standardized chemical substances prevailed for most of the nineteenth century, whereas Berthollet's ideal was eclipsed. The Normal School was globally regarded as a failure because "the professors' lessons were academic courses rather than lectures appropriated for instruction."[42] Under Napoléon's empire, teachers were trained at the university: they had to attend courses at the *Collège de France*, the *École polytechnique* and the *Muséum d'histoire naturelle*. When the Normal School was fully re-established in 1830 by King Louis-Philippe the program of the chemistry course mainly dealt with stoichiometric compounds although it maintained a first lesson dedicated to "molecular attraction" and affinity.[43] It included one laboratory class per week but virtually no practical recipes for preparing useful compounds. For many decades, only "daltonide" compounds were investigated, whereas "berthollide" non-stoichiometric compounds and equilibrium reactions were ignored until the 1870s. Only under the Third Republic did chemists face the challenge of normalizing the berthollides, of ruling chemical equilibrium with the mass action law and Henry Le Chatelier's law. However, in advocating the application of Taylorism to laboratory research, Le Chatelier extended the ideal of normalization to all practices of chemistry.[44] Definitely, Berthollet's lessons of chemistry failed to shape the future generations of French chemists!

---

42 Victor Cousin, in *École normale. Règlements, programmes et rapports* (Paris: Hachette, 1837), p. 4.

43 Programme du cours de chimie, 1st et 2nd semester in ibid., p. 50-56.

44 Henry Le Chatelier, *Science et industrie* (Paris: Flammarion, 1925).

Acknowledgments

I am indebted to Patrice Bret and Pere Grapi who collaborated with me for the edition of Berthollet's lecture in Etienne Guyon ed. *L'École normale de l'An III. Leçons de physique, de chimie, et d'histoire naturelle. Haüy, Berthollet, Daubenton.* (Paris, éditions ENS Rue d'Ulm, 2006). I also wish to thank Lissa Roberts for her critical comments on earlier versions of this essay.

CHAPTER 11

The Subversive Humphry Davy: Aristocracy and Establishing Chemical Research Laboratories in Late Eighteenth- and Early Nineteenth-Century England

Frank A.J.L. James

As other essays in this volume testify, chemistry, in all its multifarious guises throughout Europe, could be undertaken within a wide range of institutional, organisational and physical settings. This essay discusses the beginnings in England at the end of the eighteenth century of one type of setting, namely institutional laboratories funded by subscription that were either envisioned or came to be devoted to chemical research. Such laboratories required a number of material things before they could start to produce scientific knowledge: a building, building services (heating, lighting, water, and so on) and apparatus. Furthermore, appropriately trained staff including researchers, laboratory assistants and servants, also needed to be found. Less tangibly, it was important to have available retrievable recording methods and some access to disseminating the knowledge produced. These components did not appear of their own volition, but required substantial financial support, to be organized and governed, all through human agency. Even today none of this is straightforward, but when no appropriate models were available that could be followed to achieve the desired aims, the difficulties became manifold. Inevitably under such circumstances, a strong tendency existed to underdefine and underspecify what a laboratory required, particularly in terms of management, and to start with the familiar. That left ample scope for unintended consequences, the idiosyncrasies of individual agency and the subversion of original aims.

This essay explores these themes through examining the founding and development of the first two subscription funded research laboratories in England, located in the Medical Pneumatic Institution (MPI) in Bristol and the Royal Institution of Great Britain (RIGB) in London, both established in the late 1790s. This is not to suggest that research laboratories did not previously exist in Great Britain. Two of the most significant were funded privately by individual wealthy aristocrats, one in Clapham, Surrey, built by and for Henry Cavendish (1731-1810) and the other by the Second Earl of Shelburne (1737–

DOI 10.1163/9789004325562_013

1805) at his Bowood seat in Wiltshire where Joseph Priestley (1733-1804) and Jan Ingen-Housz (1730-1799) worked.[1] Both laboratories were entirely dependent on aristocratic interest and whim, with no mechanism for them to be sustained beyond their patrons' lives. It is perhaps no coincidence that much of Cavendish's apparatus later wound up at the RIGB, where he was a leading figure during its early years.[2]

It might be assumed *a priori* that subscription-funded laboratories would be organized differently from those supported aristocratically and command longer term support, since their funding base was broader and seemingly not dependent on any specific individual. In practice, it had to be learned how such novel organisations should be managed. That meant, to some extent at least, beginning with familiar organisational methods, thus potentially subjecting the new institutions to the same issues surrounding individual domination, continuity and legacy that arose with aristocratic patronage. Aristocratic laboratories provided one of the few models available in Great Britain about how a research laboratory might be managed; furthermore, a large proportion of subscriptions for both the MPI and RIGB came from aristocrats. These new laboratories were not *ex nihilo* creations, linked distinctively, according to some accounts, to notions of increasing industrialization or a rising middle class, but were also closely connected to the traditional British ruling class whose power was still enormous.

The MPI and the RIGB were also linked together closely by a single individual, Humphry Davy (1778-1829). He was the former's superintendent for nearly two and a half years from October 1798 until March 1801, when he moved to the latter. While the considerable literatures on Davy, the MPI and the RIGB do, of course, note the connection, its significance for the development of research

1 Christa Jungnickel and Russell McCormmach, *Cavendish: The experimental life* ([Lewisburg]: Bucknell, 1998), 329-30. Robert E. Schofield, *The Enlightened Joseph Priestley: A study of his life and work from 1773 to 1804* (University Park: Pennsylvania State University Press, 2004), 3-143; Norman Beale and Elaine Beale, *Echoes of Ingen Housz: The long lost story of the genius who rescued the Habsburgs from smallpox and became the father of photosynthesis* (East Knoyle: Hobnob Press, 2011).

2 The minutes of the meetings of the Royal Institution's Managers are in RI MS AD/2/B/2/A followed by volume number. This will be cited here as RI MM followed by date of meeting, volume and page numbers. The minutes of nineteenth-century meetings were published in facsimile as *Archives of the Royal Institution, Minutes of the Managers' Meetings, 1799-1903*, 15 volumes in 7 (London: Scholar Press, 1971-1976). Cavendish's apparatus is referred to in RI MM, 9 July 1810, 5: 126.

laboratories in Great Britain has not really been appreciated.[3] This essay argues that Davy's unique career trajectory from provincial obscurity in Penzance, in the far west of England, to the MPI, provided him with the skills, experience and commitment to negotiate and persuade the RIGB to add research to its activities, something that had never been intended or even envisaged by its founders. Davy brought about this significant transformation in a manner reminiscent of the way aristocratic laboratories were organized. As a consequence he established at the RIGB what became for much of the nineteenth century the best-equipped chemical and natural philosophical research laboratory in England. Many fundamental discoveries would be made there by Davy and his successors, such as Michael Faraday (1791-1867), John Tyndall (*c.*1822-1893) and James Dewar (1842-1923).[4] Their successes contributed to ensuring that the RIGB came to serve as a model, or at least a starting point, for the organisation of science in other places, most notably the Smithsonian Institution in Washington in 1846.[5]

The Medical Pneumatic Institution

The MPI stemmed from the work of Thomas Beddoes (1760-1808), the son of a reasonably wealthy tanner with significant land holdings in Shifnal, Shrop-

3 June Fullmer, *Young Humphry Davy: The making of an experimental chemist* (Philadelphia: American Philosophical Society, 2000); David Knight, *Humphry Davy: Science and power* (Oxford: Blackwell, 1992; Second Edition, Cambridge: Cambridge University Press, 1996); Sophie Forgan, ed., *Science and the Sons of Genius: Studies on Humphry Davy* (London: Science Reviews, 1980); Trevor H. Levere, "Dr. Thomas Beddoes and the Establishment of His Pneumatic Institution: A tale of three presidents," *Notes and Records of the Royal Society of London* 32 (1977): 41-9; Mike Jay, *The Atmosphere of Heaven: The unnatural experiments of Dr. Beddoes and his sons of genius* (New Haven: Yale University Press, 2009); Henry Bence Jones, *The Royal Institution: Its founders, and its first professors* (London: Longman, 1871); Morris Berman, *Social Change and Scientific Organization: The Royal Institution, 1799-1844* (Ithaca: Cornell University Press, 1978); Frank A.J.L. James, ed., *'The Common Purposes of Life': Science and society at the Royal Institution of Great Britain* (Aldershot: Ashgate, 2002).

4 Frank A.J.L. James, *Michael Faraday: A very short introduction* (Oxford: Oxford University Press, 2010); Roland Jackson, "John Tyndall and the Early History of Diamagnetism," *Annals of Science* 72 (2015): 435-89; John Rowlinson, *Sir James Dewar, 1842-1923: A ruthless chemist* (Farnham: Ashgate, 2012).

5 Heather Ewing, *The Lost World of James Smithson: Science, revolution, and the birth of the Smithsonian* (New York: Bloomsbury, 2007).

shire.[6] He attended Pembroke College, Oxford, before studying chemistry with Bryan Higgins (*c.*1741-1818) in London and then with Joseph Black (1728-1799) at Edinburgh University. Returning to Oxford in 1786, he took his MD, followed by a visit to France. There he became acquainted with many leading chemists, including Antoine Lavoisier (1743-1794), following which he adopted the new chemical nomenclature. From 1787 until his final lecture series delivered in the spring of 1792, he held the non-stipendiary Readership in Chemistry at Oxford University. He then resigned following disagreements in Oxford centering on his support for the French Revolution and his general position as a democrat, which led him to be subjected to government harassment.[7]

Needing an income, Beddoes decided to practice medicine and by early April 1793 had acquired the lease on 11 Hope Square, Hotwells.[8] Located near Bristol, the fifth largest city in England, Hotwells, a small spa village on the banks of the river Avon across from Somerset, had a long tradition of wealthy and aristocratic visitors coming to take the waters to benefit their health since the seventeenth century.[9] The Midlands engineer and industrialist James Watt sr. (1736-1819) had no doubt that Beddoes chose Bristol "for the greater [medical] practice," suggesting he believed that the choice of location was entirely deliberate, enabling Beddoes to be near wealth and potential patronage.[10]

Furthermore, Bristol was also geographically convenient for Bowood, and evidence suggests that Beddoes had links there, especially with Ingen-Housz.[11] The main influence, however, for Beddoes's choosing Hotwells was probably the presence of the Irish landowner and educational writer Richard Edgeworth

---

6 John Stock, *Memoirs of T. Beddoes, M.D., with an analytical account of his writings* (London: John Murray, 1811); Dorothy Stansfield, *Thomas Beddoes M.D., 1760-1808: Chemist, physician, democrat* (Dordrecht: Reidel, 1984).

7 Kenneth Johnston, *Unusual Suspects: Pitt's reign of alarm and the lost generation of the 1790s* (Oxford: Oxford University Press, 2013) devoted an entire chapter (96-110) to Beddoes. See also Trevor H. Levere, "Dr. Thomas Beddoes at Oxford: Radical politics in 1788-1793 and the fate of the Regius Chair in Chemistry," *Ambix* 28 (1981): 61-9.

8 Thomas Beddoes to Davies Giddy, 7 April 1793, CRO MS DG/41/2, told him to send his next letter to Hope Square.

9 Phyllis Hembry, *The English Spa 1560-1815: A social history* (London: Athlone, 1990), 245-50.

10 James Watt sr. to Joseph Black, 17 July 1793, in R.G.W. Anderson and Jean Jones, eds., *The Correspondence of Joseph Black*, 2 vols. (Farnham: Ashgate, 2012), vol. 2, 703.

11 See Jan Ingen-Housz to Thomas Beddoes, 4 August 1794, quoted in Thomas Beddoes and James Watt, *Considerations on the Medicinal Use of Factitious Airs, and on the manner of obtaining them in large quantities* (parts 1 and 2, London: Johnson, [1794]), part 1, 31. Beale and Beale, *Ingen Housz*, pp. 452-4, 481-3 (see note 1).

(1744-1817), whom he knew via his links with the Midlands industrialists. Edgeworth had been living in Clifton (north of Hotwells) with his third wife and large family since late 1791 for the sake of of his son's health. Beddoes regarded Edgeworth as being in "the highest rank of the untitled Aristocracy" (again suggesting Beddoes's concern with status and position).[12] Furthermore, Beddoes had fallen in love with Edgeworth's daughter, Anna Edgeworth (1773-1824); they married in Ireland on 17 April 1794.[13] Approving the match, Edgeworth described his future son-in-law as "a little fat Democrat of considerable abilities" and thought that if he concentrated on medicine he would make his fortune.[14] Beddoes did indeed build up a considerable and lucrative medical practice in the ensuing years.[15]

But Beddoes also spent much time developing his ideas about the possible therapeutic value of pneumatic chemistry, first mooted while still at Oxford.[16] In the preceding decades Priestley, Cavendish and others had discovered various new airs, elastic fluids or gases, and Beddoes wanted to determine whether they could be used to cure diseases (particularly consumption) or at least mitigate them. Knowledge about Beddoes's work, especially its significance if successful, circulated fairly widely and attracted considerable attention including from no less a figure than the Whig grandee Georgiana Cavendish, Duchess of Devonshire (1757-1806), wife of the fifth Duke of Devonshire (1748-1811). She twice visited Beddoes in Hotwells, once just before Christmas 1793 and again in mid-January, evincing a strong interest in his ideas.[17] On both occasions Beddoes demonstrated to her that if animals such as dogs or rabbits breathed oxygen beforehand they could survive freezing or emersion in nitrogen.[18]

---

12 Thomas Beddoes to Davies Giddy, 25 or 26 May 1793, CRO MS DG/41/21. A misreading of this passage may be the source of the mistake in Jay, *Atmosphere*, pp. 80, 289 (see note 3), in incorrectly awarding a knighthood to Edgeworth.

13 Thomas Beddoes to Davies Giddy, 25 or 26 May 1793, CRO MS DG/41/27, referred to becoming intimately acquainted with her during the previous three months.

14 Postscript by Edgeworth in Maria Edgeworth to Margaret Ruxton, 21 July 1793, National Library of Ireland MS 10166/7/105. Jay, *Atmosphere*, p. 91 (see note 3), suggested that Beddoes was not a gentleman, being "a tanner's son," and therefore there were class issues that required resolution before the marriage could take place. It is not clear on what basis Jay thought that Edgeworth would not have viewed an Oxford educated physician as anything other than a gentleman.

15 Thomas Beddoes to James Watt sr., 30 May 1797, LoB MS 3219/4/29/14.

16 Thomas Beddoes to Davies Giddy, 2 March 1795, CRO DG/42/35.

17 Duchess of Devonshire to Dowager Countess Spencer, 1 January 1794 and 16 January 1794, Chatsworth MS CS5/1201 and 1204 respectively.

18 Duchess of Devonshire to Charles Blagden, 13 and 14 January 1794, Royal Society of London MS CB/1/3/278; Duchess of Devonshire to Joseph Banks, 1 December 1794, in Neil

It was not a coincidence that, following these aristocratic visits, Beddoes began to develop plans for a pneumatic hospital. This idea appeared first in a letter to the chemist Joseph Black, written on Christmas Eve 1793.[19] In another letter written to an unidentified correspondent "immediately" after Devonshire's second visit, Beddoes wrote: "it would be more practical to determine the medical effects of elastic fluids in one year, if we had six to twelve patients in a house with apparatus, than in twelve years of private practice." He estimated that such a hospital dedicated solely to this area of research could be established with six or seven hundred pounds, though he soon revised this to "not less than £3000 & not more than £5000."[20] In a folded broadsheet entitled *A proposal towards the improvement of Medicine*, printed at the end of July 1794, Beddoes argued that he had "abundantly proved, that the application of elastic fluids to the cure of diseases, is both practical and promising." He continued that a funded "Medical Pneumatic Institution" would much more effectively establish the benefits or otherwise of pneumatic medicine than "twenty years of private practice." A successful MPI, Beddoes argued, "ought to render itself useless, by so far simplifying methods and ascertaining facts, that every practitioner of medicine, at least, may both know how to procure and how to apply the different elastic fluids." If unsuccessful, then at least it had been tried. He thought such an institution should be able to settle the matter after operating for two or three years. Money would be required to rent a building accommodating a dozen patients, apparatus, furniture, a medical superintendent to run the institution, three servants, contingent expenses and medicines.[21] Beddoes clearly understood a laboratory's material and staffing requirements, although not necessarily how they fitted together.

For nearly four years, from the autumn of 1794 until the summer of 1798, Beddoes worked with his major supporters, including Devonshire, James Watt jr. (1769-1848), Tom Wedgwood (1771-1805) and Erasmus Darwin (1731-1802), undertaking a public fundraising campaign. In the end around 200 donors subscribed just over £2000, significantly below the minimum sum that Beddoes

Chambers, ed., *The Scientific Correspondence of Joseph Banks*, 6 vols. (London: Pickering, 2007), vol. 4, 1290.

19 Thomas Beddoes to Joseph Black, 24 December 1793, *Black Correspondence*, vol. 2, p. 724 (see note 10).

20 Thomas Beddoes to unidentified correspondent, mid-January 1794, in Stock, *Beddoes*, pp. 100-1 (see note 6). Thomas Beddoes to Thomas Wedgwood, mid-March 1794, WM MS MC 35. This letter is dated on the basis that Beddoes mentioned that he was about to go to Ireland.

21 Thomas Beddoes, *A proposal towards the improvement of Medicine* (Bristol: np, 1794).

believed necessary.[22] Nearly a third of this money came from just nineteen donations greater than £20 (mostly from members of the Devonshire circle including seven aristocrats). Additionally, Wedgwood gifted £500 and Beddoes's deceased patient William Lambton (1764-1797) bequeathed £300. The reason why the general public appeal failed to raise the required amount, thus forcing Beddoes to rely heavily on his wealthy and aristocratic supporters, was due to his radical political views, which told against him. For example, despite her best efforts, the Dutchess of Devonshire (who disliked Beddoes's politics) could not obtain the support of the President of the Royal Society of London, Joseph Banks (1743-1820).[23] He regarded Beddoes's opposition "to the present arrangement of the order of Society in this Country" as disqualifying him from support.[24] Even Beddoes's strong supporters became frustrated with him; James Watt sr. told him that his political activities "will do more hurt to Pneumatics than you can possibly do good to the nation – amend your ways."[25]

During 1797 Beddoes and some subscribers became convinced that sufficient funds had been raised to ensure the practicality of imminently establishing the MPI. Discussions began about its possible location (some were not convinced of Bristol's appropriateness) and to search for a suitable superintendent.[26] So far as the latter was concerned, Beddoes noted in October, that despite "many applications" none were suitable, which "disappointed" him.[27] However, whilst staying in Penzance during the winter of 1797/8 for the sake of their health, Wedgwood and Watt's son from his second marriage, Gregory Watt (1777-1804), had formed a friendship with an apprentice apothecary, the nineteen-year-old Humphry Davy.

22 This is detailed in Frank A.J.L. James, *'The first example … of an extensive scheme of pure scientific medical investigation': Thomas Beddoes and the Medical Pneumatic Institution in Bristol, 1794 to 1799* (London: Royal Society of Chemistry Historical Group Occasional Publication, 2016).

23 Duchess of Devonshire to Earl Spencer, 30 May 1796, British Library MS add 75923 (no foliation).

24 Joseph Banks to Duchess of Devonshire, 30 November 1794, Natural History Museum Dawson Turner Collection, 9, f. 125.

25 James Watt sr. to Thomas Beddoes, 28 November 1795, LoB MS 3219/4/124/414.

26 Thomas Beddoes to James Watt sr., 26 May 1797, LoB MS 3219/4/29/13.

27 Thomas Beddoes to Thomas Girdlestone, 25 July 1797, private possession; Thomas Beddoes to James Watt sr., 24 October 1797, LoB MS 3219/4/29/23.

Humphry Davy

Davy was born in Penzance in 1778, the eldest of five children. Many anecdotes testify to his being a bright child and after attending local schools he studied at Truro Grammar School in 1793.[28] Leaving serious debts due to unwise mining investments, his father died in December 1794, a week before Davy's sixteenth birthday. To help overcome the family's financial problems, his mother, Grace Davy (1752-1826), apprenticed him to the Penzance surgeon and apothecary John Borlase (1753-1813).[29] She also opened a milliners shop and took in lodgers, including at the end of 1797 Gregory Watt to whom Davy became close. Wedgwood, also in Penzance, and Watt were both interested in chemistry and it is from this time that Davy began his chemical studies – hitherto he had been more interested in poetry.[30] Despite the existence of a 1790 English translation he read, amongst other texts, Lavoisier's *Traité élémentaire de chimie* (1789; second edition, 1793) in French, which he had learned from an émigré, and undertook some chemical experiments.[31] All this brought him to the attention of Beddoes's former student and minor member of the Cornish gentry, Davies Giddy (1767-1839). His support secured Davy access to the well-equipped laboratory in Riviere House, Hayle, owned by John Edwards (1731-1807), manager of the Cornish Copper Company.[32] Either there or possibly through an instrument maker in Penzance, Davy conducted experiments with an air pump demonstrating that a flintlock fired in a vacuum did not, contra Lavoisier, produce light. Furthermore, he also experimented on rubbing ice pieces together, from which he concluded that heat was a mode of motion rather than a fluid, a result he appears to have reached before hearing about the similar theory proposed around the same time by Benjamin Thompson, Count Rumford (1753-1814).[33]

---

28 John Paris, *The Life of Sir Humphry Davy*, 2 vols. (London: Colburn and Bentley, 1831); John Davy, *Memoirs of the Life of Sir Humphry Davy, Bart*, 2 vols. (London: Longman, 1836).

29 Davy's indenture of apprenticeship in RI MS HD/5/3.

30 Wahida Amin, "The Poetry and Science of Humphry Davy," PhD thesis, University of Salford, 2013; Sharon Ruston, "From "The Life of the Spinosist" to "Life": Humphry Davy, Chemist and Poet," Margaret Hagen and Margery Skagen, eds., *Literature and Chemistry: Elective affinities* (Aarhus: Aarhus University Press, 2013), 77-97.

31 Davy, *Davy*, vol. 1, p. 21 (see note 28).

32 Paris, *Davy*, vol. 1, p. 47 (see note 28). W.H. Pascoe, *CCC: The history of the Cornish Copper Company* (Hayle: Haylebooks, 1981), 158.

33 Rumford, "An Inquiry concerning the Source of the Heat which is Excited by Friction," *Philosophical Transactions* 88 (1798): 80-102.

Giddy suggested to Davy that he should send an account of his experiments to Beddoes which he did in April 1798, and which Beddoes later published.[34] In July Beddoes wrote to Watt sr. asking for his opinion about appointing Davy ("concerning whom apply to Gregory") to be superintendent, to which he presumably consented.[35] On 1 October, after extended negotiations, Borlase canceled the final sixteen and a half months of Davy's apprenticeship. Leaving Penzance the following day, Davy arrived in Bristol five days later. Beddoes had just moved into a large house in Rodney Place, commenting that the "houses in it are the best at Clifton & I have bought the best".[36] But, "above all," Davy told his mother soon after arriving, Rodney Place possessed "an excellent laboratory."[37] Davy did not disappoint and Beddoes reacted to him as did everyone else who met him at this time. In *The Monthly Magazine*, Beddoes described Davy as "A young man, endowed with talents for experimental researches at least equal to any person I have ever known." His patient, the poet Robert Southey (1774-1843), wrote, "We have an extraordinary young man lately settled here."[38]

Beddoes entrusted Davy with the task of spending the money that he had taken so long to raise and arranged for him to meet some of the MPI's major subscribers. For instance, Davy quickly visited the Midlands to see Watt sr. and his fellow industrialist James Keir (1735-1820). During this period Davy also began negotiations to acquire a building for the MPI in Dowry Square.[39] In March 1799 *The Bristol Gazette* carried an advertisement announcing the opening there of the "New Medical Institution" which would be attended, presumably daily, by Beddoes and Davy between 11am and 1pm.[40] Shortly afterwards, Davy drafted a "Prospectus" for the MPI (though it is not clear if this was ever printed) in which he wrote that "upwards of forty are become outpatients [at the MPI] within this fortnight & we could immediately fill the house with in-patients."[41] The following month the number of out-patients had

34 Thomas Beddoes to Davies Giddy, 14 April 1798, CRO DG/42/2. Humphry Davy, "An Essay on Heat, Light, and the Combinations of Light," Thomas Beddoes, ed., *Contributions to physical and medical knowledge, principally from the West of England* (London: Longman, 1799), 5-147.

35 Thomas Beddoes to James Watt sr., 15 July 1798, LoB MS 3219/4/29/32.

36 Thomas Beddoes to James Watt jr., October 1798, LoB MS 3219/6/2/B/72.

37 Humphry Davy to Grace Davy, 11 October 1798, RI MS HD/26/A/1.

38 *The Monthly Magazine* 6 (1798): 238. Robert Southey to William Taylor, 24 February 1799.

39 Humphry Davy to Grace Davy, 11 October 1798, RI MS HD/26/A/1.

40 *The Bristol Gazette* (21 March 1799): 3c.

41 Humphry Davy, "Prospectus of the design of the Institution," [late March / early April 1799], RI MS HD/20/B, 11-16, quotation on 14.

increased to eighty.[42] By November, there were five lady in-patients being subjected to Beddoes's novel treatment of inhaling the breath of an Alderney cow.[43]

Following the MPI's opening, Davy began experimentation on dephlogisticated nitrous air as Priestley, its discoverer, had named it, or nitrous phosoxyd as Davy initially termed it before finally calling it, following the new nomenclature developed by Lavoisier and his associates, nitrous oxide. He discovered the gas's pleasurable physiological action and self-experimented by frequently inhaling it in large quantities – on one occasion sixteen quarts (just over eighteen liters).[44] This sustained exposure to the gas so damaged his health that he spent most of November 1799 in Cornwall, where he suffered withdrawal symptoms. After thirty-three days without the gas, he inhaled nine quarts on his return.[45] In his first book, *Researches, Chemical and Philosophical; Chiefly Concerning Nitrous Oxide, or Dephlogisticated Nitrous Air, and its Respiration*, Davy described the results of his own experimentation and the accounts of many of those in Beddoes's Bristol circle whom he persuaded to inhale.[46] Published mid-1800, *Researches* cost half a guinea and ran to nearly six hundred pages. He divided the text roughly equally between providing a detailed chemical analysis of the gas and describing its physiological properties. Towards the end, and then only very briefly, did Davy make reference to any possible therapeutic use, though Beddoes thought there might be.[47] As countless writers have pointed out, Davy did not observe the anaesthetic property of nitrous oxide.[48]

Despite Beddoes's avowed aims for the MPI, for which subscriptions had been so painfully obtained, Davy, somewhat subversively, developed and pursued a rather different research agenda. In this Davy exhibited a pattern of behavior similar to those of aristocratic laboratories where the researchers followed their own interests. (What direct knowledge he might have had of them

42 Humphry Davy to Davies Giddy, 18? April 1799, Paris, *Davy*, 1: pp. 79-82 (see note 28). Paris dated this letter 10 April 1799, but such a date contradicts other evidence. Davy's numbers can, on occasion, be confused.

43 Thomas Frankland to James Smith, 18 November 1799, Linnean Society MS JES/COR/15/6.

44 Humphry Davy to Davies Giddy, 18? April 1799, Paris, *Davy*, vol.1, pp. 79-82 (see note 28).

45 Humphry Davy, *Researches, Chemical and Philosophical; Chiefly concerning nitrous oxide, or dephlogisticated nitrous air, and its respiration* (London: Johnson, 1800), 478.

46 Ibid.

47 Thomas Beddoes, *Notice of Some Observations Made at the Medical Pneumatic Institution* (London: Longman, 1799).

48 Jan Golinski, "Humphry Davy: The experimental self," *Eighteenth-Century Studies* 45 (2011): 15-28, especially 18-19.

is not known, but he later developed a taste for staying in large aristocratic country houses.) Despite Beddoes's clear aims for the MPI, he had not found a way of ensuring that they would be fulfilled, suggesting that he had not considered how research in a laboratory should be managed or guided towards a defined aim. Alternatively Beddoes may have recognized that he had too many other commitments and interests and so gave Davy a free hand, which also illustrates similarities to aristocratic laboratories.

A similar pattern occurred when news reached Davy, just as he was completing his *Researches*, of an invention made by Alessandro Volta (1745-1827), which he called a pile producing galvanic electricity, but which Davy would shortly rename the battery.[49] Davy included a very brief reference to Volta right at the end of his *Researches* and Beddoes arranged for a pile to be built.[50] For the remainder of the year Davy experimented on galvanism, recording this research in two notebooks, beginning in August, and in a series of papers published monthly (apart from January) in *A Journal of Natural Philosophy, Chemistry and the Arts* from September 1800 through to February 1801.[51] In these papers Davy announced, amongst other discoveries, that electricity would pass through organic tissue, that charcoal could be used as an electric pole and came to the overall conclusion, contra Volta, that "Galvanism ... [was] a process purely chemical."[52] Such conclusions confirmed Davy's view, probably written in his notebook during mid-1799, that "Chemistry must no longer [be] considered as a science important because it is connected with our artificial wants; but because it promises to unfold to us the laws of our own existence."[53] Right from the start, Davy believed studying galvanism would "acquaint us with some of the laws of life!," a view stemming ultimately from the origin of galvanism in animal electricity.[54] Once again Davy displayed little concern for whether his work had any therapeutic value or relevance to the MPI.

49 Humphry Davy, "An Account of some *Galvanic* Combinations, formed by the Arrangements of single metallic Plates and Fluids, analogous to the new *Galvanic* Apparatus of Mr. *Volta*," *Philosophical Transactions* 91 (1801): 397-402, 400.

50 Davy, *Researches*, p. 568 (see note 45). *A Journal of Natural Philosophy, Chemistry and the Arts* 4 (1800): 275.

51 RI MS HD/20/C and /22/B. Papers listed in June Fullmer, *Sir Humphry Davy's Published Work* (Cambridge, MA: Harvard University Press, 1969).

52 Humphry Davy to Davies Giddy, 20 October 1800, private possession, but also Paris, *Davy*, vol. 1, pp. 108-11 (see note 28).

53 RI MS HD/20/A, p. 266.

54 Humphry Davy to Davies Giddy, 3 July 1800, Paris, *Davy*, vol. 1, pp. 85-8 (see note 28).

Beddoes, who also thought they were close to understanding living systems, sought to take advantage of Davy's straying, by arguing that to exploit his work (and thus continue pursuing the MPI's stated aims) would require "the most extensive application of chemistry to physiology."[55] To achieve this he would need to appoint an expert anatomist as well as an instrument maker and establish a manufactory on the scale of Boulton and Watt's recently completed foundry for constructing steam engines at Soho, then one of the largest industrial facilities in the world.[56] Beddoes pointed out, somewhat unnecessarily, that additional funding would be required.[57] One does have to wonder how far Beddoes, or indeed anyone else, believed his unrealistic rhetoric bordering on fantasy. He essentially used the success of Davy's discovery to make a case for significant extra financial support for the MPI. In this he saw Davy as a valuable resource, writing that he considered that the effect of his presence "as more than virtually doubling the fund."[58]

Despite his value to Beddoes, Davy realized that he had major problems with his career due to "the political odium attached to its [the MPI's] founder."[59] Owing to the non-survival of papers detailing the MPI's operation, we have no knowledge of its expenditure pattern, only that Davy earned £200 annually.[60] With only finite resources, it would have been apparent to him that the MPI's duration would be limited – as Beddoes had originally intended in his 1794 *Proposal*. Davy's continuing connection with a known political radical would be an increasing liability in finding a new job, so perhaps it is not too surprizing that about two years after his arrival in Bristol, he began considering his future prospects. In two letters written in September and November 1800, he hinted to his mother that he was looking for alternative employment.[61] In the latter month writing from Lisbon, Southey opined that "Davy will not always remain at the Pneumatic Institution."[62] Though it is not clear what options might then have been open to him, by early January he had started negotiations to move to the RIGB in London.[63]

55 Beddoes, *Notice*, p. 34 (see note 47).

56 Peter Jones, *Industrial Enlightenment: Science, technology and culture in Birmingham and the West Midlands, 1760-1820* (Manchester: Manchester University Press, 2008), 55.

57 Beddoes, *Notice*, pp. 35-6 (see note 47).

58 Ibid., p. 6.

59 Humphry Davy to John Tonkin, 12 January 1801, Davy, *Davy*, vol. 1, pp. 107-9 (see note 28).

60 Humphry Davy to Grace Davy, 19 January 1799, RI MS HD/26/A/2.

61 Humphry Davy to Grace Davy, 27 September 1800, SM MS 333/2 and 19 November 1800, Davy, *Davy*, vol. 1, pp. 105-7 (see note 28).

62 Robert Southey to William Taylor, 26 November 1800.

63 Humphry Davy to Grace Davy, 31 January 1801, SM MS 333/4.

The Royal Institution of Great Britain

At a meeting held on 7 March 1799 in Joseph Banks's Soho Square house, fifty-eight men (around a quarter of whom were aristocrats), each pledged fifty guineas, a substantial sum, to become Proprietors of a new "Institution for Diffusing the Knowledge and Facilitating the General Introduction of Useful Mechanical Inventions and Improvements, and for Teaching, By Courses of Philosophical Lectures and Experiments, the Application of Science to the Common Purposes of Life."[64] They also elected a committee of Managers charged with establishing and then running the new institution.[65] In addition to Banks and Rumford, a third of this committee were aristocrats and in May George Finch, ninth Earl of Winchilsea (1752-1826) became President. The RIGB founders' original intention was to communicate scientific knowledge to its Proprietors, all of whom came from the wealthy echelons of society, as well as Life and Annual Subscribers who contributed ten and two guineas respectively. To deliver its program the RIGB needed a building and mid-1799 it moved into 21 Albemarle Street, off Piccadilly. This had been a gentleman's town-house, built during the eighteenth century, and thus needed conversion to house a scientific institution that included a well-equipped laboratory in which to prepare lecture demonstrations.[66] Rumford was charged with overseeing the necessary work and, just a year after its foundation, a temporary lecture room had been constructed. The first lecture in a course on various natural philosophical and chemical topics was delivered on 11 March 1800 by Thomas Garnett (1766-1802).[67] He had enjoyed a reasonably successful career as an itinerant lecturer during the latter part of the eighteenth century, but immediately before moving to the RIGB had been employed to lecture at the Andersonian Institution in Glasgow.

In April 1800 the Managers authorized the construction of a large lecture theater at the building's northern end, aiming for its completion by the start of

64 *Proposals for forming by subscription, in the Metropolis of the British Empire, a Public Institution for Diffusing the Knowledge and Facilitating the General Introduction of Useful Mechanical Inventions and Improvements, and for Teaching, By Courses of Philosophical Lectures and Experiments, the Application of Science to the Common Purposes of Life* ([London: np, 1799]), 43.

65 RI MM, 9 March 1799, 1: 1.

66 Frank A.J.L. James and Anthony Peers, "Constructing Space for Science at the Royal Institution of Great Britain," *Physics in Perspective* 9 (2007): 130-85.

67 *Gentleman's Magazine*, 70(1) (1800): 382. For the content see *Journals of the Royal Institution of Great Britain* 1 (5 April 1800): 15-16. Note this text was omitted from volume 1 of the bound edition of the *Journals*.

1801.[68] They contracted with the Pimlico builder Thomas Hancock to undertake the extensive work involved which meant that the building became unusable until the theater was completed.[69] This prompted the Managers (based on Banks's proposal) to dismiss five servants, including the lecture assistant, a decision that did not commend itself to Garnett who thought that the RIGB should have a permanent assistant.[70] The project overran and, at their meeting in early February 1801, the Managers decided that the second season of lectures would commence when the lecture theater was completed. Furthermore, they agreed that Banks, Rumford and Cavendish (now a Manager) would form a committee to supervise the preparation of the syllabi and that none should be published without their authorisation.[71] The reason for this latter decision was Garnett's imminent publication of the *Outlines* of his courses both on chemistry and on natural and experimental philosophy.[72] Indeed the prefaces (where he stated the times that he would be delivering the lectures) were both dated 2 February, though the chemistry volume did not appear until mid-March.[73] Garnett, behaving as one might expect of a former itinerant lecturer, published these without the Managers' permission, suggesting a lack of skill on their part about how to manage an experienced lecturer. Two weeks later they ordered that copies of the minutes recording their decisions made on 2 February should be sent to him.[74] He thus began his second season under something of a cloud so far as his employers were concerned. Furthermore, Garnett had written to Rumford requesting an increase in salary in accordance with what he believed were the terms of his original appointment.[75] The Managers deferred making a decision until the annual accounts

68 RI MM, 14 April 1800, 2: 56. Rumford to Marc-Auguste Pictet, 5 July 1800, David Bickerton and René Sigrist, eds., *Marc-Auguste Pictet 1752-1825 Correspondance sciences et techniques tome III: Les correspondants britanniques* (Geneva: Slatkine, 2000), 558-9.

69 RI MS RI/1/G/G/1.

70 RI MM, 12 June 1800, 2: 104. Rumford had proposed earlier in the day that Banks did this. Rumford to Joseph Banks, 12 June 1800, *Banks' Scientific Correspondence*, vol. 5, p. 1560 (see note 18). Thomas Garnett to Thomas Webster, 27 September 1800, Bence Jones, *Royal Institution*, pp. 172-4 (see note 3).

71 RI MM, 2 February 1801, 2: 126-7.

72 Thomas Garnett, *Outlines of a Course of Lectures on Chemistry* (London: Cadell and Davies, 1801) and *Outlines of a Course of Lectures on Natural and Experimental Philosophy: Delivered at the Royal Institution of Great Britain* (London: Cadell and Davies, 1801).

73 *Morning Chronicle* (18 March 1801): 2a.

74 RI MM, 16 February 1801, 2: 134.

75 Thomas Garnett to Rumford, 22 February 1801, copied in RI MM, 23 February 1801, 2: 136-8.

had been prepared and the RIGB's financial position ascertained.[76] When Garnett again pressed his case in mid-May, a specially convened Managers' meeting, held on the 25th refused the increase; Garnett's subsequent resignation offer was accepted.[77]

It is possible that as early as their meeting held on 5 January 1801, the deteriorating relations between Garnett and the RIGB may have prompted the Managers to ask Rumford to seek a replacement.[78] Rumford had spent much of the previous September and October in Edinburgh with Black's successor at the University, Thomas Hope (1766-1844). He had known Beddoes when they were students there and, having visited the MPI at some point during 1800, sung Davy's praises to Rumford.[79] Around 10 January, Rumford must have approached Davy, offering him by the end of the month, that should he move to the RIGB he would shortly become the "sole Professor of Chemistry."[80] Davy visited London during mid-February where he met Rumford, Banks and Cavendish.[81] These meetings resulted in the Managers appointing Davy "Assistant Lecturer in Chemistry, Director of the Chemical Laboratory, and Assistant Editor of the Journals of the Institution" on 16 February at an annual salary of 100 guineas plus accommodation.[82] This seems to represent a diminution in income, suggesting some keenness on Davy's part to leave Bristol. The same day Rumford wrote Davy's appointment letter in which he copied the Managers' minutes, adding, what was not noted there, their additional agreement that, provided he proved his fitness, he would be appointed Professor of Chemistry in the next two or three years with an annual salary of £300.[83] That promise would make up for any immediate financial loss and Davy returned to Bristol, probably for a couple of weeks, to settle his affairs. Beddoes, despite his

76 RI MM, 23 February 1801, 2: 138.

77 RI MM, 25 May 1801, 2: 180. RI MM, 15 June 1801, 2: 189-90.

78 Paris, *Davy*, vol. 1, p. 115 (see note 28), based on an account written thirty years later; there is no reference to such authorisation in RI MM, 5 January 1801, 2: 118-21.

79 Humphry Davy to Thomas Hope, 28 June 1801, transcript in RI MS HD/28/D/38. Humphry Davy to John Tokin, 12 January 1801, Davy, *Davy*, vol. 1, pp. 107-9 (see note 28). Thomas Traill, "Memoir of Dr Thomas Charles Hope, late Professor of Chemistry in the University of Edinburgh," *Transactions of the Royal Society of Edinburgh* 16 (1848): 419-34, 432.

80 Humphry Davy to Grace Davy, 31 January 1801, SM MS 333/4, noted that he had been in discussions with the RIGB for three weeks.

81 Humphry Davy to Davies Giddy, 8 March 1801, Paris, *Davy*, 1: pp. 116-9 (see note 28).

82 RI MM, 16 February 1801, 2: 134.

83 Rumford to Davy, 16 February 1801, Bence Jones, *Royal Institution*, pp. 317-9 (see note 3). Davy was appointed Professor the following year, RI MM, 31 May 1802, 3: 43.

view of Davy's key role in the MPI, approved his move to London "with great liberality."[84] On Wednesday 11 March 1801 Davy returned to the RIGB to commence the next stage of his career.[85]

At the RIGB Davy, despite no previous experience, very quickly established himself as an immensely attractive and engaging lecturer. Possibly his not having any previous experience as an itinerant lecturer meant that he could, unlike Garnett, conform more easily to the Managers' expectations. His initial two courses illustrated his immediate impact. The first starting on 25 April 1801 were evening lectures, while the second was an afternoon series "attended not only by men of science but by numbers of people of rank and fashion."[86] The pharmaceutical chemist William Allen (1770-1843) recorded in his diary the success of Davy's first lecture, describing it as "A most capital one. He bids fair to rise high in the philosophical world," a view shared by the writer in the *Philosophical Magazine*, who noted Banks's presence in the audience.[87] Davy's course on pneumatic chemistry concluded on 20 June with a lecture attended by nearly 500 people, which included a practical demonstration of nitrous oxide's physiological effects.[88] Davy was as nearly carried away with his own success as a lecturer, as anyone who had come under the influence of nitrous oxide: "I have been nobly treated by the managers, God bless us I am about 1.000.000 times as much a being of my own volition as at Bristol. My time is too much at my own disposal – So much for egotism – for weak glorious, pitiful, sublime, conceited egotism."[89] He would retain the RIGB's fashionable audience until his retirement in 1812 (at the age of thirty-three) on marriage to a widow – a wealthy heiress who had attended his 1811 lectures.[90]

During June 1801, the Managers, presumably led by Rumford, and possibly mindful that Davy was being left too much to his own volition, proposed, in line with the RIGB's utilitarian strand, that Davy should "examine the state of the arts and to begin with the process of tanning" during the autumn.[91] By the end of the month Davy had agreed to deliver a course on tanning during

84 Humphry Davy to Davies Giddy, 8 March 1801, Paris, *Davy*, vol. 1, pp. 116-9 (see note 28).

85 RI MM, 16 March 1801, 2: 150-1.

86 "Royal Institution of Great Britain," *Philosophical Magazine* 10 (1801): 86-7, 86.

87 *Life of William Allen, with selections from his Correspondence*, 3 vols. (London: Gilpin, 1846-7), vol. 1, 54 (entry for 25 April 1801). "Royal Institute [sic] of Great Britain," *Philosophical Magazine* 9 (1801): 281-2.

88 Humphry Davy to John King, 22 June 1801, Bristol Record Office, 32688/31. "Royal Institution of Great Britain," *Philosophical Magazine* 10 (1801): 86-7, 86.

89 Humphry Davy to John King, 22 June 1801, Bristol Record Office, 32688/31.

90 RI MM, 28 January 1811, 5: 178.

91 Thomas Poole to Josiah Wedgwood jr., 25 and 26 June 1801, WM MS MC 55.

November, but in exchange received three months leave, starting in July, "for the purpose of making himself more intimately acquainted with the practical part of the business of tanning." Furthermore, Davy was also "instructed" to prepare lectures for delivery in December on dying, staining and printing various cloths.[92] None of these lectures were ever delivered. This suggests that because of his success attracting audiences to the RIGB Davy could subvert the Managers' intentions without, unlike Garnett, suffering any penalty. Nevertheless, in his famous 1802 lecture *Discourse Introductory to a Course of Lectures on Chemistry* Davy went out of his way to emphasise the utilitarian value of chemistry.

At this time Davy's most significant contribution to utilitarian chemistry was his appointment, at Banks's instigation, as Professor of Chemistry to the Board of Agriculture. Until retirement he delivered annually a successful series of six lectures on agricultural chemistry to the Board.[93] Furthermore, despite the proposal in May 1805 that the Board should possess its own laboratory in its Sackville Street building, it was eventually agreed that the RIGB's laboratory would undertake this function. This had been facilitated earlier in the year by the Managers appointing Davy Director of the Laboratory and specifying its public remit.[94] This gave Davy the authority to undertake "analysis of such Substances as … the Professor of Chemistry shall deem of Scientific or Public Importance."[95] This new function took material form when Davy began a formal laboratory notebook in October. Following his death the Royal Institution had to apply to Davy's executors for the return of the first two of these notebooks which "had several years ago been taken away by Sir H Davy", illustrating his almost aristocratic sense of ownership of the laboratory.[96]

Davy did fulfill the RIGB's stated purposes to provide scientific lectures and practical scientific advice. But during his period there Davy *de facto* added scientific research to them, which had never been intended by the founders. It was the RIGB's well-equipped laboratory that allowed him also to continue the experimental work that he had started in Penzance and continued at the MPI. In particular he concentrated at the RIGB on electrical research, which formed the topic of his first paper read to the Royal Society of London in June 1801.[97] In the ensuing years he developed the first coherent theory of electro-chemical

92 RI MM, 29 June 1801, 2: 198.

93 Frank A.J.L. James, ""Agricultural Chymistry is at present in it's infancy": The Board of Agriculture, The Royal Institution and Humphry Davy," *Ambix* 62 (2015): 363-85.

94 RI MM, 28 January 1805, 4: 17. The Managers seemed to have forgotten that this was part of Davy's original job title.

95 RI MM, 11 March 1805, 4: 40.

96 RI MS HD/6 and 7. RI MM, 2 October 1829, 7: 276; 7 December 1829, 7: 285.

97 Humphry Davy, "*Galvanic* Combinations." (see note 49).

action, a term he coined.[98] During his research Davy isolated for the first time a number of chemical elements such as sodium and potassium (which he so named).[99] By turning the RIGB into a site of independent chemical research Davy fundamentally subverted the institution's founding intentions. The lecture program and providing scientific advice to the state continued, which financially supported the RIGB and thus Davy's research. And when that money became insufficient Davy also copied Beddoes's fund-raising practices in persuading the RIGB to organise a subscription to pay for a mineralogical collection and, towards the end of the decade, one to pay for a giant electric battery with which he could continue his electro-chemical researches.[100]

Conclusion

Beddoes asserted that the MPI was "perhaps, the first example, since the origin of civil society, of an extensive scheme of pure scientific medical investigation."[101] One interpretation of this view is that Beddoes believed that the MPI broke with earlier models of laboratory organisation and in terms of funding this was probably so. However, initially both the MPI and also the RIGB were organisations, which possessed underdefined or unspecified features, as one might expect at the commencement of such novel projects. This allowed a great deal of space for unintended consequences and individual human agency, especially, as in Davy's case, where he had a clear idea that he wanted to pursue independent research. Davy's work in Bristol essentially continued his chemical career begun in Penzance. There using locally available resources Davy, under the influence and guidance of Giddy, Wedgwood and Gregory Watt, had developed his ability and enthusiasm for chemical theorizing and practical experimentation. Such skills were just what Beddoes needed, signified by his immediately praising Davy's experimental prowess. With his already existing experience Davy took full advantage of the resources that Beddoes provided in Bristol. Davy, single-handedly, thus made the MPI a success in areas other than

98 Humphry Davy, "The Bakerian Lecture, on some new Phenomena of chemical Changes produced by Electricity, particularly the Decomposition of the fixed Alkalies, and the Exhibition of the new substances which constitute their bases; and on the general Nature of alkaline Bodies," *Philosophical Transactions* 98 (1808): 1-44, 2.

99 Ibid., p. 32.

100 June Fullmer, "Humphry Davy: Fundraiser," Frank A.J.L. James, ed., *The Development of the Laboratory: Essays on the place of experiment in industrial civilisation* (London: Macmillan, 1989), 11-21.

101 Thomas Beddoes, *Notice*, p. 4 (see note 47).

the intended one of determining the therapeutic value of gases. Left largely to his own devices, Davy made a significant medico-chemical discovery, published his first book and commenced electrical experimentation, which he clearly believed had further potential, especially in understanding life. But this was not what Beddoes had worked, or hoped, to establish, though he willingly sought to take advantage of Davy's discoveries. Beddoes seems not to have considered how laboratory research should be managed, but had let Davy behave in ways reminiscent of Priestley and Ingen-Housz at the aristocratic laboratory set up by the second Earl of Shelburne. That this lacuna can be attributed to Beddoes rather than Davy is apparent, since, following his departure for London, research at the MPI ceased and it soon turned instead into a more conventional hospital.[102] The MPI did not develop into a sustainable institution and could not fully survive Davy's departure.

Davy's commitment to his own research and his experience in Penzance and Bristol allowed him to add research to the RIGB's original activities, a move that would have far reaching consequences. That Davy exerted such a profound impact in developing research at both the MPI and the RIGB, suggests that his existing skills and experience were crucial in shaping both those institutions. The RIGB, much better funded (though it had more than its fair share of financial crises) than the MPI, had significant and continuing support from many wealthy, aristocratic and influential individuals. But, as with the MPI, its roles and management had not been fully defined, which again provided Davy with the institutional opportunity and space to maneuver into place his own ideas about what the RIGB should do.

In some ways his influence at both the MPI and the RIGB was similar to those aristocrats who had built their own laboratories which came to an end with their deaths (or suspension of interest), in that he constructed for himself similar freedom to pursue his own interests at both institutions. But the RIGB's institutional stability was strong enough to survive his departure and it continued to attract and foster talented men for the remainder of the nineteenth and into the twentieth century. It was in such a manner that Davy's subversion illustrates how a successful research laboratory could be created and run in a society still dominated by aristocrats.

102 Mary E. Fissell, *Patients, Power, and the Poor in Eighteenth Century Bristol* (Cambridge, Cambridge University Press, 1991), 118-19.

Acknowledgments

I am grateful to the participants of the "Situating Chemistry" workshops held in Leuven, Berlin, Florence and Haarlem for discussions on papers that I presented and particularly to Lissa Roberts for comments on this paper. I thank the following institutions for access to their manuscript collections: The Royal Institution (RI), The Library of Birmingham (LoB), the Cornwall Record Office (CRO), The Wedgwood Museum (WM), The Science Museum (SM), the British Library, the Royal Society of London, the National Library of Ireland, the Linnean Society, The Natural History Museum and Chatsworth. Letters written by Humphry Davy are freely available on-line at <www.davy-letters.org.uk> as part of the project to publish them by the end of the decade. In the meantime this paper cites their archival or printed locations. Letters from Robert Southey are available at <https://www.rc.umd.edu/editions/southey_letters> and are here cited by author, recipient and date.

CHAPTER 12

Wholesale Pharmaceutical Manufacturing in London, c.1760 – c.1840: Sites, Production and Networks

Anna Simmons

The relationship between chemistry and production was central to the development of wholesale pharmaceutical manufacturing in London in the late-eighteenth and early-nineteenth centuries. Drug manufacturing took place in a diverse marketplace united by a loose, but coherent, chemical-pharmaceutical culture. Its development benefitted from the close linkage of scientific and artisanal knowledge and practice, creating businesses from which the modern pharmaceutical industry originated. Building on the themes of this volume, this chapter will contribute to a broader history of industrialization that privileges chemistry as much as mechanics, and looks beyond innovation to provide a deeper examination of the history of productivity. Simply, pharmaceutical production can be added to the list of absentees that a focus on Newtonian mechanics has overlooked.[1] Meanwhile Mokyr's un-ashamed portrayal of an economic success story does not fit with the complex picture of development in pharmaceutical manufacturing that encompassed failure, secrecy, collaboration and competition.[2] As William J. Ashworth has commented, "the key to Britain's long term economic growth was an array of factors that lie outside the entrenched literature that has grown up around the defence of Western culture and political economy," with governmental, imperial and military factors particularly applicable to this study.[3]

Arguably even more "underappreciated" than the history of chemistry, the history of pharmaceutical manufacturing, centered as it is on the investigation and use of an extensive range of materials, plants and animals, provides a rich and relatively untouched source for studying production.[4] Much of the writing

1 Margaret Jacob, *The First Knowledge Economy: Human capital and the European economy, 1750-1850* (Cambridge: Cambridge University Press, 2014).

2 Joel Mokyr, *The Enlightened Economy* (London: Penguin, 2009).

3 William J. Ashworth, "The British Industrial Revolution and the Ideological Revolution: Science, neoliberalism and history," *History of Science* 52 (2014): 178-99, on 199.

4 Lissa Roberts, "Producing (in) Europe and Asia, 1750-1850," *Isis*, 106 (2015): 857-65, on 864.

 | DOI 10.1163/9789004325562_014

about drug manufacturers operating in the period c.1760 to c.1840 is found in company histories of long-established firms. Some of these studies can downplay the extent of production at this time.[5] Furthermore, when viewed from the perspective of large, multinational firms operating today, the origins of the industry can appear "humble."[6] Analysis of the industry's development as a whole, meanwhile, tends to emphasise its retail origins or to focus on the late-nineteenth century onwards.[7] However, during the late-eighteenth and early-nineteenth centuries, manufacturers' activities stretched beyond the pharmacy and the scale and scope of production was far from humble. Moreover, an important dynamic existed between wholesale drug manufacturing in the UK and the worldwide market for drugs. It is within this much wider international trade and productivity network that the origins of the modern pharmaceutical industry can also be located.

In the context of this volume, the neglect of the period c.1760 to c.1840 is particularly significant as it means that little has been done to integrate the early history of the pharmaceutical industry within a broader history of industrial development.[8] Moreover, a narrative of innovation shapes many of the existing individual studies of firms, notably how the introduction of new drugs into pharmaceutical and medical practice directed the history of a firm and its manufacturing activities.[9] As David Edgerton has highlighted in calling for a use-centered history of technology, one should not focus unreflectively on

---

5 Geoffrey Tweedale, *At the Sign of the Plough: Allen and Hanburys and the British pharmaceutical industry, 1715-1990* (London: John Murray, 1990), 56.

6 John P. Swann, "The Pharmaceutical Industries," Peter J. Bowler and John V. Pickstone, eds., *The Cambridge History of Science Volume 6: Modern Life and Earth Sciences* (Cambridge: Cambridge University Press, 2009), 126-40, on 127.

7 J. Burnby, "The Early Years of the Pharmaceutical Industry," Lesley Richmond, Julie Stevenson and Alison Turton, eds., *The Pharmaceutical Industry: A guide to historical records* (Aldershot: Ashgate, 2003), 1-13; Judy Slinn, "Research and Development in the UK Pharmaceutical Industry from the Nineteenth Century to the 1960s," Mikuláš Teich and Roy Porter, eds., *Drugs and Narcotics in History* (Cambridge: Cambridge University Press, 1995), 168-86. For pharmacy in general see Stuart Anderson, ed., *Making Medicines: A brief history of pharmacy and pharmaceuticals* (London: Pharmaceutical Press, 2005).

8 There is one major exception. Roy and Dorothy Porter emphasise druggists' role as manufacturers and distributors of medicines and suggest they could be "the authentic progenitors of the pharmaceutical industry," in that they are "integral to that surge of large-scale manufacturing and marketing which we call the Industrial Revolution." Roy Porter and Dorothy Porter, "The Rise of the English Drugs Industry: The role of Thomas Corbyn," *Medical History* 33 (1989): 277-95, on 282.

9 For example, A.F.P. Morson, *Operative Chymist,* Clio Medica no. 45 (Amsterdam: Rodophi, 1997).

innovation as the motor of historical development.[10] In conjunction with a stimulus from innovation, the demands of local and international markets and existing networks of supply all shaped production. Meanwhile for the actors involved, innovation was conceptualized in terms of adapting established technological processes, utilizing existing knowledge and improving technical efficiency.

Practitioners and Production

For much of the period under discussion, the boundaries surrounding manufacturing, wholesale and retail pharmacy were very fluid, whilst the substances produced often had utility beyond pharmacy.[11] Various terms were used to describe the practitioners who made drugs, for example chymist, apothecary, chemist, druggist, operative chemist, fine chemical manufacturer and so on. What these roles meant also evolved over time as professional boundaries shifted.[12] Many of these individuals were not just retailers, they were manufacturers and distributers of medicines, often engaged in overseas trade, and sometimes acting as government contractors. This chapter's primary concern is not with distinctions between the various actors' categories that are used or with changing professional and institutional regulation in this period.[13] Furthermore, the differences in the contexts in which these categories were used in Britain compared to other European countries also lie outside the scope of this study.[14] Instead this chapter's objective is to provide an insight into a world of production and commerce by focusing on the manufacture of medical drugs for sale in bulk; that is not medical drugs sold to the individual

10 David Edgerton, *Shock of the Old: Technology and global history since 1900* (London: Profile Books, 2006).

11 For the retailing perspective, see Louise Hill Curth, ed., *From Physick to Pharmacology: Five hundred years of British drug retailing* (Aldershot: Ashgate, 2006).

12 Colin A. Russell, Noel G. Coley and Gerrylynn K. Roberts, *Chemists by Profession. The origins and rise of the Royal Institute of Chemistry* (Milton Keynes: Open University Press, 1977), 14-54.

13 S.W.F. Holloway, *Royal Pharmaceutical Society of Great Britain, 1841-1991: A political and social history* (London: The Pharmaceutical Press, 1991).

14 Ursula Klein, "Blending Technical Innovation and Learned Natural Knowledge: The making of ethers," Ursula Klein and Emma Spary, eds., *Materials and Expertise in Early Modern Europe: Between market and laboratory* (Chicago: University of Chicago Press, 2010), 125-57, on 151-4.

consumer, but those supplied wholesale to hospitals, institutions, merchants and government departments, and also exported overseas.

Given the diverse range of practitioners who made drugs, the examples in this chapter include firms that described themselves as fine chemical manufacturers; individuals or frequently changing partnerships who used the term chemist or druggist or both; and also a livery company undertaking collective manufacture on behalf of its members. As a result of participating in the "Situating Chemistry, 1760-1840" international network for collaborative research, information on all of these sites of pharmaceutical manufacture and the individuals linked to them is gradually being added to the project's database, currently found at <http://situatingchemistry.org/>.[15] What is common to all these sites and individuals is that, through the medium of large-scale manufacturing, they participated in and considered themselves part of the pharmaceutical marketplace.[16] Large-scale manufacturing is obviously a relative term, relative not only to the standards of the time but also to the specific industry. Given that drugs could be prescribed to patients in quantities of grains and minims, pharmaceutical production was necessarily on a smaller scale than in other chemical industries, for example the manufacture of bleaching powder, as discussed in John Christie's chapter in this volume. Multiple operations on one site could also be supervized by a relatively small workforce, in contrast to the large numbers engaged in manufacturing for military purposes in dockyards and munitions works.[17] However, as subsequent examples of the apparatus used; the quantity of raw materials consumed; and the size of the market supplied show; for pharmacy this was production on the largest possible scale for the standards of the time.

---

15 The Business Archives Council's survey of records of the British Pharmaceutical Industry and Derek Oddy, John Perkins and John Stewart's assistance have been invaluable in this aspect of my research.

16 The concept of a generalized "medical marketplace" can be seen as outdated, with a preference for considering "markets for medical goods and services" instead. Mark Jenner and Patrick Wallis, "The Medical Marketplace," Mark Jenner and Patrick Wallis, eds., *Medicine and the Market in England and its Colonies, c.1450-c.1850* (London: Palgrave Macmillan, 2007), 1-17, on 16. However, given the complexities and fluid boundaries of the drug trade, for this chapter "pharmaceutical marketplace" is a useful way of bringing together different aspects of the "markets" for drugs.

17 Jan Lucassen, "Working at the Ichapur Gunpowder Factory in the 1790s," Parts I and II, *Indian Historical Review* 39 (2012): 19-56, 251-71. With thanks to Andreas Weber for this reference.

From Dockside to Drug Auction

Patrick Wallis' work on the massive growth in the use of commercial drugs in the early modern period clearly shows the importance of the drug trade at this time. Using Port Books listing ships' cargoes and, from 1696, annual ledgers of the Inspector–General of the Customs, he has presented new evidence on the scale, origins and content of English imports of medical drugs between 1567 and 1774. Wallis shows that the volume of imported medical drugs exploded in the seventeenth century and continued growing, but on a more gradual scale, in the eighteenth century. Many of these drugs were re-exported, as England's position as a leading international entrepôt developed. However, given the dosages in use in that period, Wallis demonstrates that common drugs such as senna and Jesuits' bark were available to the majority of the population in the eighteenth century.[18] This provides further evidence of the expansion of medical consumption at this time, with subsequent work by Wallis and Pirohakul underlining the growing centrality of therapeutics in patients' expectations of medical treatment.[19] However this poses the question: how did those involved in the pharmaceutical marketplace meet this substantial increase in demand for drugs? Whilst drug imports provide the first part of the answer, a focus on production provides the second part. The route from dockside to consumer was varied and could involve many sites, actors and networks. Plant-based drugs had to be processed in different ways to make them suitable for administration in various formats, whilst chemical medicines had to be either made from raw materials or refined to medicinal grade quality.

London provides the focus for this study as it was a major center for the international trade in crude drugs and the key location for large-scale drug production in the UK in the period c.1760 to c.1840.[20] By the late seventeenth century, over ninety-five percent of drug imports into the UK came through London, and the city possessed significant commercial advantages in terms of shipping, banking and insurance, as well as a reputation for reliability in terms

18 Patrick Wallis, "Exotic Drugs and English Medicine: England's drug trade, c.1550 – c.1800," *Social History of Medicine* 25 (2012): 20-46.

19 Patrick Wallis and Teerapa Pirohakul, "Medical Revolutions? The growth of medicine in England, 1660-1800," *Journal of Social History*, 49 (2016): 510-31, on 523.

20 It was not until the 1830s that T. & H. Smith, Duncan Flockhart & Co. and J.F. Macfarlan & Co. commenced alkaloid manufacture in Edinburgh. Morson, *Operative Chymist*, pp. 104-21 (see note 9).

of financing and the quality of goods.[21] As its sphere of influence expanded in the eighteenth century, Britain gained control over the sources of many drugs and other raw materials, such as camphor, niter, quicksilver and tincal, whilst trade restrictions forced these goods to be exported through London. These factors not only promoted the trade in raw drugs, but also the activities of the manufacturers who processed them. The broader history of the import and export of drugs into London lies outside the scope of this chapter, but this examination of sites, production and networks is part of a much wider history of globally situated interconnections, exchanges and translations in the drug trade.[22]

The exact route from ship to saleroom depended on who had imported the goods. Many of the commodities discussed in this chapter were imported by the East India Company. Until the East India Docks were opened in 1806, their ships were unloaded at Blackwall. Goods were then carried by lighters to the legal quays or sufferance wharves of the Pool of London.[23] From there, drugs such as aloes, cassia and nux vomica were sent to its Crutched Friars Warehouse. Here they were classified and sorted, with samples prepared for sale at East India House in Leadenhall Street.[24] Private trading networks, operating in tandem with monopoly companies, were particularly significant in the context of drug imports, as Timothy Davies has highlighted for the London drug merchants, Gammon and Chaloner.[25] Mincing Lane was known as the heart of London's drug trade in the nineteenth century, being subsequently described in

21 Anon., "London's Drug Market and the Romance of Mincing Lane," *Chemist and Druggist*, 30 June 1928, 850-67; Terry M. Parssinen, *Secret Passions, Secret Remedies: Narcotic drugs in British society, 1820-1930* (Manchester: Manchester University Press, 1983), 15-16; R.S. Roberts, "The Early History of the Import of Drugs into Britain," F.N.L. Poynter, ed., *The Evolution of Pharmacy in Britain* (London: Pitman, 1965), 165-85.

22 Harold J. Cook and Timothy D. Walker, "Circulation of Medicine in the Early Modern Atlantic World," *Social History of Medicine* 26 (2013): 337-51.

23 "The East India Docks: Historical development," Hermione Hobhouse, ed., *Survey of London: Volumes 43 and 44, Poplar, Blackwall and Isle of Dogs* (London: London County Council, 1994), 575-82, accessed 24 September 2015, <http://www.british-history.ac.uk/survey-london/vols43-4/pp575-582>. For changes to trading rights see John Keay, *The Honorable Company: A history of the English East India Company* (London: HarperCollins, 1991).

24 Anon., "A New Drug Showroom," *Chemist and Druggist*, 1 February 1913, 52-3; Anon., "London's Drug Market," p. 858 (see note 21).

25 Timothy Davies, "British Private Trade Networks and Metropolitan Connections in the Eighteenth Century," Maxine Berg, ed., *Goods from the East, 1600-1800: Trading Eurasia* (Basingstoke: Palgrave Macmillan, 2015), 154-67. For the broader context see Emily Erikson, *Between Monopoly and Free Trade: The English East India Company, 1600-1757* (Princeton: Princeton University Press, 2014).

the weekly trade periodical, *The Chemist and Druggist*, as "once the undoubted centre of exchange for the world's botanical drugs and essential oils."[26] Drug sales, meanwhile, were held at Garraway's coffee house in Exchange Alley, near Cornhill, until shortly before it was demolished in 1866.[27] Here brokers auctioned the lots in their catalog to an audience of wholesale druggists, export merchants and dealers. Although individual firms' buying arrangements differed, it was from this audience that significant quantities of raw materials were purchased and then processed by wholesale manufacturers.[28]

Expanding Markets

Prior to exploring the development of sites of bulk drug manufacturing in London, it is important to discuss what drove the early development of the industry in the UK. British pharmaceutical manufacturers had specific strengths in terms of their access to worldwide markets and the economic, imperial and social networks they belonged to. In this context, the industry's expansion was not only driven by increasing demand from customers for medical drugs but also facilitated by improved access to resources from Britain's empire. The growth in demand for medical drugs to supply the Army, Navy, and East India Company was particularly important, as this expenditure provided a significant stimulus for growth in production – a point that provides further evidence against explanations of British economic growth at this time with reference to 'the free market'.[29] The Garnier Family, which secured a patent in 1715 to hold the post of Apothecary General to the Army in perpetuity, was rumoured to have earned profits of £10,000 a year in the late-eighteenth century.[30] These profits derived from the lucrative terms of the post, an

26 Anon., "A Century of Commerce in Drugs," *Chemist and Druggist*, 10 November 1959, 160-6, on 160.

27 Anon., "London's Drug Market," pp. 862-3 (see note 21); Anon., "The Drug Sales," *Chemist and Druggist*, 21 August 1886, 230-2.

28 Corbyns bought direct from the auctions. The Society of Apothecaries posted a list of drugs required at Apothecaries' Hall and any merchant or druggist could offer samples to be viewed by the Society's buying committee. Howards purchased through the wholesale druggists David Taylor and Sons.

29 John Brewer, *The Sinews of Power: War, money, and the English state, 1688-1783* (Cambridge: Harvard University Press, 1988); Ashworth, "The British Industrial Revolution" (see note 3).

30 Arthur Edmund Garnier, *The Chronicles of the Garniers of Hampshire during Four Centuries, 1530-1900* (Norwich and London: Jarrold and Sons, 1900), 21.

appointment that paid ten shillings a day, to which was added a sum equal to ten percent of the value of medicines supplied. Between 1795 and 1806, the money spent by the Army on medicines amounted to over £800,000, including £70,000 on surgical instruments. The orders were placed via the Apothecary General with a range of "civil firms."[31]

More is known about drug supply to the Navy and the East India Company. The Society of Apothecaries, a city of London livery company with responsibilities for examining apprentices and regulating apothecaries' activities, primarily supplied these institutions.[32] The Society began manufacturing drugs at its premises at Apothecaries' Hall, Blackfriars, in 1672.[33] Its chemical laboratory was soon described as "the largest and the best," with supply to the Navy starting in 1703 and to the East India Company in 1766.[34] In the eighteenth century, the Society benefitted enormously from its unique position as a livery company, with a role as an arbiter of quality, situated between the trade and government spheres. This position helped to open up lucrative contracts of drug supply and such trading relationships were strengthened when, as William Ashworth has highlighted, "the events of the 1790s temporarily halted Britain's move to reform and, in fact, reinforced its existing institutions."[35]

During the Napoleonic Wars demand for the Society's drugs grew further, with an estimated 120,000 men engaged in the Royal Navy in 1801. Parallels can be drawn between the advantages for the Navy of purchasing drugs from the Society and the strengths of the contractor system of government supply that Roger Knight and Malcolm Wilcox have described in their study of the

31 Neil Cantlie, *A History of the Army Medical Department, vol. 1* (London and Edinburgh: Churchill Livingstone, 1974), 61, 187; Unfortunately I have not yet been able to consult the Garnier family papers to see if any of these "civil firms" are named.

32 E.A. Underwood, ed., Cecil Wall and H.C. Cameron, *A History of the Worshipful Society of Apothecaries of London, vol. 1, 1617-1815* (London: Oxford University Press, 1963), 8-22; Patrick Wallis, "Medicines for London: The trade, regulation and lifecycle of London apothecaries c.1610-1670" (D.Phil. Thesis, Oxford University, 2002), 23-50.

33 Anna Simmons, "Medicines, Monopolies and Mortars: The chemical laboratory and the pharmaceutical trade at the Society of Apothecaries in the eighteenth century," *Ambix* 53 (2006): 221-36.

34 W.H. Quarrell and Margaret Mare, eds., *London in 1710 from the travels of Zacharius Conrad von Uffenbach* (London: Faber and Faber, 1934), 111.

35 William J. Ashworth, "Quality and the Roots of Manufacturing 'Expertise' in Eighteenth-Century Britain," *Osiris* 25 (2010): 231-54, on 234. In the 1810s the Society supplied the army in Ireland and hoped to supply the main army as well. Cantlie, *Army Medical Department*, p. 449 (see note 31).

Victualling Board.[36] For the decade up to 1810, the Navy spent an average of £18,072 per annum with the Society for the supply of drugs, chemicals and galenical medicines, in addition to bottles, phials and mortars. For the same period, the East India Company spent an average of £20,160 per annum on medical supplies for its substantial army, plus ships, hospitals and trading posts, even though some medicines were sourced locally.[37] As the Society of Apothecaries held a monopoly of supply for all of the drugs purchased in Britain by the East India Company until its demise in 1858, some raw materials such as quicksilver were shipped from areas under the Company's control to London, processed at Apothecaries' Hall and then re-exported to South Asia.[38] Similar circular trading networks via London manufacturers existed across the Atlantic with drugs such as Barbados aloes.[39] Other destinations for the Society's medicines included hospitals in Ceylon, Malta and Mauritius, a convict establishment in Australia and the Hudson's Bay Company, in addition to numerous hospitals and institutions in London.[40] Not all of the drugs manufactured at Apothecaries' Hall were supplied direct. A great number of the Society's preparations were sent via merchants to the West Indies.[41] In addition, individual apothecaries built up extensive Transatlantic trading activities, supplying drugs, including some purchased from the Hall, to contacts in New England and the West Indies.[42]

Chemists and druggists also developed impressive overseas export markets, which drove the expansion of their businesses. William Jones traded as a drug-

---

36 Roger Knight and Malcolm Wilcox, *Sustaining the Fleet, 1793-1815: War, the British Navy and the contractor state* (Woodbridge: Boydell Press, 2010), 10-11, 29.

37 Apothecaries' Hall Archive (hereafter AHA), MS 8200/1-18, 1617-1926, Court of Assistants Minute Books (hereafter CM) CM 29 March 1811. For local sourcing see Pratik Chakrabarti, *Materials and Medicine: Trade, conquest and therapeutics in the eighteenth century* (Manchester: Manchester University Press, 2010), 33-44.

38 AHA, MS 8261, India Orders, 1827-8, state 3,011 lbs. of calomel, was sent to Bengal, Madras, Canton and Prince of Wales Island.

39 S. Stander, "Transatlantic Trade in Pharmaceuticals during the Industrial Revolution," *Bulletin of the History of Medicine* 43 (1969): 326-43, on 340-2.

40 AHA, Annotated *Pharmacopoeia Collegii Regalis Medicorum Londinensis* (London: Longman, 1809); United Stock Account Books, MS 8224, vol. 1 (1812-30), vol. 2 (1831-46); Penelope Hunting, *A History of the Society of Apothecaries* (London: Society of Apothecaries, 1998), 164-87.

41 J.F.A. Göttling, "Einige Bermerkungen über Chemie und Pharmazie in England," *Almamanach oder Taschenbuch für Scheidekünstler und Apotheker,* 1789, 128-44, on 129. With thanks to Ursula Klein for this reference.

42 I.K. Steele, *Atlantic Merchant Apothecary: Letters of Joseph Cruttenden, 1710-1717* (Toronto: University of Toronto Press, 1977).

gist in Bloomsbury in the mid-eighteenth century and had a UK-wide wholesale trade, whilst also exporting drugs to Nova Scotia, Gibraltar, and the West Indies.[43] Thomas Corbyn, who traded from 300 High Holborn, had, in addition to a substantial provincial market, a significant overseas trade, predominantly with Quakers in North America and also in the West Indies.[44] Ledgers and letter books dating from 1776 to 1780 demonstrate that another Quaker, Joseph Gurney Bevan of Plough Court (the business that ultimately became Allen and Hanburys), had forty regular customers across the Atlantic, mostly in Jamaica and Barbados, plus a secondary trade shipping bales of textiles to Europe.[45] Such activity was undoubtedly fostered by close ties between the Quaker merchant community on both sides of the Atlantic and its established networks of commerce and credit.[46] It also underlines how this story of production feeds into a broader history linking therapeutics with colonial expansion and international trade.[47]

Sites of Bulk Drug Manufacturing

It is now necessary to return to the question of how those involved in the pharmaceutical marketplace met the substantial increase in demand for medical drugs. This was achieved by firms expanding their premises (initially on site and later elsewhere) and by scaling up production. Thomas Corbyn, in addition to his premises in Holborn, had a separate laboratory, and owned a large warehouse at Cold Bath Fields. His warehouse stock book or inventory dated December 1761 included 2,500 items of materia medica, some of which were stored in very large quantities.[48] A surviving recipe book consisted of over 650 preparations and contained instructions for large-scale pharmaceutical production.[49] Samuel Towers commenced manufacturing at a laboratory in

43 G.M. Watson, "Some Eighteenth Century Trading Accounts," F.N.L. Poynter, ed., *The Evolution of Pharmacy in Britain* (London: Pitman, 1965), 45-78.

44 Porter and Porter, "The Rise of the English Drugs Industry," pp. 290-1 (see note 8).

45 Simon S. Stander, "A History of the Pharmaceutical Industry with Particular Reference to Allen and Hanbury, 1775-1843" (M.Sc. Econ. Thesis, London University, 1965), 55, 125.

46 Margaret Stiles, "The Quakers in Pharmacy," F.N.L. Poynter, ed., *The Evolution of Pharmacy in Britain* (London: Pitman, 1965), 113-30; Renate Wilson, "Trading in Drugs through Philadelphia in the Eighteenth Century: A transatlantic enterprise," *Social History of Medicine* 26 (2013): 352-63.

47 Chakrabarti, *Materials and Medicine*, pp. 19-51 (see note 37).

48 Porter and Porter, "The Rise of the English Drugs Industry," p. 288 (see note 8).

49 Wellcome Library, Corbyn and Co., Manufacturing Recipe Book, 1748-1847, MS 5446.

Oxford Street in the late 1600s. In the eighteenth century, this business moved to more extensive premises at Mount Pleasant, and sites were subsequently added in Cold Bath Fields and Maiden Lane, Battle Bridge. The latter location was used for manufacturing chemicals including ammonia (then known as hartshorn as it was obtained from distilling stags' horns and bones) and oxalic acid.[50]

However it was the Society of Apothecaries that had the greatest capacity to process and manufacture huge quantities of drugs. Their premises at Blackfriars housed the largest pharmaceutical manufacturing laboratories in London in the late-eighteenth and early-nineteenth centuries, with plans dating from 1771 and 1823 illustrating the extent of expansion during this period (see figures 12.1 and 12.2). The German apothecary Johann Göttling visited Apothecaries' Hall whilst in England in 1787 and 1788, shortly after a major extension to the trading premises had been completed. He praised the Society's manufacturing capabilities, describing two large laboratories, a still house and hand mill room, and highlighted how "all chemical preparations are prepared in large quantities."[51] His description of the apparatus for making calx of mercury is indicative of this. The brick furnace was six to seven feet long and four feet wide, with its upper part containing a sand bath, where twenty to twenty-five phials were buried. Each phial held about two pounds of water and was half-filled with quicksilver.[52] Göttling remarked that certain processes operated more efficiently when carried out in bulk. For example, he noted that the large-scale purification of ammonium carbonate was less arduous than when performed with smaller distillations.[53] A device for distilling stag horn in order to obtain the spirit (aqueous solution of ammonia) also impressed him. This used two large upturned pots, about three and half feet high, placed on top of each other to serve as a distillation receiver. The device overcame a number of the problems associated with the distillation and Göttling commented that he was surprized that a similar arrangement was not yet found in German laboratories.[54] His remarks suggest that the Hall's production method was not widely known in England either, as Göttling noted that Robert Dossie's *The Elaboratory*

50 Gustave L.M. Strauss, Charles W. Quin, John C. Brough, Thomas Archer, William B. Tegetmeier, and William J. Prowse, *England's Workshops* (London: Groombridge and Sons, 1864), 160.

51 J.F.A. Göttling, "Einige Bermerkungen," p. 129 (see note 41); See also Ursula Klein, "Apothecary-Chemists in Eighteenth Century Germany," Lawrence Principe, ed., *New Narratives in Eighteenth Century Chemistry* (Dordrecht: Springer, 2007), 97-137, on 115-16.

52 Göttling, "Einige Bermerkungen," pp. 131-2 (see note 41).

53 Göttling, "Einige Bermerkungen," pp. 136-7 (see note 41).

54 Göttling, "Einige Bermerkungen," pp. 132-6 (see note 41).

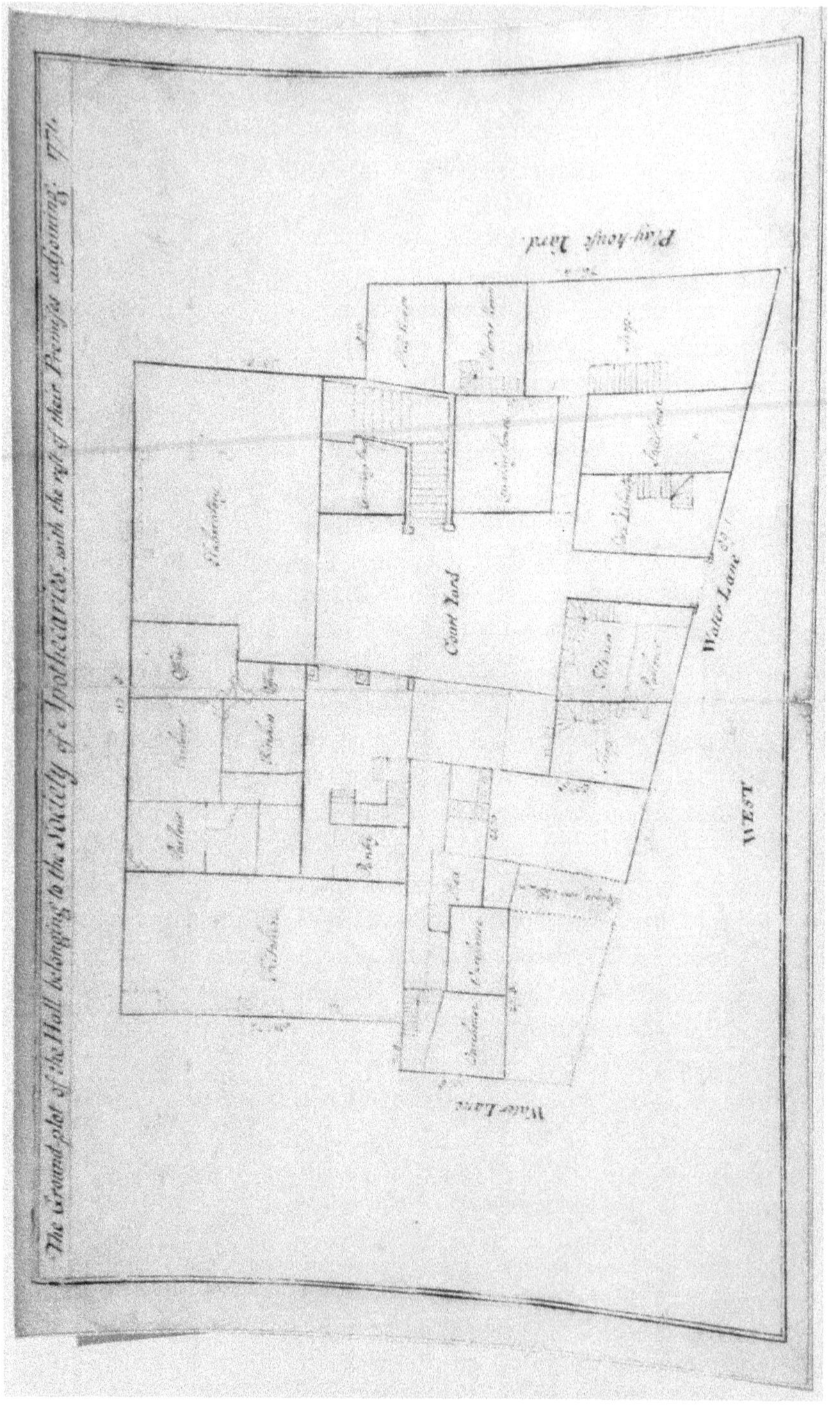

FIGURE 12.1 *Ground Plan of Apothecaries' Hall, including the Trading Premises, 1771.* IMAGE USED BY KIND PERMISSION OF THE WORSHIPFUL SOCIETY OF APOTHECARIES.

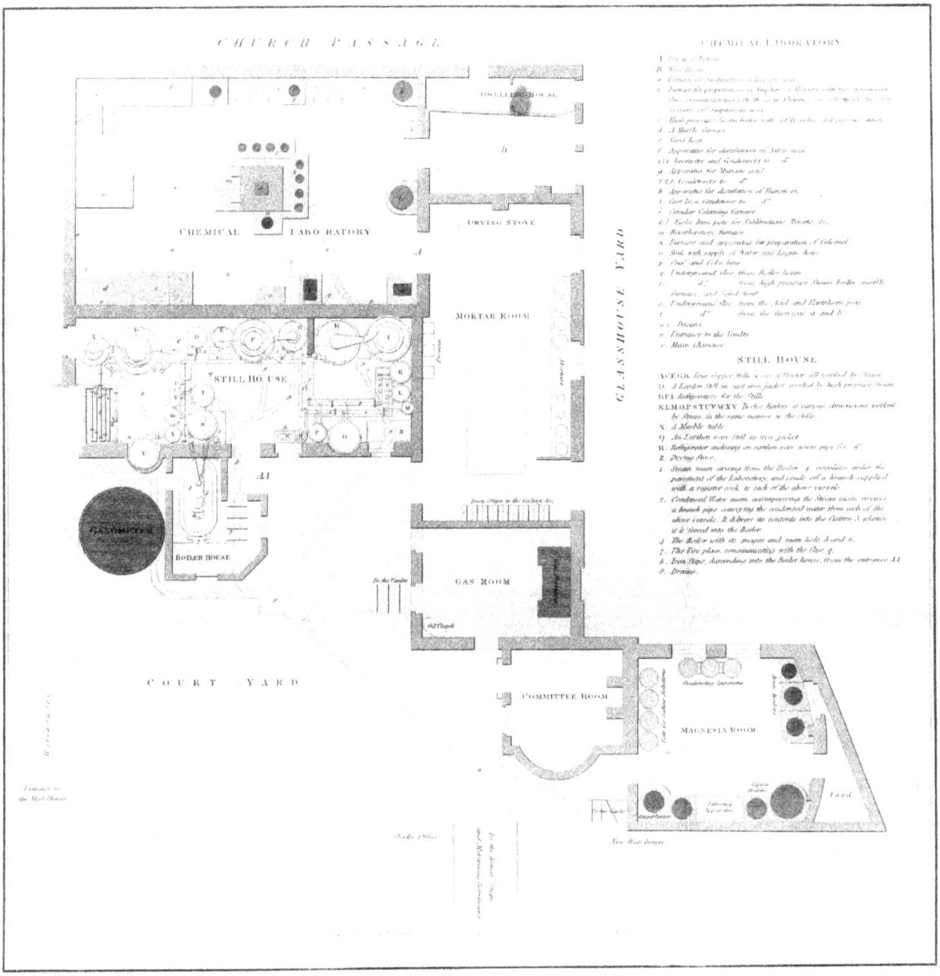

FIGURE 12.2 *Plan of Hall laboratories taken from* The Origin, Progress, and Present State of the Various Establishments for Conducting Chemical Processes, and Other Medicinal Preparations, at Apothecaries' Hall *(London: R. Gilbert, 1823).*

Laid Open also referred to the problems experienced when carrying out the distillation.[55] It seems likely that in contrast to the Enlightenment ideals of openness and freedom of knowledge, the Apothecaries' Hall laboratories at this point were a closed environment.[56] Göttling observed that it was "very difficult to gain entry here without a special recommendation."[57]

In the early nineteenth century, further development occurred to the manufacturing premises at Apothecaries' Hall, with the construction of a mill house, initially horse-powered, which enabled large quantities of drugs to be ground on site.[58] This was followed by a state of the art still house, which was significant for the novel steam technology it incorporated; a new laboratory with furnaces; new warehouses; and an eight horse-power steam engine, which powered machinery for grinding, sifting, triturating and pounding drugs.[59] As the stove for making calx of mercury illustrates, the laboratories contained existing chemical-pharmaceutical apparatus scaled up for bulk production and multiplied in number. When this was combined with mechanized drug mills and a larger workforce, consisting of a chemical operator, a galenical operator (until 1826), a foreman and around eight to ten laboratory workmen, the Society was able to manufacture and process huge quantities of drugs for a non-local market. The speed of production was such that in 1810 the Society claimed that "medicines for an Army of 30,000 men could be provided in the course of ten days in the case of an emergency."[60] The scale of production, meanwhile, is illustrated by an East India Company indent from 1827-8. The orders included 879 pounds and 5 ounces of mercury pills shipped to Bengal and Madras; 3,112 pounds of powdered cinchona *lancifolia* bark sent to Bengal, Bombay and Canton; and 36,962 pounds of magnesium sulphate dispatched to Bengal, Bombay, Madras, Canton and Prince of Wales Island.[61] Ursula Klein has highlighted how similar conditions elsewhere enabled "a continuous transition from small-scale pharmaceutical manufacture to large-scale pharma-

55 Robert Dossie, *The Elaboratory Laid Open: Or the secrets of modern chemistry and pharmacy revealed* (London: J. Nourse, 1756), 85-93.

56 Contrastingly, the Hall laboratories were publicized in an attempt to boost the Society's scientific status in the 1810s. Anna Simmons, "Stills, Status, Stocks and Science: The laboratories at Apothecaries' Hall in the nineteenth century," *Ambix* 61 (2014): 141-61.

57 Göttling, "Einige Bermerkungen," p. 129 (see note 41).

58 AHA, CM 23 October 1801, 16 September 1803.

59 Anon., *The Origin, Progress and Present State of the Various Establishments for Conducting Chemical Processes, and Other Medicinal Preparations, at Apothecaries Hall* (London: R. Gilbert, 1823).

60 AHA, CM 24 October 1810.

61 AHA, MS 8261, India Orders, 1827-8.

ceutical industry," with comparable development occurring at other sites of wholesale pharmaceutical manufacturing in London.[62]

The majority of apothecaries, chemists and druggists who began manufacturing drugs in the capital in the seventeenth and eighteenth centuries were located in the City of London, particularly around its boundaries, with clusters also situated around Oxford Street, Covent Garden and Holborn.[63] However, this changed in the period c.1760 to c.1840. As London expanded westwards and the City of London became a business rather than a residential area, the distribution of manufacturing sites began to shift.[64] Many businesses that had been founded in or near the City of London (with combined production and retail facilities on one site) constructed separate manufacturing premises outside of this area. For example, in 1795 the Quaker, Joseph Jewell, began manufacturing chemicals on a larger scale than was possible at the Plough Court pharmacy (where he was employed as an "elaboratory man"), at a new laboratory in Plaistow, Essex.[65] Luke Howard joined fellow Quaker, William Allen, in partnership at Plough Court in 1797 and together with Jewell concentrated on developing the Plaistow laboratory. A laboratory journal from the turn of the nineteenth century provides an insight into its daily operation, as well as highlighting the range and extent of production.[66] It indexes seventy-one principal products, including ammonia, borax, nitric and citric acids, camphor, ether, mercurials and potassium salts.[67] The entry for 7 February 1800 records that 3,403 ½ pounds of rough camphor were treated yielding 3,120 ½ pounds of the sublimed. Costings of labor, glassware and fuel, alongside yields and technical details are all noted, illustrating how a paper record of commercial factors and experimental observation was increasingly used to manage workers and the processes under their supervision.

62 Ursula Klein, "Apothecary's Shops, Laboratories and Chemical Manufacture in Eighteenth Century Germany," Lissa Roberts, Simon Schaffer and Peter Dear, eds., *The Mindful Hand: Inquiry and invention from the late renaissance to early industrialisation* (Chicago: University of Chicago Press, 2007), 247-76, on 275.

63 A detailed analysis of the location of sites for wholesale pharmaceutical manufacturing in London will form the focus of a separate article.

64 Michael Ball and David Sunderland, *An Economic History of London, 1800-1914* (London: Routledge, 2001) 121, 171-3, 182, 361-2.

65 A.W. Slater, ed., "Autobiographical Memoir of Joseph Jewell, 1763-1846," *Camden Miscellany* 22 (1964): 113-78, on 115.

66 London Metropolitan Archives, Records of Howards and Sons, Laboratory Journal, ACC 1037 291/1.

67 Anon., "Howards of Stratford and Ilford," *Chemist and Druggist*, 25 April 1914, 115-23, on 116.

Allen and Howard's partnership was amicably dissolved in 1807 and around this time the laboratory was relocated to larger and more accessible premises at Stratford.[68] Here, with Jewell as a junior partner, Howard specialized in fine chemicals. At Stratford the large-scale refining of crude Tibetan tincal, niter and camphor imported by the East India Company expanded further, with, for example, five tons of saltpeter purchased on 7 September 1819.[69] By 1821 Howards employed over thirty workmen, increasing to forty-three by the 1830s. Despite fluctuating economic conditions, with a boom during the Napoleonic wars, followed by a post-war slump, sales grew to a peak of £44,916 in 1825. A severe economic downturn followed and Howard and Jewell retired at the end of 1830, but sales only once dipped slightly below £30,000 in the years 1826-37.[70] Allen and his successors, meanwhile, continued manufacturing at Plough Court, with a remarkably consistent turnover of around £15,000 per annum for much of the period 1816-40.[71] They undertook cod liver oil production on site from the 1840s, before later establishing processing plants in Norway, with refining carried out at Plough Court.[72] It was not until 1878 that a new factory was opened at Bethnal Green.

For many businesses founded in the early nineteenth century, a shift from shop-based to factory-based manufacture tended to happen more rapidly, with the introduction of new product ranges often driving expansion. Having gained experience in Paris, Thomas Morson started his business in Fleet Market in 1821, and was the first to manufacture quinine sulphate and morphine salts on a commercial scale in England. His price list from 1821 featured seventeen "new chemical preparations employed as medicine" and included morphine, strychnine, emetine and quinine sulphate.[73] Demand was such that he moved to bigger premises in Southampton Row in 1826, where a 300-square-foot laboratory was built at the rear of his shop.[74] Morson erected works in the Hornsey Road shortly afterwards. There he began manufacturing creosote, which had

---

68 A.W. Slater, "Howards, Chemical Manufacturers, 1797-1837: A study in business history" (M.Sc. Econ. Thesis, London University, 1956), 179, states that the Stratford laboratory opened in 1807. However most other sources give 1805.

69 Ibid., pp. 56-7 (see note 68).

70 Slater, "Autobiographical Memoir," pp. 121-2 (see note 65); Slater, "Howards," pp. 320-1 (see note 68).

71 Stander, "A History of the Pharmaceutical Industry," pp. 164-5 (see note 45).

72 A plant founded in Newfoundland in 1860 was short-lived. Tweedale, *At the Sign of the Plough*, p. 75, p. 78 (see note 5).

73 Wellcome Library, Records of Thomas N.R. Morson Ltd., Product and Price Lists, 1821-1900, SA/MORH 1.

74 Morson, *Operative Chymist*, p. 44 (see note 9).

again been recently discovered. In the longer-term, Morson failed to exploit his early entry into the quinine market. Production continued until the middle of the nineteenth century, but in 1866 the German firm, Böhringer, supplied Morson with significant quantities of quinine, suggesting that manufacture had ceased.[75] However, this did not affect the firm's growth in other areas. By the 1860s, Morson was producing over 500 different chemical substances, made in all grades of purity. There were also more than 250 extracts, essences and tinctures, in addition to proprietary preparations and gelatine.[76]

The acquisition of new sites did not only signal an expansion of laboratory premises or the manufacture of new products. Larger premises might also be needed for preparing and packaging orders; bigger warehouses were required for storage; or new partnership agreements meant different properties were leased or owned. George Maw started in the London pharmaceutical trade by entering into a partnership in 1807 with his cousin, William Hornby, who was already established as wholesale druggist at 20 Fenchurch Street. Maw left this partnership in 1814 to purchase a surgical plaster factory in Whitecross Street, near Shoreditch. This factory later expanded to produce druggists' sundries, toiletries and pharmaceutical products. Maw then acquired larger premises in Aldermanbury in 1820 and at Aldersgate Street in 1834, as various relatives joined the firm and its range of activities diversified.[77]

As the industry grew, a new sort of manufacturer emerged, of which Maw was an example. Although the need to refine chemicals to medicinal grade quality was not new in the pharmaceutical trade, manufacturers had commonly carried out these steps themselves to guarantee purity.[78] However, increasingly firms specialized in fine chemicals or manufactured semi-prepared products to supply the pharmaceutical trade. In 1833, Stafford Allen, a miller at Amersham and the nephew of William Allen of Allen, Hanburys and Barry (as the firm was then styled), went into partnership with Charles May, a druggist and herb grower from Ampthill, Bedfordshire. They opened drug mills in Cowper Street, City Road, London and the site at Ampthill was used to provide the London business with raw materials. The firm processed these materials into semi-manufactured products, such as powders, distilled oils,

75 Morson, *Operative Chymist*, p. 79 (see note 9).

76 Morson, *Operative Chymist*, p. 225 (see note 9).

77 Anon., "More Historic London Wholesale Houses," *Chemist and Druggist*, 31 January 1914, 155-7, on 156; Anon., "London Pharmaceutical Industry," *Chemist and Druggist*, 24 June 1933, 667-95, on 677.

78 For the Society of Apothecaries' purification of magnesium sulphate see AHA, E/7 Loose Papers, Box 3. Evidence for similar practices exists at William Jones' Bloomsbury business.

extracts and emulsions, which were supplied to major London manufacturers and wholesalers.[79] Similarly, a cost price book from Allen, Hanburys and Barry illustrates how supplies were purchased from various manufacturers and drug merchants, with substances used in the firm's own pharmaceutical production or sold on directly to customers.[80]

Networks of Supply

As increasing specialization developed within wholesale pharmaceutical manufacturing, firms adapted their business methods to utilize various networks of supply based on established familial, social, religious and economic connections. In this context, the price and purchasing arrangements between manufacturers that regulated the availability and cost of bulk chemicals became more significant. Howards had private arrangements with firms nationwide, which helped the newly formed firm survive price-cutting in the 1810s. For example, Thomas and William Henry of Manchester bought large quantities of tartaric acid from Howards, whilst Howards purchased most of its magnesium sulphate from the Henrys. It is not surprizing that in some cases, price and purchasing arrangements led ultimately to 'mergers' or 'acquisitions.' Luke Howard had come to a price agreement with John Towers of Cold Bath Fields, regarding potassium salts in 1808, but Towers subsequently sold out to Howards in 1816.[81] Quaker ties were also important in this respect. In the 1830s Howards began to produce iodine compounds on a large scale and purchased iodine (made from kelp) from their "'respected Friend' Patrick Miller, in the North." In 1832 Howards agreed to take 2,500 ounces of iodine every three months on condition that Miller "agreed not to sell to any house at a lower price than that charged to the Messrs Howards, or to send any 'hydriodate of potash' to London while the contract ran."[82]

Changes to duties and existing mercantilist laws also had a major impact on what was manufactured by a firm or purchased from other suppliers.[83]

79 Anon., "Centenary of Stafford Allen and Sons," *Chemist and Druggist*, 1 July 1933, 22-3; Entry for Stafford Allen & Sons Ltd in Richmond et al., eds., *The Pharmaceutical Industry*, p. 83 (see note 7).

80 Royal Pharmaceutical Society, Allen, Hanburys and Barry Cost Price Book, 1822-44, IRA 1997.008.

81 J. Burnby, "The Early Years," pp. 8-9 (see note 7).

82 Anon., "Iodine," *Chemist and Druggist*, 19 June 1897, 974.

83 Most drugs and chemicals were dutiable in the UK until 1845 (Anon., "The Good Old Times," *The Chemist and Druggist*, 19 June 1897, 967), with quinine sulphate bearing duty

A broader examination of the trade in cinchona bark and quinine sulphate production in Europe lies outside the scope of this study.[84] However quinine sulphate provides an interesting example of firms' different responses to the introduction of a new drug into the marketplace. It illustrates how innovation, in terms of the adoption of new chemical knowledge and practice, was not exclusively the motor of pharmaceutical development. Pierre-Joseph Pelletier and Joseph Caventou had first isolated quinine from cinchona in Paris in 1820.[85] John Warrick, an importer of foreign drugs and chemicals based in Blackfriars, reputedly sold the first ounce of quinine in England.[86] The purchaser was Thomas Morson, who then manufactured and sold quinine sulphate from Fleet Market in 1821. However it was Howards, not Morson, who went on to dominate quinine production in Britain. Howards first manufactured quinine sulphate on a commercial scale in 1823,[87] but faced strong competition from French producers, such as Pelletier and Levaillant, selling through London drug merchants. Luke Howard and Joseph Jewell evolved an extraction process which produced quinine sulphate of reasonable purity from cinchona bark at their Stratford factory. However, Howards' production capabilities were hampered by the relatively high import duties charged on crude drugs compared to those incurred on imports of the finished product.[88] In such circumstances, it is not surprizing that other manufacturers decided against developing production themselves. Some production was undertaken at Apothecaries' Hall in the early 1820s, but for reasons of quality, price, and the nominal import duty of one penny per ounce on quinine sulphate, by 1834 the Society of Apothecaries preferred to import it directly from France.[89] Howards lobbied HM Privy Council for Affairs of Trade and Commerce regarding the discrepancy in duty

up to 1870. For the complications arising from the Navigation Acts for Howards' importation of camphor see Anon., "Quinine and Camphor," *Chemist and Druggist*, 19 June 1897, 974.

84 I am very grateful to Laurence Brockliss, John Cardwell and Michael Moss and for sharing their research on this trade, which will be published in due course as part of a project on the health of the Navy.

85 Marcel Delépine, "Joseph Pelletier and Joseph Caventou," *Journal of Chemical Education* 28 (1951): 454-61.

86 Anon., "London Wholesalers in 1863 and Now," *Chemist and Druggist*, 26 July 1913, 143-9, on 148.

87 Anon., *Howards, 1797-1947* (Ilford: Howards & Sons, 1947), 7.

88 Similar problems existed with opium duty. Anon., "Cinchona and Opium Duties," *The Chemist and Druggist*, 19 June 1897, 975.

89 *Report of the Select Committee on Medical Education, Society of Apothecaries, part III* (602), P.P. 1834, XIII, 64-5.

levels in 1836 and until 1860 British quinine manufacturers were protected by an import duty of six pence per ounce.[90] This mirrors aspects of the complex history of British industrialization abetted by government protection and regulation that William J. Ashworth has described for the eighteenth century.[91] It also underlines the importance of strategic political action to support the development of chemical production, as John Christie has highlighted in his chapter in relation to Charles Tennant's efforts to abolish salt duty. Howards' production of quinine increased steadily: in 1836 it was 6,000 ounces, rising to 15,000 ounces in 1838 and never falling below 100,000 ounces a year after 1847.[92] By the 1860s over 200 workmen were employed at Howards' Stratford factory and more than a ton of bark was processed each day – the transition to large-scale pharmaceutical industry had taken place.

Borax refining had been a key part of production for Howards since the Plaistow laboratory opened at the turn of the century. Both Tibetan tincal imported via the East India Company, and borate of lime from Peruvian coastal saline deposits had been used as raw materials.[93] In the late 1820s a new source from Italy, exploited by Count Lardarel, came onto the market and almost wiped out the Tibetan tincal trade in Europe.[94] John Eliot Howard obtained a sample of this Tuscan boracic acid in 1830 from the wholesale druggist David Taylor and Sons of Mincing Lane. Analysis showed the sample was of high quality and free from muriatic acid.[95] The firm subsequently inquired how it would be imported. Taylors' reply underlines the value, scale and complexity of the networks of supply involved in the international borax trade:

> We can now tell you [write Grant & Co. on March 14, 1836] how the operation stands. Larderell is bound to deliver to Hepburn, Pullars & Co. 21,000,000 lbs. of boracic acid at a price somewhat above Liv. 41 (per ton) in seven lots from 1st January 1837 to the end of June, 1839. Six months before the end of that period H., P. & Co. can denounce the contract and pay down Liv. 200,000 as *Caparra*, [deposit] which is to be discounted

90 Anon., "Quinine and Camphor," p. 974 (see note 83); Strauss et al., *England's Workshops*, p. 146 (see note 50).

91 William J. Ashworth, "The Intersection of Industry and the State in Eighteenth Century Britain," in Roberts et al., eds., *The Mindful Hand*, 348-77 (see note 62).

92 Redbridge Information and Heritage Service, Archives of Howards and Sons, B.F. Howard, *Howards 1847-1947: A Treatise*, 1956, 3.

93 Strauss et al., *England's Workshops*, p. 146 (see note 50).

94 N.J. Travis and E.J. Cocks, *The Tincal Trail: A history of borax* (London: Harrap, 1984), 24-6. With thanks to Andreas Weber for this reference.

95 Anon., "An Analysis by John Eliot Howard," *Chemist and Druggist*, 19 June 1897, 975.

> gradually at each delivery as usual when advances are made. Besides these advances H., P. & Co. lend Larderell Liv. 400,000 for 10 years at 5 p.c. Larderell issues notes of Liv. 5,000, payable 10 years hence and paying 5 p.c. interest. H., P. & Co. expect to negotiate these notes and be freed actual disbursements. The first three millions of acid they have resold to W. Lloyd (the refiner) at about 50 Liv.[96]

Such arrangements ensured that prices remained high for those in possession of the raw material after the original supply began to be exploited. They also underline the relationship between the worldwide commodity market and bulk manufacturing in the UK at this time. In the 1860s, Howards were amongst the largest consumers of Tuscan boracic acid worldwide.[97] However while Howards' consumption was at a globally significant level, the firm did not deal directly with overseas customers. By the late 1820s, Allen, Hanburys and Barry also had few transatlantic contacts.[98] Difficulties with shipping and obtaining payment for goods meant that agents played an increasingly important role in the networks of supply. When Howards received an enquiry from Mr H.J. Esszingh in Cologne for refined borax in 1841, they quoted a price of seventy-two shillings per hundredweight and requested that he place his order via an agent in London.[99]

Conclusion

The economic drivers of growth in the London pharmaceutical marketplace in the late-eighteenth and early-nineteenth centuries stemmed from the industry's location at the center of an international network of drug supply and processing, with productivity at its heart.[100] As Britain's empire increased in size, the balance of the push and pull relationship between production and supply became central: sources of raw materials for production expanded, but the markets for processed drugs and chemicals also grew. Rather than inventing new technologies, it was the scaling up of existing chemical-pharmaceutical

96 Anon., "Borax Reminiscences," *Chemist and Druggist*, 19 June 1897, 975.

97 Strauss et al., *England's Workshops*, p. 146 (see note 50).

98 Stander, "Transatlantic Trade," p. 333 (see note 39).

99 Anon., "Borax Reminiscences," p. 975 (see note 96).

100 On London as an industrial center, see Thomas Misa, *From Leonardo to the Internet: Technology and culture from the renaissance to the present* (Baltimore: Johns Hopkins University Press, 2004), pp. 59-73.

apparatus and expansion in terms of workforce, site and product range that were significant factors in the development of bulk drug production in London, as a transition from shop-based to factory-based manufacture occurred. Instead of the Enlightenment ideals of openness and the free dissemination of knowledge, tension between the relative values of publicity and secrecy persisted and pricing agreements between rival manufacturers were commonplace. Long-standing networks of supply based on colonial, economic, social, familial, and religious connections provided a strong framework for industrial development and drove the expansion of the industry. Such continuity provided a context for gradual change and allowed incremental innovations in practices, techniques and processes to occur.

Wholesale pharmaceutical manufacturing operated in an environment characterized not by clichés of a British 'free' market, but instead characterized by a market 'organized' around interactions between sites through networked exchanges and circulation. In this market, both cooperation and competition between producers were significant; British governmental policies, contracts and expenditure provided a major stimulus for growth; and the ability to utilize the resources of empire whilst also responding to its demands was paramount. Although this story has a London base, its reach was global as wholesale drug manufacturing functioned in an intricate productivity network of empire and international trade.

Acknowledgments

I am very grateful to the editors of this volume and all the members of the Situating Chemistry Research Network for the feedback and comments they have provided on this article. I am particularly indebted to Joe Cain and the Science and Technology Studies Department at University College London for continuing to support me as an Honorary Research Associate. The Society of Apothecaries has been most generous in giving permission to use and cite its archives and Janet Payne, from the Society's Friends of the Archives, has been extremely helpful with my research. The London Metropolitan Archives, Redbridge Information and Heritage Service, the Royal Pharmaceutical Society and the Wellcome Library have also kindly allowed access to their collections.

CHAPTER 13

Chemical Glasgow and its Entrepreneurs, 1760-1860

John R.R. Christie

The principal focus of this essay, the town of Glasgow and the chemical works of St. Rollox, is local, but has a general resonance, for St. Rollox may be regarded as a paradigmatic case of industrialized chemical production within the encompassing orbit of the industrial revolution. Here, inarguably it seems, are found the kinds of research-based, knowledge-induced technical innovation, entrepreneurship, growth rates, scale transformations, employment and wage patterns, which allow assimilation to the historiographical normativity of industrialization, at least in British terms. The most recent and conceptually sophisticated treatment of St. Rollox informed by the history of chemistry is found in Hasok Chang and Catherine Jackson's edited volume on *The Life of Chlorine*.[1] The authors critique the technologically determinist, linear approaches of older writers on the history of chemical technique, which explained the history of chemical and industrial development in chlorine bleaching through a narrative structured by progressive innovation moving from science to technological advance to industrial production. In its stead they recommend an approach emphasizing the complex, contingent and feed-back looping elements characteristic of interpretation based upon 'social shaping of technology'.

The approach adopted in this essay has some kinship with this advocacy, but has additional characteristics. Its most acute focus is upon 'situation', the town of Glasgow, and the industrial site of the St. Rollox chemical works within it.[2] It equally emphasizes contemporary Glaswegian sites of chemical produc-

1 Manichi Chung, Saber Farooqi, Jacob Soper and Olympia Brown, "Obstacles in the Establishment of Chlorine Bleaching," Hasok Chang and Catherine Jackson, eds., *An Element of Controversy: The life of chlorine in science, medicine, technology and war* (London: British Society for the History of Science, 2007), 153-178, for chlorine bleaching in this period, and for St. Rollox in particular 168-74; First published nearly fifty years ago, Alfred Musson and Eric Robinson, *Science and Technology in the Industrial Revolution* (Manchester: Manchester University Press, 1969), 231-371, remains the most informatively detailed treatment of British chemical technology and manufacture during the period.

2 There exist few archival sources for St. Rollox and no substantial history of the company. The Mitchell Library Glasgow holds site-development plans 1830-1900, and legal depositions. I have located an early letter-book, useful primarily for business organization, and have additionally

DOI 10.1163/9789004325562_015

tion other than St. Rollox, and the infrastructural development of educational and collective commercial institutions within the town. It further focuses upon the chemical-entrepreneurial figures, Charles Tennant and Charles Macintosh, at the center of St Rollox chemistry and industrial expansion. As young men, they worked in a pre-industrial manufacturing setting. Their careers therefore allow us to track practical chemistry's transition from pre-industrial to thoroughly industrialized settings, a period coinciding with the rapid sequence of development in chemical science during this period. Because some narratives of industrial revolution emphasize qualitative, ultimately discontinuous change, these two careers thus also provide an opportunity for critical attention to that historiographical reflex, with respect both to practical chemistry, and to industrialization. Further, as 'entrepreneurs of the Industrial Revolution', Tennant and Macintosh, in their very different ways, repay analysis of the varied forms of chemical-entrepreneurial activity they exhibited. This affords critical reflection upon the vocabulary of technical and economic 'innovation', and more particularly, upon models of entrepreneurship which currently preoccupy economic historians. Are chemical entrepreneurs easily absorbed by such models as further exemplars, or might they induce attention to a more nuanced understanding of entrepreneurship?[3]

A Chemical Behemoth

The St. Rollox Chemical Works, owned and operated by Charles Tennant and Co. in Glasgow, was often described as being, in its mid-Victorian heyday, the largest in Europe, if not the world. Across numerous measures (physical extension, employment, product diversification and output, fuel consumption, cost reduction) the company exhibited exceptionally impressive growth. Growth and size are however often matters of relative judgment, and parameter-dependent. The economic historian may wish to fix upon financial elements, such as annual financial turnover, rates of profitability and the like, as opposed to physical extension of site and other physical detail. There is a point to such

used contemporary accounts, family histories and biographies, and informal local and parish histories.

3 For recent work on entrepreneurship in this period, see François Crouzet, *The First Industrialists: The problem of origins* (Cambridge: Cambridge University Press, 1985); Joel Mokyr, "Entrepreneurship in the Industrial Revolution," David Landes, Joel Mokyr and William Baumol eds., *The Invention of Enterprise* (Princeton, NJ: Princeton University Press, 2012), 183-210.

selectivity. Site-wise a large bleachfield might compare in acreage with early St. Rollox, indeed there was one such in Glasgow, but its financial parameters would have nonetheless stood in stark contrast. One might alternatively think of employment size, for a large site does not of necessity employ large numbers (think again of the bleachfield). There is also interesting international comparison to be made on the question of size. The Gunpowder Manufactory run by the British military in Ichapur, India, employed more than two thousand workers, at least triple the number of St. Rollox workers in the 1820s, and at a considerably earlier date, and this example serves also remind us that, during the eighteenth century, and into the nineteenth, the employer with the highest number of workers engaged in manufacturing, on the largest sites, was the government, in particular the Army and Navy offices, with their extensive dockyards and munitions works.[4]

St. Rollox originated as a chlorine bleaching manufacture in 1797, sited in a semi-rural location, north of Glasgow, close to the newly completed Monkland Canal. Tennant used a Berthollet chlorine bleaching liquor, then modified with the addition of lime (as opposed to potash). The move to the Tennant-Macintosh patented chlorine bleaching powder in 1799 was the key element in St. Rollox' early expansion, the powder remaining an ongoing profitable staple of production throughout our period. The site diversified internally, producing sulfuric acid and soap, and after the abolition of salt importation duty in the early 1820s, Leblanc process soda, both ash and a lesser amount of crystal. Its physical plant grew, and was improved, with expensive installation of platina vats (instead of lead) for concentrated vitriol, more furnaces, chimneys, warehouse storage, canal basin, and railway terminal in 1831. By the 1850s the company had its own on-site cooperage, a foundry for equipment-making, and interests in local coal mines. The central chemical works alone had come to occupy thirteen acres. Tennant also began purchasing sea-going schooners, the basis by 1848 of a large mercantile fleet on the river Clyde, with steam vessels for the London and Baltic trades, and coastal sailing vessels for limestone importation (Ireland) and sulfur (Italy).[5] St. Rollox site-expansion was spectacular, but, through Tennant's relentless application, it became in local, national and international terms also geographically tentacular, by rail, and by water – thus Leviathan accompanied Behemoth.

---

4 Jan Lucassen, "Working at the Ichapur Gunpowder Factory in the 1790s," *Indian Historical Review* 39 (2012): 19-56 (part 1), 251-71 (part 2); With thanks to Andreas Weber, who brought this work to my attention.

5 James Dawson Burn, *Commercial Enterprise and Social Progress: Or, gleanings in London, Sheffield, Glasgow and Dublin* (London: Piper, Stephenson & Spence, 1858), 118.

Contemporary perceptions of St. Rollox recorded its massive productive presence in the north of Glasgow. J.D. Burn was struck by its 'stupendous' size overall, the enormous bulk of the vitriol vats and the chlorination chambers, the mountainous tonnage of heaped chemical raw materials, the hundreds of furnaces, the heat, and the staggering amounts of materials consumed and produced. St. Rollox's impact upon visitors, dutifully noted in quantitatively numerical terms, was equally registered in physical, material terms other than simple tonnages consumed and produced. To Burn's awed perception, and in his laudatory account of a progressive manufacture, there was nonetheless consciousness of an infernal working environment, an 'extraordinary labyrinth', a 'devil's den', dominated by the 'monstrous' dimensions of an immense chimney, 'Tennant's Stalk'.[6] Its material penetration of the surrounding environment became as noticeable. George Dodd's description a decade earlier was comparable with Burn's, noting St. Rollox gigantism, monstrosity, and the bewildered impressions of the novice observer, but unlike Burn, it also emphasized the deleterious working environment, both hazardous and unhealthy.[7]

This progressive narrative had already received cultural registration in educated, popular bourgeois and upper-class reading, whereas one needs to look to demotic and musical working class expression for registration of dissent from the progressive narrative, and contrasting emphasis upon working conditions. Tennant and Co. appeared in Walter Scott's *The Antiquary* as an item of common conversation in a provincial east-coast town, its thriving business noted, and its commercial prospects speculated upon, a readily intelligible sketching of common communicative life, and to that extent a witness to the company's rising and pervasive presence in Scottish commercial culture.[8] A contrasting view of St Rollox is found in verses recited by Hugh Aitken Dow at a Royston School reunion in 1875, a school founded by Tennant's for the children of their workers, so the likelihood is that Dow was a local resident, and had been a St.

6 Ibid., pp. 114-119.

7 George Dodd, Charles Knight, James Thorne, Harriet Martineau, William Harvey and William Wylie, *The Land We Live In: A pictorial, historical and literary sketch-book of the British Islands*, vol. III, Scotland, Ireland and the Devonshire Coast (London: William Orr and Co., 1846), 213.

8 Walter Scott, *The Antiquary*, original publication 1816, edition referenced is *The Antiquary*, vol.1 (New York: Van Winkle and Wiley, 1816), ch.15, 154, 160; *The Antiquary*'s representation may be somewhat anachronistic. The novel is set in the 1790s, at which time Tennant's had existed at most for four years. It is possible that the company's extended network of business agents and volume of sales at that time was such as Scott depicted, but it is at least as likely that it represents the state of affairs around Scott's time of writing, 1815-16.

Rollox chemical worker, perhaps as early as the 1840s.[9] Invaluable as a direct comment by a chemical worker, Dow's verses portray the oppressive, sooty and sulfurous atmosphere of the St. Rollox Works and its surroundings, and also its physiological effects, 'smarting eyes', and 'muffled noses'. A nether–world is again invoked, with a concluding image of Vulcan, but, the verses insist, the god of fire, forge and metallurgy can no longer compete effectively with St Rollox. Later in the century, the nose of St. Rollox' chief chemist, nicknamed Sniffer Crystal, had been eroded, by his constant olfactory occupation of gauging chemicals.[10] If St. Rollox was comparable with contemporary English chemical works, then indoor workers, in the vitriol rooms, the soda sheds, or at the chlorination chambers, would after a number of years be moved to work outside in the yards, so deleterious were the effects of working the chemical processes.[11] Chemical workers were of course acutely aware of such specific occupational consequences, and of occupational life expectancy.[12]

It is difficult to avoid the conclusion that St. Rollox' general environmental and particular physiological impacts were pervasive, oppressive and unhealthy. The chimneys poured smoke, soot and chemical fumes into the sky, a subject of sarcastic pictorial comment. In terms of an Aristotelian material cosmos, St. Rollox' coverage was absolute. Earth and water absorbed its chemical waste. The air was increasingly contaminated by its carbon and gaseous exhalation, the fire of the furnaces flamed incessantly, and the ethereal sky was occluded by the smoke, rising to join Glasgow's often low cloud ceiling, whose frequent rain returned fractions of the furnace carbon and acidic fumes to earth.

None of this is of course particularly surprising. We are narratologically accustomed to accounts of economically progressive, massive and continuous output expansion, technological innovation, factory labor concentration and exploitation, with the downside consequences of urban pollution, disease and crowded, squalid housing, and we are comparably accustomed to an accompanying, progressive 'science and technology' narrative, old-fashioned

9 Text available at the Royston Road Project, <http://www.roystonroadproject.org.>, section "Garngad & Royston"; For music, listen to Ron Angel or Big Big Sea, "The Chemical Worker's Song", both available on YouTube. This is a twentieth century song, also known as "The Process Worker's Song", sometimes "The ICI Song", and features life expectancy, caustic burn, gypsum and cyanide.

10 Simon Blow, *Broken Blood: The rise and fall of the Tennant family* (London: Faber and Faber, 1987), 59.

11 W.A. Campbell, "The Alkali Industry," Colin A. Russell ed., *Chemistry, Society and Environment* (Cambridge: The Royal Society of Chemistry, 2000), 75-106, on 82.

12 Anon., "Glasgow Sketches," (1889), a compilation of nineteenth century excerpts from local newspapers, Mitchell Library (Glasgow Room), ref. G330.193.01444 MIL.

perhaps, but still exerting influence, and which may be briefly sketched as follows. In Sweden, the chemist Scheele observed chlorine's bleaching properties, with experimentation on plant materials. In France, Claude-Louis Berthollet undertook further experimentation, producing the innovative technique of chlorine-based liquor bleaching, which he publicized to French bleachers. James Watt learned of his new process, which, although difficult and dangerous to manage, possessed potential advantages over existing bleaching techniques, particularly the reducing of cloth bleaching to direct chlorine treatment, eliminating the lengthy techniques of alternative alkalization and acidification followed by time-and-space consuming exposure to sunlight. It could all be moved indoors, and done with the newly discovered, liquefied chlorine gas, with cost reduction and labor-process reconfiguration possessed of developmental potential. Berthollet was close to the heart of the chemical revolution in France, and its new chemistry of gases. This scientific context had thus produced dramatically innovative technique, publicized through Berthollet's benevolent, philosophical commitment to human betterment, and James Watt informed colleagues and compatriots of it when he returned to Britain. Amongst them were the progenitors of St. Rollox. Tennant and Macintosh adopted the liquor technique with some modification, and commenced the new works with it. Within a few years, they effected a radical improvement in the new chemical technology with a process which produced a powder, and this produced significant further cost savings, particularly in relation to packaging, transport and distribution, the liquor being much heavier, bulkier, and prone to lose its efficacy, than the powder. The technical revolution originating in and accompanying the chemical revolution was now essentially complete, and the result in Scotland was St. Rollox and its economically expansive and culturally iconic, if ambivalent history, forged by the practical chemical acuity and exceptional entrepreneurial abilities of the Tennant and Co. partnership. Such a 'linear' account, neatly enough, conjoins a history of pure science terminology of experimentation and discovery and applied science-driven technological innovation, with the equally significant terminology of innovative industrial entrepreneurship – a tale of two revolutions, discontinuously instituting modern chemistry and modern urban industry, whose integration proved to be an unstoppably powerful and transformative historical force, whatever evaluative attitudes historians may hold towards it. Without denying the factual basis for the progressivist narratives of chemistry and industry, it is nonetheless possible to use further historical detail in ways which refuse simple endorsement of them, and introduce further resources for conceptualizing and narrating the origins and development of key aspects of chemical manufacture and industry, practical chemical technique, chemical science and education, chemical entrepreneurship, and Glasgow itself, within the period 1760-1860.

Behemoth Deconstructed

We have long known of the very considerable difficulties, of chemical technique, experimental and practical development, of competition and cost and hazard to health, which attended attempts to introduce Berthollet's oxymuriatic liquor into commercial production in the late 1780s and 1790s. The extensive and detailed research of Musson and Robinson revealed a lengthy series of efforts, in Lancashire, Nottinghamshire and the west of Scotland, to introduce chlorine bleaching on a commercially viable scale.[13] From Berthollet's original work until the patent application for the Tennant's powder took fourteen years. Those years were filled with further practical chemistry experimentation, numerous trials involving a variety of different technical apparatus (containment vessels for acid and gas, complicated flask arrangements for managing the gas, a measurement device for gauging bleaching strengths), direct gas-to-cloth treatment, and a number of different chemical additives to liquors to manage both the degree of caustic strength and the respiratory and ocular hazards of working with chlorine.

Processes which worked with linen did not necessarily work with cotton. Printed calicos offered further problems. Alkaline addition diluted bleaching efficacy, and transport of bulky, heavy liquor was not propitious. Some became convinced that even with workable and reliable liquor, the whole process would be just too costly for commercial viability. The competitive environment induced levels of secrecy, and of betrayal, among the groups trialling the varieties of new processes. The new processes themselves, if successful, would, it was feared, face further competition from combinations of old-method bleachers, who would co-operate to undercut new-method prices, a particular vulnerability given the anxious cost estimates of new-method trialists. Several competing groups of manufacturing chemists devoted sizeable investment of time and money, convinced of the chlorine route to a radically reformed commercial bleaching. At least three of these groups, each containing experienced bleachers and reputable practical chemists, nonetheless ended in failure, and were obliged to give up on this particular chemical quest. Then when some reliably usable liquors were formulated, and the Tennant's powder emerged as

13 Musson and Robinson, *Science and Technology*, pp. 251-337 (see note 1).

a further reliable option, this by no means signaled the effacement of the older techniques.

The process of 'innovation' was not simply 'difficult' and 'bumpy'; this really does not capture the reality of the case. It was long-drawn-out, populated with serious financial hazard and a substantial business failure rate; at any given moment within the period, it appeared as entirely contingent upon a large number of chemical, technical and business variables; and as an historical process it continued to remain incomplete, as the persistence of older methods in tandem with new tended to demonstrate. Any account of the chlorine innovation which focuses simply on Berthollet's discovery at one end and successful powder manufacture at the other is thus prone to ignore the actual historical process of what gets called, and singularized, as 'innovation'. The process was multiple in its attempted novelties, and most of them failed over the medium term.

Further investigation, details of which now follow, tends to indicate the misleading insufficiency of solely 'revolutionary' narratives, insofar as their underlying concept of change is one confined to radical innovation inducing discontinuous development. It also indicates the inadequacy of accounts of chemical Glasgow's patterns of development focused solely on the pre-eminent case of St. Rollox, and questions the limiting nature of entrepreneurial modelling derived from the paradigm industries of the period, cotton, steam, coal, mechanics and the like.

Chemical Glasgow

The analysis which follows is geographical in the first instance, based as it is on an informative Glasgow map of 1828.[14] Its particular virtue is clear identification of Glasgow's sites of manufacture with precise location, and calculable area occupied by works sites. It is firstly possible to form an accurate impression of the range and number of manufacturing concerns. Cotton Works (twenty-two) and Foundries (eighteen) predominate numerically, and there are only two strictly denominated 'Chemical Works'. To form a more accurate impression of chemical manufacture, we should however add to these the other producers of chemicals, the four Soda Works, two Gas Works, and the

14 David Smith, *Plan of the City of Glasgow and its Environs* (Glasgow: Wardlaw & Co., 1828); For zoom facility, legibility and close inspection navigation, see University of Glasgow Library's site: <http://www.gla.ac.uk/services/library/collections/virtualdisplays/mapsofglasgowhistoricaltodigital/davidsmith1828planofthecityofglasgowanditsenvirons/>

Acid and Vinegar Works; and also the eight Dye Works, the two Crystal (leaded glass) Works and the Coal Tar Works, making a total of twenty-one manufactures where chemical products, crucial chemical processes, manipulation and practical chemical expertise were fundamentally involved.[15]

This total still excludes the highly chemical potteries, the refineries, breweries and distilleries, but for chemical completism with reference to Glasgow, note two further sites, those of the University of Glasgow and the Andersonian Institution, where chemical science was taught. In this overall picture of chemical Glasgow, the sheer size of St. Rollox, (the dark section at the top of the north-west quadrant of map section, Fig. 13.1) still predominates as notably the largest manufacturing site in the city; but we should note too that the largest sites tend to be chemical, exemplified in addition to St. Rollox by the main gasworks (south-west quadrant), and the Cudbear Works (south-east quadrant). The latter two will also prove to be of particular significance in the developmental pathways we will shortly follow. For the time being, the considerable presence of both cotton and iron manufactures, each of them probably outnumbering chemical manufactures in employment terms, may be noted, along with the extensive areas occupied by chemical production.

The textile sector and the chemical cannot of course be functionally separated in local or national terms, in that the expansion of bleaching and dyeing materials which is registered on David Smith's map is directly ascribable to coeval expansion of cotton textile production, witnessed by the number of Glasgow cotton works. The contrast between Glasgow textiles and chemicals at this time is between the large number of smaller units of production in cotton, and the smaller number of large units of production in chemicals. The proliferation of four soda works in addition to St. Rollox is explicable in terms of the recent abolition of salt importation duty, reminding us of the crucial role played by government policy in directing the course of industrial development.

To this picture of the comparative placement of chemical manufacture within Glasgow's industrial setting we can add further relevant detail from academic, educational chemistry. Anderson's Institute numbered in its trustees and professors men with chemical interests. Macintosh was a trustee, as was his business associate John Wilson, whilst Willian Couper, another friend and business associate, became a teacher of chemistry there.[16] The University of

15 Figures compiled from Smith, *Plan of Glasgow* (see note 14).

16 Roger Emerson and Paul Wood, "Science and Enlightenment in Glasgow, 1690-1802," Charles Withers and Paul Wood, eds., *Science and Medicine in the Scottish Enlightenment* (East Linton: Tuckwell Press, 2002), 79-143, on 97-99.

FIGURE 13.1 *Section of David Smith,* Plan of the City of Glasgow and its Environs, *1828. Showing (1) St. Rollox Chemical Works, Charles Tennant & Co. (2) Glasgow Gas Works (3) Cudbear Dye Works, George and Charles Macintosh.* USED WITH THE KIND PERMISSION OF UNIVERSITY OF GLASGOW LIBRARY.

Glasgow by this time possessed a distinguished genealogy of teachers of chemistry, including William Cullen, who worked on both salt and bleaching process manufacture in the 1750s, Joseph Black who worked on bleaching agents and pursued artificial soda with James Watt, and William Irvine, with his particularly relevant and meticulously quantified analysis of the properties of sulfur in heating and cooling.[17] By the time of this survey of the late 1820s, the University's professor of chemistry was Thomas Thomson, an early supporter of Daltonian atomism, with interests also in pursuing the gains made by German analytical chemistry, and making these gains tell with regard to chemical teaching and research.[18] The work of these academic chemists had a variety of bearings upon the practical chemistry of manufacturers, in particular the early attention (1750s-1770s) to alkalization in bleaching and the potential of lime, and Thomson's enthusiasm for Hermann Klaproth's ruthless analytics, which left no residue unexamined.

Some of William Irvine's pupils and associates formed a small Chemical Society whilst in Glasgow, in 1785-7. Whatever the chemical preoccupations of the Glasgow students, the society helped cement relations between two young men who would continue their association momentously, as original, founding partners of Tennant and Co., namely Charles Macintosh and the above-mentioned William Couper.[19] This noteworthy group also contained two more sometime professors of chemistry, Adair Crawford of the Woolwich Military Academy, and John McLean, professor at the College of New Jersey; also Alexander Tilloch, publisher of *The Philosophical Magazine*, John Finlay, a close chemical correspondent of Macintosh, and John Wilson of Hurlet, where he and Macintosh would initiate Scotland's first alum works. Macintosh contributed papers on dyeing, crystallization and alcohol, this last exhibiting familiarity with Lavoisian chemistry.[20]

The Chemical Society thus offered opportunity for young chemists to pursue their particular, practical chemical interests, but at least as important for the future of manufacturing chemistry were the emergent associative bonds which it fostered. From this associational point of view, two further institutional venues have relevance. The Glasgow Chamber of Commerce, the first of

17 William Irvine, *Essays, Chiefly on Chemical Subjects* (London: J. Mawman, 1805), 475-90.

18 For Thomson, see J.B. Morrell, "Thomas Thomson: Professor of chemistry and university reformer," *British Journal for the History of Science* 4 (1969): 245-286; and idem. "The Chemist Breeders: The research schools of Liebig and Thomas Thomson," *Ambix* 9 (1972): 1-46.

19 Emerson and Wood, "Science and Enlightenment," pp. 128-129 (see note 16) fn.63 has detail on this society; An earlier useful source is George Macintosh, *Biographical Memoir of the Late Charles Macintosh* (Glasgow: Blackie & Co., 1847), 6-8.

20 Ibid., pp. 6-8.

its kind in Britain, was started in 1783, immediately in the wake of the peace ending the American War of Independence. The 1780s marked something of a shift of commercial focus, away from colonial trade monopolization of tobacco and toward domestic manufacture. This shift is noticeable in the Chamber of Commerce's early focus upon improving the quality of manufactures, and influencing government on tax and tariff. Charles Macintosh became a member, and his father, George, was an early president of the institution. Thus, when Macintosh joined the founding partnership of St. Rollox, it meant that Tennant, who tends to be regarded as senior partner (he took over financial control early on), was forging a relationship with the pre-eminent chemical manufacturers of Glasgow, the Macintosh family. One might begin to think therefore, that Tennant joined the Macintoshes, rather than the reverse; in the *Memoir* of his father, George jnr. mentions that Macintosh and Tennant were, "for several years previous", business partners by the time of *both* St. Rollox bleaching patents (that is, previous to 1797).[21] The thought is reinforced by consideration of the fact that two of Charles Macintosh's business associates in the Hurlet alum works, John Wilson and James Knox, were also partners in Tennant and Co. The Hurlet alum works was virtually contemporary in origin with St. Rollox, and in associational business terms, the firms had overlapping partnerships. This circumstance, and its timing, serves to emphasize the way in which, by the second half of the 1790s, this forceful grouping of chemically-inflected manufacturers had attained the chemical and commercial confidence to undertake not a single but a double initiative with reference to the foundation of major new enterprises.

In the 1780s and 1790s then, Glasgow clearly exhibited a set of educational and collective commercial institutions, voluntarist in nature other than the University. From our viewpoint, these institutions did not simply teach and promote chemical science and its manufacturing uses and potentials. They also generated and sustained a chemico-commercial culture, visible nodes of association for commercially-oriented chemists whose ties of friendship, family and manufacturing business would provide social cohesion to the development of Glasgow's chemical manufactures in the coming decades.

Chemical Entrepreneurship. The Staple Highway: Charles Tennant

The narrative of St. Rollox' origin, foundation, spectacular growth and commercial success, whilst acknowledging the significance of Charles Macintosh,

21 Macintosh, *Biographical Memoir*, p. 37 (see note 19).

tends to be understandably dominated, in terms of human agency, by the figure of Charles Tennant, a compelling story of transformation of a young rural silk weaver into the industrial Colossus of St. Rollox, an epic socio-economic trajectory of the industrial revolution. Young Tennant was a silk weaver, working in this highly-skilled occupation in Ayrshire. Aware perhaps of the increasing demand for textile bleaching, from long-established linen and recently growing cotton textile production, he switched occupation to bleaching, and ran a rural bleachfield in Darnley, Ayrshire, from 1788. There, hearing of Berthollet's chlorine technique, most likely *via* James Watt or Watt's father-in-law the bleacher John McGrigor, both of whom were working on improvements to the Berthollet technique, Tennant commenced, after some experimentation, using the Berthollet potash-chlorine liquor with success, eventually modifying it by substituting lime for the potash. His neighbour at Darnley was William Wilson, whose daughter Tennant would later marry in 1795. Wilson, a merchant and factor to the Earl of Glasgow, also had a son, name of John, whom we have already met, a fellow member of the University Chemical Society with Charles Macintosh, and a business associate of Macintosh in the nearby Hurlet Alum Works, sited on land owned by the Earl. Three years later, shortly after the foundation of Tennant and Co., the firm took out a patent for the chlorine liquor (eventually lost in an informative court case of 1802), but further experimentation, using slaked lime and chlorine gas, produced the solid bleaching powder which replaced the liquor, and provided the chemical basis of St. Rollox' early ascent, and some financial basis for its expansion. The local communicative networks of family, friends, Chemical Society and business association were clearly instrumental in forming the genesis of St. Rollox and its new bleaching chemistry.

Thus far we have Tennant as the alert, opportunistic, young entrepreneur, moving occupation riskily from a skilled to a less skilled occupation with more expansive potential, pursuing new processes, recognizing talented expertise and incorporating it by business association into ambitious commercial novelty. This does not, however, suffice as a characterization of the way he developed St. Rollox, which, after the new bleaching powder, may be considered as chemically conservative. That is, it concentrated on long existent methods, not chemical innovation, for producing basic staples, such as soap and sulfuric acid, rather than new methods for new products. When Leblanc process soda was adopted in the 1820s, it was certainly new in Glaswegian and Scottish terms, but it was by no means the first artificial soda, and all the pre-installation start-up costs, experimental research, up-scaling, equipment design, were absent, for by then it was a known and tried process of chemical production, with a waiting market, relatively riskless. Similarly, the lead chamber process for sulfuric acid had been practiced in Scotland since 1749, at John Roebuck's Prestonpans manufacture. Tennant's concentration was not on radically innovative technique, other than the bleaching powder, but on using St. Rollox for radically increasing the quantities of common, basic chemical materials and products. The language of 'growth', 'expansion' and 'innovation' tends to obscure this conservative, and fundamental, feature of St. Rollox chemical production.

Other notable features of Tennant's entrepreneurial practice include the distributed network of sales agents, and his concentration on packaging, transport and product distribution. One reason the Glasgow-Garnkirk railway was an early arrival in Scottish rail development terms was Tennant's early realization of railway's advantages. He befriended George Stephenson, and worked to ensure the Caledonian Railway spur to Garnkirk. The mercantile fleet also displayed his attention to transport, now in national and international import-export terms. In all these forms of development, Tennant's entrepreneurial style exhibited the impulse to own and control as many facets of the commerce as possible. In addition to production of relatively riskless chemical staples, the on-site packaging, immediate rail and water access, and the seagoing fleet bespoke a quest for, and ability to achieve, not simply maximal control and the co-ordination advantages derived therefrom, but as much independence as possible, relative freedom from reliance upon other packagers, distributors, and raw material suppliers.

Externally to the company, Tennant also paid effective attention to politics, acting for the advance of St. Rollox with the Leblanc process by pushing for abolition of salt importation duty, and opposing one chemical vested interest, which wished to prevent the advent of artificial soda. This interest was the Kelpers, numerous on the Scottish islands and west coast, the gatherers and burners of seaweed, or at least the proprietors and managers of estates where such work occurred. They produced much of the potash used in Britain, a strategically significant group in terms of chemical manufacture. This was the underlying conflict. It was about whether or not artificial soda, which needed mineral salt, would chemically supplant the established, vegetable-derived potash, rather than simply about free trade and Treasury receipts on salt, and Tennant and St. Rollox emerged a winner in the conflict.[22]

22 Obituary of Charles Tennant, Institution of Civil Engineers Letter Books, 9 vols. 1839-49, vol. 1, Shelf Mark 624/629 (410)31G.

Secret Works and Serial Invention: Charles Macintosh

If Tennant's entrepreneurship after bleaching powder can be described as chemically conservative, staple-focused quantitative expansion, complemented by strategic political action where necessary, and by in-house development and control of as many commercial factors as possible, then consideration of Charles Macintosh's complex and varied career, marked by serial manufacturing initiatives, multiple business associations and chemical invention, offers a striking contrast in entrepreneurial character. The relevant background pre-dates his birth in 1766, for the Macintosh family were already chemical manufacturers. His father George, with the financial aid provided by the Glasgow tobacco trader John Glassford, took over an established and patented, though then failing, dye making business in the 1770s, starting manufacture at Dunchatton in east Glasgow (see Fig. 13.1 for location of cudbear dye works). Macintosh Snr., "who is a great hunter after secrets", employed the previous owners of the process as managers.[23] The dye, named cudbear, was lichen-derived, and chemically capable of manipulation of color gradations through pink and red to blue and purple, and the Macintosh manufacture proved successful and long-lived. At its inception and for decades after it possessed two very striking features. It was surrounded by high walls, to hide the production processes from the curious eyes of 'intelligencers' or spies. In this its protective design followed that of Roebuck's lead vat acid process at Prestonpans. Even more striking was the work force, composed of Highlanders who were largely monoglot Gaelic speakers. The Macintosh family came from the northern county of Ross-shire, and they knew that Scots Highlanders were already familiar with the process, at least in undeveloped terms, because they used lichen to make their own textile dyes. The work force was essentially internally imported, semi-skilled labor, and the workers numerous. They were moreover sworn to secrecy, attended a nightly roll call, and were housed on-site, inside the walls. Many rarely left the compound, nor learned any English.[24] George Macintosh had erected a quadruple security barrier, of oath, roll call, wall and language, to protect the details and practice of the cudbear process, and the security would seem to have worked insofar as no other cudbear

23 McGrigor to Watt, April, 1788, cited in Musson and Robinson, *Science and Technology*, p. 293 (see note 1).

24 George Stewart, *Curiosities of Glasgow Citizenship; As exhibited chiefly in the business career of its old commercial aristocracy* (Glasgow: Maclehose, 1881), 70.

manufacture appeared in Glasgow.[25] Dunchatton was thus a remarkable socio-historical enclave, a site of production whose fuller comprehension requires appreciation of internal migration from the *Gaidhealtachd*.[26] Industrial urbanization of such populations, and the persistence of local Gaelic cultures in such circumstances are also important.[27] This place, of chemical enclosure and jealously guarded secrecy, was where Charles lived from childhood, the Macintosh house being on site and also functioning as the company office.

Charles at a still youthful age, as well as attending the chemistry class at Glasgow University (and Joseph Black's in Edinburgh, also later Thomson's in Glasgow), participated in the family business, traveling on at least two occasions, to Germany, Holland and France, on sales business for the company, and for the Prestonpans acid works, now taken over by John Glassford. He probably also pursued chemical intelligence, experimenting on French plants for dye potential, and noting Holland's sugar of lead manufacture and its successful export business to Britain.[28] He started his own entrepreneurial career before he was twenty, a sal ammoniac plant comparable to James Hutton's in Edinburgh. This commenced an entrepreneurial career of a large number of chemical manufacture initiatives, by no means all successful, but some of which brought him further prosperity, scientific reputation (F.R.S 1823), and eponymous commercial renown (the 'Mackintosh' [*sic*], a waterproof coat). He commenced sugar of lead manufacture in 1786, apparently lowering costs enough to export to Rotterdam, successfully undercutting prices of the Dutch sugar of lead manufacture which had so impressed him on his continental tour. He and his father learned of efforts in trials of the new chemical bleaching, and in 1788 Charles was "making experiments at home in order to find out the secret", then traveling to Manchester to learn more.[29] He had also launched mineral acetate production, substituting lime for alumina, a British novelty, and of use particularly in calico printing. He would start the Hurlet Alum works with John Wilson, and initiated a further alum site north east of Glasgow

25 Though see ibid., p. 71, where he records that one Highlander absconded to London, providing technical information to a company which set up short-lived cudbear production.

26 The Gaelic speaking areas of Scotland.

27 For informative analysis of such processes, see Charles Withers, *Urban Highlanders: Highland-lowland migration and urban Gaelic culture, 1700-1900* (East Linton: Tuckwell Press, 1998).

28 This and following detail on Macintosh's chemical career are taken from Macintosh, *Biographical Memoir*, particularly the "Introduction," pp. xii-xix, and the additionally narrated accounts of Macintosh's activities throughout the main text (see note 19).

29 Musson and Robinson, *Science and Technology*, pp. 293-4 (see note 1).

in the Campsie hills. The production of the chlorine bleaching powder for the new Tennant Co. in 1798, again highly novel in its gas-to-solid (chlorine-to-slaked lime) reaction, occasionally ascribed to Tennant, was the foundational experimental work of Macintosh.

The torrent of manufacturing processes did not cease with the establishment of St. Rollox, where he remained as a partner until 1814. For the East India Co. he produced a fused saltpeter process with considerable weight and space saving for stowage capacity for sea transport from India, the process demonstrated to the Directors' satisfaction, but not adopted. On the dyeing front, in addition to continuing the Cudbear Works, and the Turkey Red dyeworks started by his father in 1787 with the aid of Papillon's expertise in the "secret", he established a Prussian Blue process, and prussiate of potash for calico printing.[30] Prospecting at the Glasgow Gas Works he found naphtha (a coal tar derivative), a by-product of the coal-gas production. Either detecting its India rubber-dissolving property himself, or more likely knowing of James Syme's discovery of it in Edinburgh in 1818, he formulated the production of waterproofed cotton by using sheets of rubberized naphtha sandwiched between layers of cotton, an eventually successful manufacture which he relocated to Manchester. His last significant cooperative venture was with James Nielson, an eventually successful patenting, in 1828, for Nielson's hot- blast furnace metal-smelting process. His initial acquaintance with Nielson was probably a result of his prospecting of the coal tar by-products at the Glasgow Gas Works, where Nielson was a foreman. Prior to that, he had produced a new preparation of iodine, further innovations in textile treatment for calico printing, started a yeast factory in London (eventually failing), and patented an iron-steel conversion process with carburetted hydrogen.

Overall, his activity differed substantially from that of Charles Tennant. Less single-minded and more diversified than Tennant, his operations nonetheless possessed an identifiable internal coherence, for instance in the number of projects focused upon textile treatment and dyeing. The alum works and acetate process were relevant from this viewpoint, the potential application among others being their specific properties as mordants. He also used Glasgow Gas Works ammonia by-product in the Cudbear processes. We tend perhaps to over-individuate in our focus on sites, and this can misdescribe their functionality. Instead of individualized accounts of Macintosh's apparently diverse manufactures, an account which recognizes a chain of enterprises, succes-

30 Macintosh, *Biographical Memoir*, p. 22 (see note 19); George Macintosh Snr., writing to Charles, claimed to have thoroughly improved upon what they had from Papillon, improving color and shortening dyeing time.

sively connecting alum, ammonia and acetate with cudbear, Turkey Red and thence outward to calico printing, is more realistic for appreciating the overall coherence of these aspects of Macintosh's entrepreneurship.

Further characteristics reinforce the contrast with Tennant. Macintosh was far more mobile, a well-traveled and well-educated practical chemist, not solely Glasgow-sited, and willing to divest himself of existing enterprises (the Turkey Red Works and St. Rollox) to free his capacity for further initiative in chemical production. His multiple projects tended to remain of medium size, unlike the singular gigantism of St. Rollox, and a thread of persistent chemical novelty, chlorine powder, mineral acetates, dyes and mordants for calico printing, rubberized cotton, ran through them. Stated thus, the Macintosh successive chemical coverage of the key chemical processes prior to actual textile treatment by bleach and dye becomes readily apparent.

Just as indicative of his entrepreneurial character was his already indicated habit of chemical prospecting and scavenging. His first effort, sal ammoniac manufacture, relied, like Hutton's, on the free waste product of soot. The alum works were firstly based on cast off schist from local coal mines. The naphtha was an unused gas production by-product, like the ammonia. The point of scavenging in this sense is not just the finding of new materials. The materials were available, unintentionally as it were, and in the first instance, as the result of the labor of others. First-phase production thus came if not for free, at least for reduced cost. Nowadays we might understandably call him a chemical recycler, or re-purposer, but perhaps that does not quite capture the prospecting and scavenging habit quite spectacularly displayed by Macintosh. Rather, he gave chemical purpose and commercial value to the purposeless and valueless cast-offs of others' labor.[31] Mobile, prospective, scavenging, qualitatively diversifying rather than quantitatively accreting, this increasingly expert and chemically innovative entrepreneurial practice thus provides a thorough and instructive contrast with Tennant's and St. Rollox.

These chemical entrepreneurs were indubitably successful in chemico-economic terms, but the routes to success, the staple chemical highway of St. Rollox, the longer, twisting trail of multiple chemical and manufacturing initiatives followed by Macintosh, show that in this period, no singular entrepreneurial mode was definitive of chemical manufacturing success. Nor are these entrepreneurs straightforwardly assimilated by current modelling of industrial revolution owner-industrialists and of entrepreneurship, although a degree of comparative light is thrown upon them by recent work.[32] They may,

31 See Simon Werrett's and Lissa Roberts and Joppe van Driel's essays in this volume.

32 Crouzet, *The First Industrialists*, and Mokyr, "Entrepreneurship," (see note 3).

for instance, be described as broadly middling rank in origin (Tennant specifically a skilled, independent artisan), and both had extensive prior experience in the manufactures they industrialized. However, unlike some other significant industrial-scale works owners, Tennant did not undertake sector diversification that we know of into forms of finance, nor into non-chemicals within the manufacturing sector, nor enterprise outside manufacturing other than his in-house mercantile shipping.[33] Judgment of Macintosh's entrepreneurial characteristics should not allow his diverse set of medium scale operations to obscure the sense of coherent functional interaction of the textile-directed chain of the various operations, nor ignore his prospecting and scavenging proclivities, modes of entrepreneurship not easily recognizable in available models. As partner-owner of one large-scale enterprise, and with reference particularly to his range of inventive abilities, technical expertise and scientific cultivation, he is comparable especially with fellow-Scot James Watt, perhaps also with James Keir in the Midlands, and Thomas Henry in Manchester. The work of Peter Jones has rendered this a type more categorically recognisable than hitherto, the distinctive '*Savant-Fabricant*' figures of early industrialization, men who successfully combined scientific knowledge, technical expertise and manufacturing experience.[34]

Joel Mokyr's essay on entrepreneurs in the industrial revolution continues his strategic interpretive stress upon institutional and cultural factors.[35] He picks out the behavioral normativity of gentlemanly politeness and trust, the latter particularly significant with respect to contract and credit; trade associations and informational networks; and the strategic importance of choice of partners in new industrial ventures. This latter point is dramatically endorsed by the St. Rollox partnership of Tennant and Macintosh. With respect to local Glasgow commercial culture, we might add two sorts of relevant institution not covered by Mokyr, the new formal and quickly chartered collective institution, the Chamber of Commerce, with its focus on manufacture; and, of genetic significance, the University, not simply for its chemical teaching, nor yet for any commitment to techno-scientific public communication, but for the voluntarist association of the Chemical Society, and the combination of chemical enthusiasms, personal and commercial associations at whose origins it lay. Overall then, we may at least start to think of Glaswegian chemical entrepreneurship as partially intelligible within some recent analytical perspectives of

33 For such diversification see Crouzet, *First Industrialists*, pp. 16-17 (see note 3).

34 Peter Jones, *Industrial Enlightenment: Science, technology and culture, in Birmingham and the West Midlands, 1760-1820* (Manchester: Manchester University Press, 2009), 116-29.

35 Mokyr, "Entrepreneurship," (see note 3).

economic history, but at the same time, as having additional and different features to extend and qualify such historical understandings.

Conclusion

By way of conclusion, it may be worth reflecting on the broader issue of the nature of the historical developments involved, as premised in discontinuous conceptions of chemical and industrial revolutionary change, and as gesturally represented above (in the section "Chemical Behemoth"). Certainly the expanding St. Rollox site represented a visibly dramatic step-change, both for chemical manufacture generally, and for bleaching powder, then acid and soda production specifically. Also certainly, the chemical knowledge of Macintosh was up-to-date, informed by the new chemistry of Lavoisier and his French colleagues, and the new chemical technique of St. Rollox is directly relatable to a key component of revolutionary chemistry, namely the multiplication of gas discovery, the properties and compositional implications thence investigated, and the provocation all this provided for a systematic reformulation of the science, together with its practical methods of analytical and synthetic procedure. Macintosh's chemical circle was clearly aware of this. One friend wrote, in 1786, four years before the English translation of Lavoisier's *Traité élémentaire,* "From the perusal of Fourcroy's chemistry, I have become a perfect convert to the aerial system, although I think most of the disputes on this subject, and the doctrine of phlogiston, are mere playing on words."[36] This evidenced the chemical attitudes of the young practical chemists, keenly interested in the very latest chemistry, their focus particularly on the systemic role of gas components, the 'aerial system', but not overly distracted by intractable theoretical dispute. To this not inconsiderable extent, the changes in question, chemical, technical and economically productive, do not require violent shoe-horning into a narrative stressing intensive, discontinuous change.

There are however other detailed features of change which further complicate any straightforwardly discontinuous emphasis. Remaining for the moment with the chemical dimension of manufacture, the key role of chlorine needs a fundamentally qualifying addition. The first Tennant & Co. patent, for bleaching liquor was lost in 1802, because it also specified lime in suspension, which Tennant had used to replace Berthollet's potash additive, and lime was already in comparable bleaching use, a point sufficient in the legal specifications of

36 John Finlay to Charles Macintosh, February 1786, in Macintosh, *Biographical Memoir*, p. 19 (see note 19).

the time to invalidate the patent. The second St. Rollox patent was for the dry bleaching powder process, chlorine-based but crucially dependent on the reaction with slaked lime to produce not liquor but powder, the great commercial advantage of St. Rollox. Whence and why the lime?

Since 1750, bleaching had chemically modernized with sulfuric acid, produced by Roebuck's works, and considerably shortening bleaching time. In the 1760s Edinburgh's academic chemists Joseph Black and Francis Home had argued chemically for the introduction of legally banned lime, convinced of its relevant property, under appropriately focused quantitative management.[37] Tennant and Macintosh were also lime enthusiasts in this genealogy of Scottish bleaching technique, Macintosh in particular likely to have known of the work of Black and Home. From this viewpoint, first Tennant, then Macintosh, might be considered not as adding lime to chlorine, but adding French chlorine to chemically established Scottish lime. If that may be thought of as tendentiously overstating the case, then consider also Macintosh's statement to a correspondent: "Lime has long been a favourite nostrum of mine, having first used it many years ago."[38] He used it with reference to sal ammoniac, alum and elsewhere, and had thorough familiarity with its properties in its mild, caustic, liquid and slaked states. It was his familiar chemical standby, a first port of reactive chemical call. It was thus utterly unsurprizing that he should investigate its potentials in chlorine combinations. Lime, so to speak, does the business, in its itinerary through sal ammoniac, acetate, alum and chlorine processes of production. In this sense, Macintosh was not simply a 'modern' chemist, but an educated and knowledgeable inheritor of the previous generation of pre-revolutionary chemists' practical and theoretical knowledge. Chemically, therefore, we require an equal stress upon pre-revolutionary chemistry to understand the genealogy of St. Rollox bleaching. That in turn induces an authentically continuous dimension to the chemical history of St. Rollox.

Distinctive features of the Cudbear Works also bear more extended examination for their presence elsewhere. Its early modern, pre-industrial retentive secrecy, reminiscent of Prestonpans, was not Macintosh's only secretive process. The Campsie alum works was also known as the 'secret works', as, in imitation of cudbear manufacture in Glasgow, Macintosh attempted to sequester details of his alum process. The Turkey Red Works and 'secret' were also

37 Joseph Black, "An Explanation of the Effect of Lime upon Alkaline Salts," Francis Home, *Experiments on Bleaching* (Dublin: T. Ewing, 1771), 265-282.

38 Macintosh to his Tennant business partner, James Knox, Glasgow, 20th January, 1800, in Macintosh, *Biographical Memoir*, p. 25 (see note 19).

sequestered. This surviving ethos of pre-industrial manufacture can be contrasted with Tennant's far more communicative attitude, for example his willingness to instruct Irish bleachers. Macintosh worked for a time with Lancashire bleachers, but retained an attitude of guarded secrecy toward aspects of development of the Berthollet process, as indeed did James Watt, both men unwilling to forego any competitive edge which withheld chemical knowledge might provide.[39]

The partners' practices, both convergent and divergent at particular moments, illustrate the ways in which pre-industrial behavioral reflexes could and did persist, contemporary with the supposed liberal openness of a knowledge-based industrial economy, or more precisely, the way in which a publicly communicative ethos of accessible exchange of scientific and technical knowledge had a differential uptake with respect to manufacturing practice, even as between close entrepreneurial partners.[40] These kinds of concluding examples, emphasizing the persistence of various kinds of pre-revolutionary commercial forms and behaviors, and of chemical knowledge and practice, coeval nonetheless with consciousness of fundamental change in chemical science and with comparably fundamental economic development, reinforce the need for a more nuanced historiographical semantics, capable of registering and relating all relevant forms of change during the industrial and chemical revolutions.

39 For the introduction of Berthollet's process in Britain and the role of James Watt, see Musson and Robinson, *Science and Technology*, pp. 251-337 (see note 1).

40 The thesis of the growth of an openly communicative techno-scientific culture in this period is argued in Joel Mokyr, *The Enlightened Economy: An economic history of Britain, 1700-1850* (New Haven and London: Yale University Press, 2009).

CHAPTER 14

Relations between Industry and Academe in Scotland, and the Case of Dyeing: 1760 to 1840

Robert G.W. Anderson

What was the role of academic chemists in relation to those who were directly involved in developing Scotland's burgeoning chemical industry between 1760 and 1840? Traditionally, most historians have suggested that there was little involvement between academics and manufacturers across Europe. More recently, some historians suggest that it is difficult to disentangle their relations; it is best, they argue, to speak of hybrid identities. Neither extreme is sustainable when Scottish evidence is examined; an intermediate position, depending on the particular individual concerned, paints a more plausible picture. Some academics interacted closely with enterprising manufacturers, others less so. This essay provides a general introduction to the issues, followed by an analysis of the case of the pre-synthetic dyestuffs industry.

The traditional view holds that the academic world of the late eighteenth and early nineteenth centuries did not play a role. In a paper of 1797, Theophilus Lewis Rupp (d.1805), a German-born Manchester cotton manufacturer wrote,

> The arts, which supply the luxuries, conveniences, and necessaries of life, have derived but little advantage from philosophers [...] The chemist, in particular, if we except the pharmaceutical laboratory, has but little claim on the arts: on the contrary, he is indebted to them for the greatest discoveries and a prodigious number of facts, which form the basis of his science [...] [N]o brand of the useful arts is less indebted to him than that of changing the colours of substances. The art of dyeing has attained a high degree of perfection without the aid of the chemist, who is totally ignorant of the rationale of many of its processes, and the little he knows of this subject is of a late date.[1]

1 Theophilus Lewis Rupp, "On the Process of Bleaching with the Oxygenated Muriatic Acid; and a Description of a New Apparatus for Bleaching Cloths, with that Acid Dissolved in Water, without the Addition of Alkali," *Memoirs of the Manchester Literary and Philosophical Society* 5 (1798): 298-313.

DOI 10.1163/9789004325562_016

This view was reiterated by the American economic historian Witt Bowden (1885-1979) in 1925 and by British economic historian, Peter Matthias (b.1928), who in 1983 wrote, "innovations were not the result of the formal application of applied science, nor a formal product of the educational system [...] determination, intense curiosity, quick wits, clever fingers, luck, capital [...] and a backer [...] were more important [...] than a scientific training".[2]

A recent opposing view to this distinction between the academic world and industry has been developed by historian of science Ursula Klein, who asserts that throughout the eighteenth century,

> [...] chemical science and technology were strongly and systematically linked with one another. The interconnection of chemical science and technology was not promoted by a few individuals occupied with both scientific and technological enterprises. Rather, this interaction was firmly established on the communal and institutional level of eighteenth-century chemistry.

She stresses her point by applying the seemingly anachronistic concept of 'technoscience', the intricate entanglement of what have conventionally been considered as two different cultures, to cover both "savants experimenting at academic institutions and craftsmen and artisans producing in workshops."[3] Practitioners, it is argued, essentially merged towards a single chemical culture. It must be asked, however, how generally applicable this was throughout Europe? Klein's evidence is largely taken from the German world (particularly that described in Karl Hufbauer's *The Formation of the German Chemical*

---

2 Peter Mathias, *The First Industrial Nation: An economic history of Britain, 1700-1914*, second edition (London: Methuen, 1983), 124-25; Witt Bowden, *Industrial Society in England Towards the End of the Eighteenth Century* (New York: Macmillan, 1925); second edition (London: Frank Cass & Co., 1965), 11.

3 Ursula Klein, "Technoscience *Avant la Lettre*," *Perspectives on Science* 13 (2005): 226-66, quotations on 226-27; Klein goes so far as to use the term 'fusion' (226) to describe the degree of merging of the chemical cultures. On chemistry as a technoscience, see also Bernadette Bensaude-Vincent, "Chemistry as a Technoscience," Jean-Pierre Llored, ed., *The Philosophy of Chemistry: Practices, methodologies, and concepts* (Newcastle-upon-Tyne: Cambridge Scholars Publishing, 2013), 330-41; For an intermediate view, see Eda Kranakis, "Hybrid Careers and the Interaction of Science and Technology," Peter Kroes and Martijn Baker, eds., *Technological Development and Science in the Industrial Age* (Dordrecht: Kluwer Academic Publishers, 1992), 177-204, on 178.

Community).[4] But if we attend to the relations in Scotland between university chemists and those who employed chemistry in the fields of manufacture and commerce over the period 1760 to 1840, the more sophisticated and integrated German chemistry culture (particularly that in relation to pharmacy) cannot be assumed.[5] Similarly, the German level of organisation in the field of mining, with its academies and ceremonial, did not exist in the same way in Scotland; neither did posts associated with them. Klein also describes the field of organic chemistry, in which new carbon compounds were produced, as technoscientific. But it was a particularly German and French development from the 1820s, whose activity led to an alliance between chemical science and the synthetic dye industry from the late 1850s.

Edinburgh, situated near the Forth valley where industrial initiatives were rapidly developing, was one of Europe's major centers of chemistry teaching in the later eighteenth century. To an extent the cultures of chemical pedagogy and industrial practices overlapped there. There was a significant amount of social interaction between many academic chemists and entrepreneurs and for some this extended further to levels of co-operation and advice. Their identities, however, generally remained connected either to academe or the industrial and commercial world. While academics were in touch with industry in a variety of ways, their main concern was clearly seen as transmitting chemical knowledge to future medical practitioners.[6]

Industrialists were certainly never involved in formal teaching activities and practically all chemistry teaching was conducted by those who had graduated with medical degrees. A few entrepreneurs had undertaken formal university education and some did subscribe to a course of chemistry lectures as non-matriculated students. On occasion, academics were asked by industrialists, government agencies and publically funded bodies to offer advice on industrial and health matters. But, as said, they largely maintained their distinct identities.

4 Karl Hufbauer, *The Formation of the German Chemical Community 1720-1795* (Berkeley: University of California Press, 1982).

5 Ursula Klein, "Apothecary Shops, Laboratories and Chemical Manufacture in Eighteenth-century Germany," Lissa Roberts, Simon Schaffer and Peter Dear, eds., *The Mindful Hand: Inquiry and invention from the late Renaissance to early industrialisation* (Amsterdam: Royal Netherlands Academy of Arts and Sciences, 2007), 247-78.

6 The requirement for physicians to have knowledge of general chemistry has been little discussed by historians. The chemistry curricula taught includes a great deal of chemistry which had no bearing on their professional needs. Courses by chair holders in *materia medica* and botany which would have been much more relevant to the daily work of a general practitioner.

From the foundation of the Edinburgh medical faculty in 1726, it is possible to observe three phases in the relationship between the chemistry professors and industry up to the 1840s. During the first phase, the professor Andrew Plummer was directly involved with his medical colleagues in the production of drugs for distribution to pharmacies around Edinburgh, at least until 1742. They played a strongly entrepreneurial role: they were unsalaried and needed to earn more than their student fees brought in.

A second phase involving William Cullen and Joseph Black between 1747 and 1799, saw their close but distinctive relationship with industrialists, with whom their interests coincided. William Cullen (1710-90), lecturer in chemistry in Glasgow from 1747 and successor to Plummer in Edinburgh from 1756, was particularly damning of his teaching. He wrote in his notes, "Pharmaceutical Courses of Chemistry have not deserved the place they have hitherto held in our Schools – that they are not fitted to lead us to a general knowledge of Chemistry."[7] Furthermore, he taught, "Some persons may expect very particular Inquiries on the Subject of certain Arts. Examples of the practice of these and Essays made for their Improvement and these arts they would chiefly attend to – but I here must inform such persons that we don't pretend to shew the practice of particular Arts in the way of Trade & Business."[8] The emphasis Cullen was making here was that he would not teach *particular* Arts – his teaching, while practical, would be generally applicable.

The third phase, involving Thomas Charles Hope, 1796 to 1843, saw much more polarization. Hope (1766-1844) was initially antagonistic even to the teaching of practical chemistry, though his attitude softened later in his career. In the world outside his lecture theater, major developments such as chlorine bleaching, were being pursued by industrialists (in Scotland, particularly by Charles Tennant (1768-1838) in Glasgow). Hope's only involvement with bleaching, however, seems to have entailed a single experiment, conducted in the presence of Humphry Davy (1778-1829) and other gentlemen, which concerned theory rather than practice.[9] Hope's territory was based around the university's Old College and in the social round in Edinburgh's New Town, rather than sites of industry. His successor, William Gregory (1803-1858), did have some personal connection with the alkaloid industry in Edinburgh in that his method of mor-

7 R.G.W. Anderson, *The Playfair Collection and the Teaching of Chemistry at the University of Edinburgh 1713-1858* (Edinburgh: Royal Scottish Museum, 1978), 11; Cullen probably attended Plummer's lectures as a student.

8 Ibid.

9 Archibald Clow and Nan L. Clow, *The Chemical Revolution: A contribution to social technology* (London: Batchworth Press, 1952), 500.

phine production was taken up by the firm of J.F. Macfarlan in the 1830s, though this was more than a decade before he was selected for the Edinburgh chair.[10] Two years before his appointment in 1844, Gregory wrote a telling letter to the 4th Earl of Aberdeen (1784-1860), indicating a very different attitude from that of Hope:

> If any nation is bound to encourage and promote the study of practical chemistry it is the British nation, which has derived, and continues to derive, such vast advantages from the applications of its principles to the useful arts. Yet, if we investigate the subject, we find that the opportunities afforded in this country for the study of practical chemistry are exceedingly limited.[11]

Glasgow offers other cases that are a bit more ambiguous. Andrew Ure (1778-1857) served as professor of natural philosophy at Glasgow's Anderson's Institution, lecturing also on chemistry and mechanics, from 1804 to 1830, when he resigned his post to become what his biographer has called "the first consulting chemist in Britain."[12] While his resignation emphasized a separation between the academic world and the realm of chemical industry, Ure had, for example, worked as a consultant to the Irish Linen Board in 1814. And, reflecting on his career in his highly successful *Dictionary of Arts, Manufactures and Mines* of 1839, which he described as embodying "the results of my long experience as a Professor of Practical Science," Ure spoke of his continued interest in educating his nation's practitioners.[13]

Those who studied in Thomas Thomson's laboratory in Glasgow University (where he was professor of chemistry from 1818 to 1852) also present us with a picture that neither conforms clearly to the traditional distinction between academic and industrial chemists nor the claim of chemists as hybrid 'technoscientists'. In his study of this cohort, Jack Morrell does divide them between the categories of "academic" and "industrial" chemists. But he is quick to add

10 Anderson, *The Playfair Collection*, p. 48 (see note 7).

11 William Gregory, *Letter to the Right Honourable George, Earl of Aberdeen, K.T. [...] on the State of the Schools of Chemistry in the United Kingdom* (London: Taylor & Walton, 1842), 12.

12 Donald Cardwell, "Ure, Andrew," *Oxford Dictionary of National Biography* (Oxford: Oxford University Press, 2004).

13 Andrew Ure, *Dictionary of Arts, Manufactures and Mines* (London: Longman, Orme, Brown, Greene, & Longmans, 1839), iv-v; For further consideration of Ure's career, see the discussion on his research on dyes below.

that "the above classification is arbitrary" and specifically mentions Walter Crum, whose work in the field of dyeing is discussed below, as a "case in point".[14]

Chemistry Teaching in Scotland

Formal academic teaching of chemistry in Great Britain over the period 1760 to the late 1820s essentially entailed the courses available at Scottish universities, especially those of Glasgow and Edinburgh. The English universities, Oxford and Cambridge, did offer chemistry classes from time to time but the subject did not form part of the examined curriculum.[15] The University of London treated chemistry seriously from its beginning but that was not until after its foundation in 1826. The other two Scottish universities, St Andrews and Aberdeen, were small and did not particularly promote the subject of chemistry, though there was sporadic activity from certain professors. In contrast, Edinburgh and Glasgow were seriously involved with the subject. Edinburgh established a chair of chemistry and medicine in 1713 and its medical faculty thirteen years later.[16] The University of Glasgow set up a lectureship from 1747 and appointed William Cullen.[17] From these years onwards, a continuous series of appointments was made. In Glasgow, a Regius chair in chemistry was established in 1818. Both universities drew large classes. In the 1820s, Thomas Charles Hope's class at Edinburgh on occasion attracted more than 500 subscribers per annum.[18]

14 Jack Morrell, "The Research Breeders: the Research Schools of Liebig and Thomas Thomson," *Ambix* 19 (1972): 1-46, see particularly the lists on 21, 22; quotation on 21.

15 Mary D. Archer and Christopher D. Haley, eds., *The 1702 Chair of Chemistry at Cambridge: Transformation and change* (Cambridge: Cambridge University Press, 2005); Peter J.T. Morris, "The Eighteenth Century: Chemistry Allied to Anatomy," Robert Joseph Paton Williams, Allan Chapman and John Shipley Rowlinson, eds., *Chemistry at Oxford: A history from 1600 to 2005* (London: Royal Society of Chemistry, 2008), 52-78.

16 John C. Powers, "Leiden Chemistry in Edinburgh: Herman Boerhaave, James Crawford and Andrew Plummer," Robert G.W. Anderson, ed., *Cradle of Chemistry: The early years of chemistry at the University of Edinburgh* (Edinburgh: John Donald, 2015), 25-58.

17 Andrew Kent, ed., *An Eighteenth Century Lectureship in Chemistry: Essays and bicentenary addresses relating to the chemistry department of Glasgow University* (Glasgow: Jackson Son & Co, 1950).

18 For numbers of students registering for the annual course of chemistry lectures given by Thomas Charles Hope, see Jack Morrell, "Practical Chemistry in the University of Edinburgh, 1799-1843," *Ambix* 16 (1969): 66-80 (on 76, note 84).

It is important not to ignore Anderson's Institution in Glasgow, established in 1796 (from 1828, renamed Anderson's University) through a legacy of John Anderson (1726-96), professor of natural philosophy at the University of Glasgow.[19] Because the funds he bequeathed were insufficient to establish a fully-fledged university, it was decided to begin teaching natural philosophy and chemistry; other subjects were added subsequently. This was because the institution was established specially to teach manufacturers and artificers who had not previously undertaken mathematical studies. In some ways it can be seen as a precursor of mechanics' institutes. George Birkbeck (1776-1841), who is regarded as the prime originator of mechanics' institutes, taught at Anderson's Institution from 1799. It might thus be seen as ironic that it was this institution which Ure left to become an independent chemical consultant.

Scotland's Universities and Industry

Eighteenth-century Scottish chemists were encouraged to develop interests in industry by benefiting from premiums made available through a body which was established to promote the Scottish economy, the Board of Trustees for Fisheries and Manufactures in Scotland. This offered financial inducements, provided by the London-based British Government.[20] From when it started in the 1720s, the initial sum set aside for this purpose was £6000 per annum. Among the higher priorities was encouragement of the linen industry, for which £2650 had been ring-fenced. The traditional means of whitening cloth – leaving it laid out on bleachfields for several days in the sun, sometimes adding sour milk – was clearly inefficient. Black and his colleagues suggested innovative bleaching techniques. Francis Home (1719-1813), professor of *materia medica*, was awarded a prize of £100 for his proposal to use dilute sulfuric acid in the process rather than sour milk. But it was not always to larger-scale industries that Black offered advice. In 1783 the Board asked him to adjudicate on Mrs Gordon's method of making wine vinegar.[21]

19 James Muir, *John Anderson, Pioneer of Technical Education and the College he Founded* (Glasgow: John Smith & Son, 1950); John Butt, *John Anderson's Legacy: The University of Strathclyde and its antecedents* (East Linton: Tuckwell Press, 1996), 1-24.

20 Annette M. Smith, "State Aid to Industry – an Eighteenth Century Example," T.M. Devine, ed., *Lairds and Improvement in the Scotland of the Enlightenment* (Glasgow: 9th Scottish Historical Conference, 1978), 21-30.

21 For this example, see Robert G.W. Anderson and Jean Jones, eds., *The Correspondence of Joseph Black*, 2 vols. (Farnham: Ashgate, 2012), vol. 1, 614-15; Mrs Gordon could not be identified.

Andrew Ure reflected on this increasing input by academic chemists into manufacturing practices in the preface to his *Dictionary of Arts, Manufactures and Mines*.

> I have embodied in this work the results of my long experience as a Professor of Practical Science. Since the year 1805, when I entered at an early age upon the arduous task of conducting the school of chemistry and manufactures in the Andersonian Institution, up to the present day, I have been assiduously engaged in the study and improvement of most of the chemical and many of the mechanical arts. Consulted professionally by proprietors of factories, workshops and mines, of various descriptions, both in this country and abroad, concerning derangement in their operations, or defects in their products, I have enjoyed peculiar opportunities of becoming familiar with their minutest details, and have frequently had the good fortune to rectify what was amiss, or to supply what was wanting.[22]

This judgment was by no means unique. In 1814, Sir John Sinclair (1754-1835), who promoted the compilation of the *Statistical Account of Scotland*, declared: "At present there are a greater number of intelligent practical chemists in Scotland, in proportion to the population, than perhaps in any other country of the world."[23]

Not much is known about laboratories situated in Scottish universities up to 1820, and what went on in them. Occasional details are indicated when funds were applied for either from the University or (in Edinburgh's case) its patron, the Town Council. A certain amount can be deduced about extra-mural chemical laboratories, which were developed by freelance chemists to teach students who preferred to learn the subject outside the confines of a university.[24] Other laboratories which were set up throughout Britain after 1821 included those attached to mechanics' institutes, bodies which initially were dedicated to providing education for the working class. The first one of these was the Edinburgh School of Arts.[25] Its Glasgow counterpart, the Glasgow Mechanics' Institution,

---

22 Ure, *Dictionary of Arts*, p. iv (see note 13).

23 John Sinclair, *General Report of the Agricultural State and Political Circumstances of Scotland* (Edinburgh, 1814), appendix 2, 307.

24 Robert G.W. Anderson, "Chemistry Beyond the Academy: Diversity in Scotland in the early nineteenth century," *Ambix* 57 (2010): 84-103.

25 P.N. O'Farrell, *Heriot Watt University. An illustrated history* (Harlow: Pearson Educational, 2004).

emerged two years later from a split in the mechanics class of Anderson's Institution. These various bodies were mainly concerned with teaching workers science. Much less is known about private laboratories (as opposed to teaching lecture theaters) used for experimental activity by the professors where personal investigation and research could be carried out. Black must have had private facilities for conducting his many analyses.[26]

Knowing about social relationships between academic chemists and entrepreneurs is useful to determine the role of each. Very little can be gleaned from published books and papers; more is available from correspondence. It is likely that much of the interaction was informal. The loosely structured Lunar Society in Birmingham was a locus for the exchange of ideas, but no formal minutes were kept which could provide evidence.[27] In Edinburgh and Glasgow there were plenty of social clubs, but these, by their nature, left very little trace of what went on in them. Occasional light is shed in social letters. In one written in May 1784, John Hope (1725-86), professor of botany at Edinburgh, revealed to Matthew Boulton, "Last night we had a full meeting at the Oyster Cellar, Mr Cort, Lord Dundonald, Hutton, Black, MacGowan, etc. Dr Hutton whispered to me, what a number of projectors, and Black said I was a fool of a one myself. We had as usual a great deal of pleasantry, and every now and then some useful and interesting conversation."[28] This makes it clear that the Oyster Club brought together a group whose interests were complementary. The landowning aristocracy was represented by Lord Dundonald (1749-1831) whose estate at Culross was used for various industrial purposes; John McGowan (d.1803) was solicitor to the Customs and Excise; Henry Cort (1741-1800) was an innovative ironmaster; and James Hutton (1726-1797) could be described as an Edinburgh chemical manufacturer (that was where his income was made; he held no post in geology).[29] These men were carousing on equal terms with the university professoriate represented by Joseph Black and John Hope (1725-86), whilst having "every now and then some useful and interesting conversation."

This indicates the importance of letters as evidence. Black's correspondence of about 830 items is the key to revealing his relationship with entrepreneurs

26 Black certainly had a private room: Adam Ferguson made a comment about its tidiness, it "never being lumbered with [...] the apparatus of experiments", see Anderson and Jones, *Correspondence*, vol. 1, p. 49 (see note 21).

27 What evidence there is has been assembled by Peter M. Jones, see his *Industrial Enlightenment: Science, technology and culture in Birmingham and the West Midlands* (Manchester: Manchester University Press, 2009), 82-94, especially Table 3.0 on 92-3.

28 Clow and Clow, *The Chemical Revolution*, p. 415 (see note 9).

29 On Hutton, see the essay by Lissa Roberts and Joppe van Driel in this volume.

and industrialists.[30] Nearly seventy of his correspondents sought direct advice on industrial issues. Through this evidence, there can be no doubt that Black contributed a great deal to the developing chemical industry in Scotland and elsewhere. It is possible to categorize his assistance under five heads. He corresponded with landowners who wanted to improve agricultural yields. For this cohort he also conducted mineral analyses in his laboratory to determine how to extract metals profitably. He was frequently contacted by the Board of Fisheries and Manufactures to help them make judgments on giving grants and premiums to those submitting new ideas for improving various processes. The Board of Customs asked him about issues concerning taxation. Finally, he was contacted by those seeking advice on the likely commercial viability of developing new chemical works.

Throughout all this, it is not clear what Black himself gained. Was he was simply being philanthropically responsive? One scheme certainly did occupy his time, thought and energy. This was his own attempt to produce alkali by a cheap method. Soda and potash were important in the Scottish industry, not only textiles but glass and soap as well. Traditional methods of alkali production, by burning wood or seaweed and leaching-out salts, were extremely inefficient, though Black advised on this, determining experimentally which ferns from various Scottish estates offered the best yield.[31] Statistics were also determined for kelp found on various Scottish beaches.[32] It was important because imports from overseas, especially Spanish barilla, were subject to interruption by the many wars and skirmishes of the period. There would be significant financial rewards if a new chemical process using cheap reagents could be made to work. The attempted method was to juxtapose plates of solid salt and lime, hoping to get them to react together by means of a double decomposition which, it was hoped, would result in soda and chloride of calcium. Experiments by Black were conducted in parallel with James Watt's in Glasgow and their correspondence reveals their frustration at their failure, even though over-optimistically, at the same time, they talked about patenting the process. Watt's work was on a large scale; he refers to a reagent 'house' being constructed, 36 feet in length, 18 feet wide and 15 feet in height; he estimated that after one year, nine tons of alkali would be produced. Black at a

30 Anderson and Jones, *Correspondence,* 2 vols. *passim* (see note 21).

31 Ibid., pp. 511-13, letter from Black to Ebenezer Macculloch, 9 July 1782; Mcculloch is described here as 'General Surveyor of the Linen Manufactures, under the Honble. Board of Trustees'.

32 Black's experiments on kelp are recorded by Andrew Fyfe, "On the Comparative Value of Kelp and Barilla," *Transactions of the Highland Society* 5 (1820): 10-64, on 29.

later stage spoke in general about how long his experiments took to conduct (though he did not specify which experiments these were).[33] Both men were certainly concerned that others would develop a viable scheme before they did.[34] This hands-on experimentation with an industrial process in mind was exceptional. It seems unlikely that Black's laboratory facilities resembled Klein's description: "large parts of the academic chemical laboratory instruments resembled the instruments of assayers, smelters, apothecaries, and distillers"; there was simply insufficient space.[35]

More details are known about experiments requested by others because some replies survive. Black had a particularly extensive correspondence with Henry Home, Lord Kames (1696-1782), a scientifically minded judge, agricultural improver and leading member of the Edinburgh literati.[36] Kames was attempting to drain part of his estate and turned to Black because he wanted to know about the nature of alluvial clays. He wrote in a preface to one of his books, "An imprimatur from one of the ablest chymists of the present age [that is, Black] has given me some confidence of being on the right track."[37]

The Board of Fisheries and Manufactures generally turned to the academic world for advice on practical matters. An example involving Black (and James Hutton) concerned the development of a new type of furnace designed by one William Cottrell (fl.1780s), intended for the smelting of ores to produce pig iron.[38] Black and Hutton had their doubts from the outset, but "Since so much has already been laid out upon an experiment, which seems to be conducted with ingenuity, we should be sorry not to see it brought to some conclusion."[39] A very thorough report was finally submitted on 27 October 1783.[40]

33 Anderson and Jones, *Correspondence*, vol. 2, pp. 1068-69 (see note 21).

34 See correspondence between Black and Watt, 23 January to 10 July 1769 in ibid., vol. 1, especially pp. 206-13.

35 Klein, "Technoscience," p. 240 (see note 3).

36 Anderson and Jones, *Correspondence*, vol. 1, pp. 312-322 and 323-329 (see note 21).

37 Henry Home, *The Gentleman Farmer; Being an attempt to improve Agriculture by subjecting it to the test of rational principles* (Edinburgh: W. Creech, 1776), xii-xiii.

38 Anderson and Jones, *Correspondence*, vol. 1, pp. 649-50 (see note 21). Letter from Robert Arbuthnot, Secretary to the Board, to Black and Hutton, 27 June 1783.

39 Ibid., vol. 1. pp. 650-52. Draft letters from Black and Hutton to the Board of Fisheries and Manufactures, after 27 June 1783.

40 Ibid., vol. 1, pp. 670-75. Report of Black and Hutton submitted to the Board of Fisheries and Manufactures, 27 October 1783.

The Board of Customs consulted Black about the point at which poor quality, untaxed 'culm' becomes taxable, good quality coal.[41] Another commission for Customs and Excise carried out by Black related to whether tax should be exacted on the other salts in sea water, as it was on common salt and focused on the substance bittern (the solution remaining when seawater has been concentrated by evaporation, to the point where the common salt component has crystallized out).[42] The Collector of Customs at Port Glasgow wanted to know, for a start, what it was. Although Black was willing to give advice on scientific aspects of the matter and carry out calculations, he made it clear that he did not want to be drawn into giving an opinion which would affect tax legislation.[43]

Black was astute when it came to judging financial consequences of establishing industrial processes, something which perhaps would not be expected of an eighteenth century academic scientist. This comes through very clearly from Black's correspondence with Archibald Cochrane, 9th Earl of Dundonald. Cochrane owned an estate on the north shore of the Firth of Forth at Culross. Not a wealthy man, he developed various industrial schemes which, he hoped, would ease his financial situation. Black was cautious about Cochrane's ability to bring anything he started to a successful conclusion. He wrote to Cochrane's brother-in-law in surprizingly forthright terms:

> In my preceeding conversations with his Lordship he had appeared to me to be very ingenious & acute but at the same time so overflowing with Projects & so extreamly [sic] sanguine [...] I may say extravagant in his estimation of them that I neither expected to find any thing Solid or Satisfactory nor had the least hope that my advice or opinion would be attended to.[44]

One of these Projects was tar production, which Cochrane intended to produce by distilling coal, available on his estate. Cochrane had ambitions to sell

41 Ibid., vol. 1, pp. 345-38; On culm, see also the essay by Roberts and Van Driel in this volume.

42 Ibid., vol. 2, pp. 1214-15 and 1216-23.

43 William Ramsay, *Life and Letters of Joseph Black* (London: Constable, 1918), 108; The letter from which Ramsay made this judgment no longer survives. Contrast this with the role of academic chemists as advisors to the Dutch government, see Joppe van Driel and Lissa Roberts, "Circulating Salts: Chemical governance and the bifurcation of "nature" and "society"," *Eighteenth Century Studies* 49 (2016): 233-63.

44 Anderson and Jones, *Correspondence*, vol. 2, pp. 599-602 (see note 21). Letter, Black to Andrew Stuart, Edinburgh, 25 January 1783.

his tar to the Navy to preserve ships' bottoms from becoming worm-eaten. Black's analysis was very detailed, considered under six separate headings. He performed different sets of calculations showing that the income at time of war would be greater than if the country were at peace (the weekly value would be £88.8.0 compared with £69.16.0, respectively).[45] Of Black's highly practical contribution to Cochrane's enterprise, Sir John Dalrymple, Solicitor General for Scotland, wrote that Black was, "the best judge, perhaps in Europe, of the merit of such inventions."[46]

Though it is clear that Black was a key figure in advising on the development of a variety of industrial ventures, he was cautious about getting too involved – and he was good at making excuses, even to the aristocracy. When the Earl of Hopetoun, who had asked Black to analyze samples of gravel for gold, summoned him to Hopetoun House in March 1772, a distance of less than 13 miles, Black wrote, "I intended to have taken the opportunity of the present holidays to have paid my respects to your L^d^ship at H^p^. House but the change of the weather made me delay that Ride to another time."[47] There are other similar examples of this attitude.

Thomas Charles Hope did not play the same sort of role as Black in advising those developing industries, though much less is known about his personal inclinations; very little of his correspondence survives. Hope himself declared, "I consider my vocation to be the teaching the science [chemistry]."[48] Until recently, it was thought that Hope cared little about developing up-to-date content for his chemistry course, but reference to his teaching notes show that he was well aware of innovations and he updated his teaching continually.[49] Though it is fair to say that he concentrated largely on more theoretical aspects, he did not ignore the industrial side of chemistry as seen in relation to dyes.

George Wilson (1818-59) was an Edinburgh figure who operated at the borders of university teaching and was significantly involved with industrialists.[50]

45 Ibid., vol. 1, pp. 580-89. Black's notes and calculations.

46 John Dalrymple, *Addresses and Proposals* [...] *on the Subject of Coal, Tar and Iron Branches of Trade* (Edinburgh: Thomson Gale, 1784).

47 Anderson and Jones, *Correspondence*, vol. 1, pp. 250-52. Letter, Joseph Black to the 2nd Earl of Hopetoun, 14 March 1772.

48 Thomas Stewart Traill, "Memoir of Dr Thomas Charles Hope," *Transactions of the Royal Society of Edinburgh* 16 (1849): 419-34, on 431.

49 Robert G.W. Anderson, "Thomas Charles Hope and the Limiting Legacy of Joseph Black," Anderson, ed. *Cradle of Chemistry*, pp. 147-62 (see note 16).

50 Jessie Aitken Wilson, *Memoirs of George Wilson* (Edinburgh: Edmonston and Douglas, 1860).

Trained in medicine at Edinburgh, Wilson had no intention of practising, seeing his training as preparing him for a career in chemistry. After graduation in 1838 he turned down an offer to work with Thomas Graham at University College, London, settling into a life of teaching chemistry extramurally in Edinburgh. In 1855, when a search was on for someone to direct the newly established Industrial Museum of Scotland, one of Wilson's former university colleagues, the influential Lyon Playfair (1818-98), recommended him for the job. Though it would be several years before a museum building became available, Wilson was immediately active in summoning the help of industrialists to fill his stores with teaching material.[51] The University became concerned that Wilson, a popular teacher, would usurp the role of professor of chemistry, so it created a Regius Professorship of Technology and appointed Wilson to it. It could thereby exert a degree of control over his activities. Wilson's policy for teaching industrial processes was clearly set out in two lectures given at the Philosophical Institution in 1856.[52] At least forty-eight extramural chemistry lecturers who taught courses starting between 1768 and 1855 were involved in a similar way.[53] A small proportion of these were involved in industrialization. George Dixon Longstaff (1799-1892), for example, conducted tar distillation at Leith shortly after graduating and then returned to England where he developed industrial interests in oils, colors and varnish while teaching at mechanics institutions.[54]

The Case of Dyes

The considerable importance of the linen and cotton industry in Scotland provides context for Joseph Black's keen interest in dyeing.[55] He prepared Prussian Blue in front of his class, a fact recorded by Thomas Charles Hope in his student

51 R.G.W. Anderson, "'What is Technology?': Education Through museums in the mid-nineteenth century," *British Journal for the History of Science* 25 (1992): 169-84.

52 George Wilson, *On the Objects of Technology and Industrial Museums. Two lectures addressed to the Philosophical Institution, Edinburgh, in February 1856* (Edinburgh: Sutherland & Knox, 1856).

53 Robert G.W. Anderson, "Chemistry Beyond the Academy: Diversity in Scotland in the early nineteenth century," *Ambix* 57 (2010): 84-103.

54 Anon., "George Dixon Wagstaff," *Journal of the Chemical Society* 63 (1893): 751-54.

55 Alastair J. Durie, *The Scottish Linen Industry in the Eighteenth Century* (Edinburgh: John Donald, 1979), 32-94.

lecture notes as having taken place in March 1784.[56] Some of Black's correspondence is directly concerned with dyestuffs, especially the dyes cudbear and Turkey red.[57] George Mackintosh (1739-1807), partner in a Glasgow concern which manufactured these dyestuffs wrote to Black: "I have late discovered a preparation which makes cudbear dye linen and cotton with the same ease, as it does woollens, which till now it could not do. I would cheerfully lay my process before you could I expect to be honoured with your sentiments upon it, as I want to have it fully complete before I discover it to the public."[58] Obvious from this is that Mackintosh was seeking Black's valued *imprimatur*.

In November 1782, Mackintosh wrote again to thank Black and ask for his "approbation and opinion [of] the late discovery I made of making cudbear strick [sic] on linens and cottons."[59] Later in November 1782, he sent six samples of dyed linen and five samples of dyed silk thread to Black. Perhaps Mackintosh felt that he was writing too often, "[b]ut it is an affair of some importance to individuals."[60] Black's replied encouragingly, ending his letter with a friendly promise to keep details of their correspondence secret,[61]

The final letter of the sequence is from Mackintosh to Black in 1785, when he sent Black six handkerchiefs "of a very excellent colour." The letter refers to a French dyer, probably Pierre Jacques Papillon (fl.1780s-90s), who came to work in the extensive dyeworks at Dalmarnock established by Mackintosh and David Dale. He was one of a succession of European dyers who struggled to learn the secret of making fast red dye from madder. Papillon arrived in Glasgow in 1785, via Manchester. The dye he produced was considered to be excellent, though he was an employee rather than a partner and he left in 1787. He petitioned the Board of Manufactures on three occasions, between 1787 and 1790, for a grant to set up and run his own dyeworks, to which they were sympathetic, but quickly realized that they could scarcely offer him money

56 Thomas Charles Hope's lecture notes taken in Black's class, 1783-84, Edinburgh University Library Research Collections, MS Dc.10.9[5]. Hope dated the demonstration as being done in March 1784.

57 Turkey red was used widely in Scottish works, developed in 1785 at Dalmarnock on the River Clyde near Glasgow. The excellent color was obtained from the rubia plant by a complicated process. See W.T. Johnston, *The Secret of Turkey Red* (Livingston: Officina W.T. Johnston, 1993); see also Clow and Clow, *Chemical Revolution*, pp. 217-19 (see note 9).

58 Anderson and Jones, *Correspondence*, vol. 1, pp. 522-24 (see note 21). Letter, Mackintosh to Black, 31 August 1782.

59 Ibid., vol. 1, pp. 543-45. Letter, Mackintosh to Black, 2 November 1782.

60 Ibid., vol. 1, pp. 545-57. Letter, Mackintosh to Black, 4 November 1782.

61 Ibid., vol. 1, pp. 550-1. Draft letter, Black to Mackintosh, n.d.

from public funds for a secret process.[62] A solution was eventually found: Black was asked to examine Papillon's petition, in confidence, and adjudicate it. This he did, with a positive recommendation. It was agreed that there would be a ban on publication of the process for a period of twelve years, after which it would be made publicly available.[63]

Another interaction Black had concerning dyes was with William Hamilton (1758-1807) in 1791. Hamilton was an Irishman who had been a student of Black's when working for his MD, awarded in 1779. Hamilton had been translating *Éléments de l'Art de la Teinture* by Claude Louis Berthollet (1748-1822), published in Paris in 1791. He complained to Black that someone else was publishing another translation of the same book, implying piracy, and that Black had sanctioned it.[64] Black explained in a reply that he had asked for a copy of the original French version, and having received a copy, he was asked by a publishing firm whether it should be translated.[65] Black commended it highly, saying that it would be very useful and should be published in English. Sample pages of translation received a favorable opinion from him and the book was duly produced – which was what so distressed Hamilton. Again, this incident indicates the status which Black had achieved in being sought to make judgments on industrial matters.

In 1795, Black purchased a bond for £500 sterling from Robert Graham of Gartmore (1735-1797) in the Culcreach (or Culcruech) Cotton Company, a very substantial sum.[66] Black's relationship with Graham, who was a considerable landowner, politician and poet, is not known; no correspondence between them survives. Graham was Rector of Glasgow University between 1785 and 1787, which possibly explains how they knew each other. The Culcreach Company had been established at Fintry, in the Campsie Hills, twenty miles north of Glasgow, by Alexander Spiers.[67] Black also took out a £500 sterling bond with the Edinburgh and Leith Glasshouse Company. He had developed a close friendship with its proprietor, Archibald Geddes, but the professional

62 Johnston, *Secret*, appendix 2, pp. 11-14 (see note 57).

63 Papillon's Turkey red process was published in the *Edinburgh Evening Courant* on 1 December 1803.

64 Anderson and Jones, *Correspondence*, vol. 2, p. 1151 (see note 21). Letter, Hamilton to Black, 14 January 1792.

65 Ibid., vol. 2, p. 1152. Draft letter, Black to Hamilton, n.d.

66 Ibid., vol. 2, p. 1467: "Minutes taken at the opening of the Repositories of the late Dr. Joseph Black [...] Bond by Robert Grahame of Gartmore esq & other partners of the Culcreuch Company for £500 St in favor of the said Dr Black – dated 2d. Decr. & 8 Dec. 1795."

67 Some papers are in Glasgow University Library Special Collections, MS Gen 1717.

roles of the two men appear distinct; Geddes attended two sessions of Black's chemistry class, paying his three guineas per session, like everybody else.

Hope's sparse research interests were different from Black's: he was the son of John Hope, holder of the chair of botany at Edinburgh, and a few of his published papers dealt with plant chemistry. There was quite a substantial section on inks and dyeing in his chemistry course, the evidence for this being found in the notes from which he lectured. He started the latter subject of his course by stating: "The Art of Dyeing is one of the most interesting of the Chemical Arts – It comprehends the methods of applying & fixing durably colouring matter on animal & vegetable fibre, so as to communicate on the endless variety of beautiful Tints [...]. Every process is in fact chemical one & often very intricate." He then offered definitional discussions of dyeing in general and topical dyeing, further explaining that:

> If no Chemical Attraction be exerted between the fibre of the stuff & the colouring matter, this matter having been soluble & actually dessolved in the water of the vat, continues soluble in the same menstruum – consequently when the dyed stuff is treated with water the colouring matter is compleatly carried off & the stuff again becomes colourless.[68]

Following this introduction to the subject, Hope continued with a section on fixing coloring agents with mordants – solutions of alum or tin salts. He then treated indigo: "Dr Roxburgh of Calcutta teaches [...] leaves of Nerium Tinctorum, or diffr. Species of Indigofera are mixed with cold water in a boiler wc. Is heated slowly to 150° & then the clear water is strained off – Dr Roxburgh has proved that the production of the Blue substce or Indigo is owing to the absorption of Oxyg by the Green base."[69] In 1809, Hope felt that, as usual, he was running behind with his course and he decided not to discuss lichens and cudbear.[70] He did go on to teach a substantial section on how patterns are printed on to calico.

68 Edinburgh University Library Research Collections, MS Gen 268-277, envelope 9, 'Astringents/Colouring Matter'.

69 William Roxburgh (1751-1851), a Scots surgeon working in Calcutta, had been taught by Thomas Charles Hope's father, John Hope; he experimented with sources of indigo for dyes from 1790, see Alexander Dalrymple, *Oriental Repertory*, 2 vols. (London: G. Biggs, 1793-1797), vol. 2 (London, 1793), 39-44; Roxburgh's full paper was published 18 years later: William Roxburgh, "Account of a new Species of Nerium," *Transactions of the Society* [...] *for the Encouragement of Arts* 28 (1811): 251-307.

70 Cudbear is a natural red-purple dye, extracted from lichens. Its production was patented in 1758 by George and Cuthbert Gordon, and from 1764 it was manufactured in Leith,

Glasgow academics, including Andrew Ure when he was professor of natural philosophy at the Andersonian Institution, were also interested in dyes from a chemical point of view. Relevant to the Scottish dye industry is Ure's 1824 translation of Berthollet's *Elements de l'Art de Teinture*. Acknowledging that it had previously been translated and published by Hamilton, Ure's excuse for another edition was that Berthollet had added to his original work.[71] There must be doubts about this claim. Berthollet's second edition appeared in 1804, Ure's translation was published in 1824. It seems very likely that the translation was a vehicle for publishing the considerable research into dyes which Ure was carrying out. There are, in fact, 138 pages of notes which are pure Ure. Passages from *Elements of the Art of Dyeing* indicate that he considered that laboratory experiments were necessary if dyeing were to improve; the manufacturing arts would only improve if they were guided by science. In the Introduction to his 1824 book, he wrote:

> The Arts can make but limited progress, when they are directed merely by a blind practice. Thus, they have remained for several centuries in nearly the same state in China and in India. But if artisans be guided by the knowledge of those properties which have been investigated by physics, and its complement, chemistry, there is no boundary to the perfection which they may reach. How many advantages has that nation, so powerful by its industry, derived from Watt, Wedgwood, Henry, and some other philosophers. In this respect a happy revolution has been affected among ourselves. Our manufacturers are no longer intrusted to ignorant workmen. In the greater part of them are found to be enlightened individuals, well informed philosophers, to whom indeed we must have recourse, if we wish to excite the progress of the useful arts, and remove the obstacles which stand in their way.[72]

It becomes apparent from Ure's remarks at the end of volume two, that he was conducting his own experiments on dyes, not just translating Berthollet's text. For example, when considering the relation between chemical change and color in indigo, he wrote:

---

though not very successfully. See Cuthbert Gordon, *Memorial of Mr. Cuthbert Gordon, Relative to the Discovery and Use of Cudbear, and Other Dying Wares* (London: s.n., 1784).

71 C.L. and A.B. Berthollet, *Elements of the Art of Dyeing* [...] *Translated from the French* [...] *With Notes* [...] *by Andrew Ure*, 2 vols. (London: Thomas Tegg, 1824), vol. 2, xxiv.

72 Ibid., vol. 2, p. 1.

> While engaged in these experiments, I discovered that if the action of sulphuric acid upon indigo be stopped at a certain point, a new substance, altogether different from cerulin, is produced, formed at the instant that indigo changes from yellow to blue by the action of sulphuric acid.[73]

Concerning alternative chemical substances to alum as mordants, he indicated clearly that he had been conducting tests of his own: "Although all researches hitherto made have been ineffected [sic] to find a substitute for alum, we have nevertheless made trial of a great number of substances with wool."[74] The ongoing nature of his experimental work to create a particular color is clearly understood by this statement: "We propose to determine the chemical nature of the combination formed upon wool by cochineal, tartar, and a solution of tin, and to make known the result of our enquiries upon the colour of scarlet."[75] As if to emphasise his hands-on activity, the 1841 edition of *Elements of the Art of Dyeing*, published seventeen years later, has as its subtitle: *A New Edition, Revised and Corrected by an Experienced Practical Dyer and Calico Printer.*

A number of others were making knowledgeable contributions to dyeing in or near Glasgow, one being Walter Crum (1796-1867), who bridged the academic and industrial worlds.[76] He attended Glasgow University, studying under Thomas Thomson. His university career culminated in a paper published in *Annals of Philosophy* of 1823 on the subject "Indigo".[77] This contains a good deal of detail about his laboratory experiments and refined analytical technique. Thereafter he joined his family calico printing business at Thornliebank, now

73 Ibid., vol. 2, p. 372.

74 Ibid., vol. 2, p. 324.

75 Ibid., vol. 2, p. 326.

76 Anon., "Obituary Notices of Fellows Deceased," *Proceedings of the Royal Society of London* 16 (1868): viii-x ("The intimate knowledge that Mr Crum thus acquired combined with his general scientific attainments, enabled him to introduce many useful improvements into his own business and increase the excellence of its manufactures."). For details of Crum's interest in the theory of dyeing, see Bernadette Bensaude-Vincent and Agustí Nieto-Galan, "Theories of Dyeing: A view on the long-standing controversy through the works of Jean-Francois Persoz," *Natural Dyestuffs and Industrial Chemistry in England, 1750-1880*, eds. Robert Fox and Agustí Nieto-Galan (Canton: Science History Publications, 1999), 3-24.

77 Walter Crum, "Experiments and Observations on Indigo, and on certain Substances which are produced from it by means of Sulphuric Acid," *Annals of Philosophy* (February 1823): 81-100.

a Glasgow suburb. But he continued to have scholarly urges and he published various papers pertinent to dyeing, Between 1830 and 1861 he regularly offered papers to the Philosophical Society of Glasgow. In 1844 he was appointed a Fellow of the Royal Society of London. In that same year, he proposed a theoretical interpretation of the interaction between cotton fibers and natural dyes, which brought him into contention with the French color chemist, Jean-François Persoz (1805-68), who held an alternative theory.[78] Though primarily a manufacturer, Crum maintained scientific contact with Thomas Thomson, Michael Faraday, Jean-Baptiste Dumas, Alexander von Humboldt and Justus von Liebig. As mentioned above, it is indeed difficult to categorize his position, whether as an academic or an industrialist.

Conclusion

There are large gaps in our knowledge of the contribution made by academic chemists to the Scottish industrial world, in the case here, of dyeing in Scotland in the pre-synthetic dyestuffs era. It is indisputable, however, that academe made a distinctive contribution. In this particular case, and probably others as well, it is not unreasonable to challenge the theses of Rupp, Bowden and Matthias, but equally to question the extent to which Klein's 'technoscience' label is applicable, in determining the degree of separation or interlacing of the roles of academic chemist and industrial practitioner. Joseph Black, in one of his lectures, made a most telling statement about his personal attitude to their relationships, which veers the argument somewhat more towards Klein's view than that of Matthias et al.

> I call every man a Philosopher who invents anything new or improves any business in which he is employed – even the Farmer who considers the nature of different soils or makes improvements on the ploughs he uses, I must call a Philosopher, though perhaps you can call him a Rustic one. Nor am I inclined to give much credit to those men who shut up their Closets in study and retirement have obtained the appellation of Learned Philosophers they in general puzzle more than they illustrate, they are

78 Walter Crum, "On the Manner In which Cotton Unites with Colouring Matter," *Philosophical Magazine* 24 (1844): 241-46; Bensaude-Vincent and Nieto-Galan, "Theories," pp. 8-14 (see note 76).

> wrapt in a veil of Systems and of Theories and seldom make improvements or discoveries of Use to Mankind.[79]

---

79 Quoted in Margaret C. Jacob and Larry Stewart, *Practical Matter: Newton's science in the service of industry and empire* (Cambridge, MA: Harvard University Press, 2004), 116-7; This quotation comes from lecture notes taken by Lowell Edgworth, who probably attended the 1795-96 final course given by Black, with Hope's assistance.

Bibliography of Secondary Sources

Albury, William, "The Order of Ideas: Condillac's method of analysis as a political instrument in the French Revolution," John Schuster and R.R. Yeo, eds., *The Politics and Rhetoric of Scientific Method: Historical studies* (Dordrecht: Reidel Publishing Co., 1986), 203-225.

Alder, Ken, "A Revolution to Measure: The political economy of the metric system in France," M. Norton Wise, ed., *Values of Precision* (Princeton: Princeton University Press, 1995), 39-71.

Alder, Ken, *Engineering the Revolution: Arms and enlightenment in France, 1763-1815* (Princeton: Princeton University Press, 1999).

Alder, Ken, "Innovation and Amnesia: Engineering rationality and the fate of interchangeable parts manufacturing in France," *Technology and Culture* 38 (1997): 273-311.

Algazi, Gadi, "Scholars in Households: Refiguring the learned habitus, 1480-1550," *Science in Context* 16 (2003): 9-42.

Allchin, Douglas, "James Hutton and Coal," *Cadernos IG/UNICAMP* 7 (1997): 167-183.

Allen, Tim, Mike Cotterill, Geoffrey Pike, "The Kentish Copperas Industry," *Archeologia Cantiana*, 122 (2002), 319-334.

Amin, Wahida, "The Poetry and Science of Humphry Davy," PhD Thesis, University of Salford, 2013.

Anderson, R.G.W., "A Source for Eighteenth Century Chemical Glass," Georgi Dragoni, Anita McConnell and Gerard L'E. Turner, eds., *Proceedings of the Eleventh International Scientific Commission Symposium* (Bologna: Grafis Edizioni, 1994): 47-52.

Anderson, R.G.W., "Chemistry Beyond the Academy: Diversity in Scotland in the early nineteenth century," *Ambix* 57 (2010): 84-103.

Anderson, R.G.W., *The Playfair Collection and the Teaching of Chemistry at the University of Edinburgh, 1713-1858* (Edinburgh: Royal Scottish Museum, 1978).

Anderson, R.G.W., "Thomas Charles Hope and the Limiting Legacy of Joseph Black," R.G.W., Anderson, ed., *Cradle of Chemistry: The early years of chemistry at the University of Edinburgh* (Edinburgh: John Donald, 2015), 147-162.

Anderson, R.G.W., "What is Technology?: Education through museums in the mid-nineteenth century," *British Journal for the History of Science* 25 (1992): 169-84.

Anderson, R.G.W., and Jones, Jean, eds., *The Correspondence of Joseph Black*, 2 vols., (Farnham: Ashgate, 2012).

Anderson, Stuart, ed., *Making Medicines: A brief history of pharmacy and pharmaceuticals* (London: Pharmaceutical Press, 2005).

Anes, Gonzalo, *Economía e ilustración en la España del siglo XVIII* (Barcelona: Ariel, 1969).

Anon., *Howards, 1797-1947* (Ilford: Howards & Sons, 1947).

Arber, E.A. Newell, *The Natural History of Coal* (Cambridge: Cambridge University Press, 1911).

Archer, Mary, and Christopher D. Haley, *The 1702 Chair of Chemistry at Cambridge: Transformation and change* (Cambridge: Cambridge University Press, 2005).

Arkestijn, N.L.M., "Met de bajonet op de keel. Inkopen met ongedekt papiergeld," E.H.P. Cordfunke and H. Sarfatij, eds., *Van Solidus tot euro: Geld in Nederland in economisch-historisch en politiek perspectief* (Hilversum: Verloren, 2004), 139-146.

Arnold, David, *Toxic Histories. Poison and pollution in modern India* (Cambridge: Cambridge University Press, 2016).

Ashbee, Ruth, "The Discovery of Chlorine: A window to the chemical revolution," Hasok Chang and Catherine Jackson, eds., *An Element of Controversy: The life of chlorine in science, medicine, technology, and war* (London: British Society for the History of Science, 2007), 15-40.

Ashworth, William J. "'Between the Trader and the Public': British alcohol and the proof of good governance," *Technology and Culture* 42 (2001): 27-50.

Ashworth, William J. *Customs and Excise: Trade, production and consumption in England, 1640-1845 (Oxford: Oxford University Press, 2003).*

Ashworth, William J. "Quality and the Roots of Manufacturing 'Expertise' in Eighteenth-Century Britain," *Osiris* 25 (2010): 231-254.

Ashworth, William J. "The British Industrial Revolution and the Ideological Revolution: Science, neoliberalism and history," *History of Science* 52 (2014): 178-199.

Ashworth, William J. "The Ghost of Rostow: Science, culture and the British Industrial Revolution," *History of Science* 46 (2008): 250-274.

Ashworth, William J. "The Intersection of Industry and the State in Eighteenth Century Britain," Lissa Roberts, Simon Schaffer and Peter Dear, eds., *The Mindful Hand: Inquiry and invention from the late renaissance to early industrialisation* (Amsterdam: Edita, Royal Netherlands Academy of Sciences, 2007), 348-77.

Atkins, Peter, *Liquid Materialities. A history of milk, science and the law* (Surrey: Ashgate, 2010).

Aymes, Jean René, *España y la revolución francesa* (Barcelona: Crítica, 1989).

Bada, Juan, "Don Pedro Díaz de Valdés, obispo de Barcelona (1798-1807)," *Anthologica Annua* 19 (1972): 651-674.

Baker, Keith M., *Condorcet: From natural philosophy to social mathematics* (Chicago: University of Chicago Press, 1974).

Baker, Keith M. and William A. Smeaton, "The Origins and Authorship of the Educational Proposals Published in 1793 by the *Bureau de Consultation des Arts et Métiers* and generally Ascribed to Lavoisier," *Annals of Science* 21 (1965): 33-46.

Ball, Michael and David Sunderland, *An Economic History of London, 1800-1914* (London: Routledge, 2001).

Balland, Antoine, *Les pharmaciens militaires français* (Paris: Imprimerie-librairie militaire universelle L. Fournier, 1913).

Balz, Viola, Heiko Stoff, Alexander v. Schwerin and Bettina Wahrig, eds., *Precarious Matters. The history of dangerous and endangered substances in the nineteenth and twentieth centuries* (Berlin: Max Planck Institute, 2008).

Barillé, Claire, Thomas Le Roux and Marie Thébaud-Sorger, "Grenelle, 1794. Secourir, indemniser et soigner les victimes d'une catastrophe industrielle à l'heure révolutionnaire," *Le Mouvement Social* 249 (2014): 41-71.

Barker, Richard, "French Entrepreneurship During the Restoration: The record of a single firm, the Anzin Mining Company," *The Journal of Economic History* 21 (1961): 161-178.

Barles, Sabine, *L'invention des déchets urbains: France, 1790-1970* (Seyssel: Champ Vallon, 2005).

Barles, Sabine, *La Ville Délétère. Médecins et ingénieurs dans l'espace urbain, XVIIIe-XIXe siècles* (Seyssel: Champvallon, 1999).

Barrenechea, José Manuel, *Valentín de Foronda, reformador y economista ilustrado* (Álava: Diputación Foral, 1984).

Bartrip, Peter, "A 'Pennurth of Arsenic for Rat Poison': The Arsenic Act (1851) and the prevention of secret poisoning," *Medical History* 36 (1992): 53-69.

Beale, Norman and Elaine Beale, *Echoes of Ingen Housz: The long lost story of the genius who rescued the Habsburgs from smallpox and became the father of photosynthesis* (East Knoyle: Hobnob Press, 2011).

Beek, Marcel van der, "'s Rijks Munt en de aanmuntingen door De Heus," *De Beeldenaar* 2 (1995): 357-366.

Belhoste, Bruno, *La Formation d'une technocratie. L'École polytechnique and ses élèves de la Révolution au Second Empire* (Paris: Belin, 2003).

Belhoste, Jean-François and Woronoff, Denis, "Ateliers et manufactures : Une réévaluation nécessaire," Françoise Monnier, ed., *A Paris sous la Révolution, nouvelles approches de la ville* (Paris: Publications de la Sorbonne, 2008), 79-91.

Jones, Henry Bence, *The Royal Institution: Its founders, and its first professors* (London: Longman, 1871).

Bensaude Vincent, Bernadette, "Between Chemistry and Politics: Lavoisier and the Balance," *Eighteenth Century, Essays and Interpretation* 33 (1992): 217-237.

Bensaude Vincent, Bernadette, "Chemistry as a Technoscience," Jean-Pierre Llored, ed., *The Philosophy of Chemistry: Practices, methodologies, and concepts* (Newcastle-upon-Tyne: Cambridge Scholars Publishing , 2013), 330-341.

Bensaude Vincent, Bernadette, *Lavoisier. Mémoires d'une révolution* (Paris: Flammarion, 1993).

Bensaude Vincent, Bernadette and Christine Blondel, eds., *Science and Spectacle in the European Enlightenment* (London: Ashgate Publishing, 2005).

Bensaude Vincent, Bernadette and Ferdinando Abbri, eds, *Lavoisier in European Context. Negotiating a new language for chemistry* (Sagamore Beach: Science History Publications, 1995).

Bensaude Vincent, Bernadette and Agustí Nieto Galán, "Theories of Dyeing: A view on the long-standing controversy through the works of Jean-Francois Persoz," Robert Fox and Agustí Nieto Galán, eds., *Natural Dyestuffs and Industrial Chemistry in England, 1750-1880* (Canton, MA: Science History Publications, 1999), 3-24.

Bensaude Vincent, Bernadette and Isabelle Stengers, *Histoire de la chimie* (Paris: La Découverte, 1993).

Beretta, Marco, "Pneumatic vs. 'Aerial Medicine': Salubrity and respirability of air at the end of the eighteenth century," *Nuova Voltiana: Studies on Volta and his time* 2 (2000): 49-71.

Beretta, Marco, *The Definition of Chemistry from Agricola to Lavoisier*. Uppsala Studies in History of Science, 15 (Sagamore Beach: Science History Publications, 1993).

Beretta, Marco, *The Enlightenment of Matter* (Sagamore Beach: Science History Publications, 1993).

Berman, Alex, "J.B.A. Chevallier, Pharmacist-Chemist : A major figure in nineteenth-century French public health," *Bulletin for the History of Medecine* 52 (1978): 200-213.

Berman, Morris, *Social Change and Scientific Organization: The Royal Institution, 1799-1844* (Ithaca: Cornell University Press, 1978).

Bernal, J.D., *Science in History*, 4 vols. (Harmondsworth: C.A. Watts, 1965).

Bertomeu Sánchez, José R., and Antonio García Belmar, "Pedro Gutiérrez Bueno (1745-1822) y las relaciones entre la química y la farmacia durante el último tercio del siglo XVIII," *Hispania* 208 (2001): 539-562.

Bertomeu Sánchez, José Ramón, "Animal Experiments, Vital Forces and Courtrooms: Mateu Orfila, François Magendie and the study of poisons in nineteenth-century France," *Annals of Science* 69 (2012): 1-26.

Bertomeu Sánchez, José Ramón, "Classrooms, Salons, Academies and Courts: Mateu Orfila (1787-1853) and nineteenth-century French toxicology," *Ambix* 61 (2014): 162-186.

Bertomeu Sánchez, José Ramón, "Managing Uncertainty in the Academy and the Courtroom: Normal arsenic and nineteenth-century toxicology," *Isis* 104 (2013): 197-225.

Bertomeu Sánchez, José Ramón, "Smell, Chemistry and Microscopy: Bloodstains and nineteenth-century forensic medicine," *Annals of Science* 72 (2015): 490-516.

Bertomeu Sánchez, José Ramón, *La verdad sobre el caso Lafarge. Ciencia, justicia y ley durante el siglo XIX* (Barcelona: El Serbal, 2015).

Bertomeu Sánchez, José Ramón and Agustí Nieto Galán, eds., *Chemistry, Medecine and Crime. Mateu J.B. Orfila (1787-1853) and his times* (Sagamore Beach: Science History Publications, 2006).

Bertomeu Sánchez, José Ramón and Antonio García Belmar, "Louis Jacques Thenard's Chemistry Courses at the Collège de France, 1804-1835," *Ambix* 57 (2010): 48-63.

Bickerton, David and René Sigrist, eds., *Marc-Auguste Pictet 1752-1825: Correspondance, sciences et techniques. Tome III: Les correspondants britanniques* (Geneva: Slatkine, 2000).

Blanckaert, Claude, Michel Porret, eds., *L'Encyclopédie méthodique, 1782-1832. Des Lumières au positivisme* (Genève: Librairie Droz, 2006).

Blow, Simon, *Broken Blood: The rise and fall of the Tennant Family* (London: Faber and Faber, 1987).

Blussé, Leonard, *Visible Cities. Canton, Nagasaki, and Batavia and the coming of the Americans* (Cambridge: Harvard University Press, 2008).

Bödeker, Hans Erich, "Economic Societies in Germany, 1760-1820: Organization, social structures and fields of activities, " Koen Stapelbroek and Jani Marjanen, eds., *The Rise of Economic Societies in the Eighteenth Century* (London: Palgrave MacMillan, 2012), 182-211.

Boegheim, L.M.J., "François Loriaux, de stichter van de Duitenmunt te Soerabaja," *De Beeldenaar* 3 (1996): 131-134.

Bonastra, Quim, "Los orígenes del lazareto pabellonario. La arquitectura cuarentenaria en el cambio del setecientos al ochocientos," *Asclepio* 60 (2008): 60-61.

Bonnassieux, Pierre, *Conseil de commerce et bureau du commerce 1700-1791. Inventaire analytique des procès-verbaux* (Paris: Imprimerie nationale, 1900).

Boomgaard, Peter, ed., *Empire and Science in the Making. Dutch colonial scholarship in comparative global perspective, 1760-1830* (New York: Palgrave Macmillan, 2013).

Bosma, Ulbe, *The Sugar Plantation in India and Indonesia. Industrial production, 1770-2010* (Cambridge: Cambridge University Press, 2013).

Boudia, Soraya and Nathalie Jas, eds., special issue, "Risk and Risk Society in Historical Perspective," *History and Technology* 23 (2007).

Bourquelot, Émile, *Le centenaire du Journal de pharmacie et de chimie (1809-1909)* (Paris: Octave Doin et fils, 1910).

Bouvet, Maurice, "Nicolas Vauquelin, droguiste," *Revue d'histoire de la pharmacie* 46 (1958): 246-252.

Bowden, Witt, *Industrial Society in England Towards the End of the Eighteenth Century*, second edition (London: Frank Cass & Co., 1965).

Bray, Francesca, *Technology, Gender and History in Imperial China: Great transformations reconsidered* (London; New York: Routledge, 2013).

Bret, Patrice, "Des essais de la Monnaie à la recherche et à la certification des métaux: un laboratoire modèle au service de la guerre et de l'industrie (1775-1830)," *Annales historiques de la Révolution Française* 320/2 (2000): 137-148.

Bret, Patrice, *L'Etat, l'armée, la science. L'invention de la recherche publique en France, 1763-1830* (Rennes: Presses Universitaires de Rennes, 2002).

Brewer, John, *The Sinews of Power: War, money, and the English State, 1688-1783* (Cambridge: Harvard University Press, 1988).

Brewer, John and Roy Porter, *Consumption and the World of Goods* (New York: Routledge, 1993).

Brock, William H., *Justus von Liebig: The chemical gatekeeper* (Cambridge: Cambridge University Press, 1997).

Brose, Eric Dorn, *The Politics of Technological Change in Prussia: Out of the shadow of antiquity, 1809-1848* (Princeton: Princeton University Press, 1993).

Brown, Chandos Michael, *Benjamin Silliman: A life in the young republic* (Princeton, NJ: Princeton University Press, 2014).

Bucknill, John, *The Coins of the Dutch East Indies* (London: Spink, 1931).

Burnby, J., "The Early Years of the Pharmaceutical Industry," Lesley Richmond, Julie Stevenson and Alison Turton, eds., *The Pharmaceutical Industry: A guide to historical records* (Aldershot: Ashgate, 2003), 1-13.

Burney, Ian, "Languages of the Lab: Toxicological testing and medico-legal proof," *Studies in History and Philosophy of Science* 33 (2002): 289-314.

Burney, Ian, *Poison, Detection, and the Victorian Imagination* (Manchester: University Press, 2006).

Butt, John, *John Anderson's Legacy: The University of Strathclyde and its antecedents* (East Linton: Tuckwell Press, 1996).

Callon, Michel, "What Does it Mean to Say that Economics is Performative?," *CSI Working Papers Series* 5 (2006).

Callon, Michel and Vololona Rabeharisoa, "Research "In the Wild" and the Shaping of New Social Identities," *Technology in Society* 25 (2003): 193-204.

Cameron, Rondo, "A New View of European Industrialization," *The Economic History Review* 38 (1985): 1-23.

Campbell, R.H., A.S. Skinner and W.B. Todd, eds., *The Glasgow Edition of the Works and Correspondence of Adam Smith* (Oxford: Clarendon Press, 1976).

Cannadine, David, "The Present and the Past in the English Industrial Revolution, 1880-1980," *Past and Present* 103 (1984): 131-172.

Cantlie, Neil, *A History of the Army Medical Department*, vol. 1 (London and Edinburgh: Churchill Livingstone, 1974).

Cardwell, Donald, "Ure, Andrew," *Oxford Dictionary of National Biography* (Oxford: Oxford University Press, 2004).

Carrasco Jarabo, Paula, "Vida y obra de Pedro Gutiérrez Bueno" *Boletín de la Sociedad Española de Historia de la Farmacia* 15 (1964): 154-69; 16 (1965): 10-24; 71-86; 101-18 and 153-77.

Carrillo, Juan L., *Juan Manuel de Aréjula (1755-1830). Estudio sobre la fiebre amarilla* (Madrid: Ministerio de Sanidad y Consumo, 1986).

Carrillo, Juan L., Pedro Riera Perelló and Ramón Gago, "La introducción en España de las hipótesis miasmática y prácticas fumigatorias. Historia de una polémica (J.M Aréjula – M.J Cabanellas)," *Medicina e historia* 67 (1977): 8-26.

Carrillo, Juan L., Jesús Castellanos and María Dolores Ramos, *Enfermedad y sociedad en la Málaga de comienzos del siglo XIX: El diálogo de los muertos en la epidemia de Málaga (c.1803)* (Málaga: Universidad de Málaga, 1980).

Carrillo, Juan L., and L. García Ballester, "El comportamiento de las clases y grupos sociales de Málaga en las epidemias de fiebre amarilla," *Cuadernos de la historia de la medicina española*, 11 (1972): 88-95.

Cassin, Barbara, ed., *Dictionary of Untranslatables: A philosophical lexicon* (Princeton: Princeton University Press, 2014).

Chakrabarti, Pratik, "Empire and Alternatives: Swietenia febrifuga and the cinchona substitutes," *Medical History* 54 (2010): 75-94.

Chakrabarti, Pratik, *Materials and Medicine: Trade, conquest and therapeutics in the eighteenth century* (Manchester: Manchester University Press, 2010).

Chambers, Neil, ed., *The Scientific Correspondence of Joseph Banks*, 6 volumes (London: Pickering, 2007).

Chandler, Alfred, "Anthracite Coal and the Beginnings of the Industrial Revolution in the United States," *Business History Review* 46 (1972): 141-181.

Chang, Hasok, and Catherine Jackson, eds., *An Element of Controversy: The life of chlorine in science, technology, medicine and war* (Oxford: British Society for the History of Science, 2007).

Chauvaud, Frédéric, *Experts et expertise judiciaire : France, XIXe et XXe siècles* (Rennes : PUR, 2003).

Chauvaud, Frédéric, *Les experts du crime. La médecine légale en France au XIXè siècle* (Paris: Aubier, 2000).

Clarke, Adele and Joan Fujimura, eds., *The Right Tools for the Job* (Princeton: Princeton University Press, 1992).

Claver, Alex, *Dutch Commerce and Chinese Merchants in Java. Colonial relationships in trade and finance, 1800-1942* (Brill: Leiden, 2014).

Clow, Archibald and Nan Clow, *The Chemical Revolution: A contribution to social technology* (London: Batchwork Press, 1952).

Cockayne, Emily, *Hubbub: Filth, noise and stench in England, 1600-1770* (New Haven: Yale University Press, 2007).

Coggan, Philip, *Paper Promised: Debt, money and the new world order* (New York, NY: Public Affairs, 2012).

Cohen, I. Bernard, *Revolution in Science* (Cambridge, MA: Harvard University Press, 1985).

Cook, Harold J. and Timothy D. Walker, "Circulation of Medicine in the Early Modern Atlantic World," *Social History of Medicine* 26 (2013): 337-51.

Cooper, Alix, "Homes and Households," Katharine Park and Lorraine Daston, eds., *The Cambridge History of Science*, Vol. 3: *Early Modern Science* (Cambridge: Cambridge University Press, 2006), 224-237.

Cooper, Denis R., *The Art and Craft of Coinmaking. A history of minting technology* (London: Spink & Son 1988).

Cooper, Timothy, "Peter Lund Simmonds and the Political Ecology of Waste Utilization in Victorian Britain," *Technology and Culture* 52 (2011): 21-44.

Cooter, Roger and Stephen Pumfrey, "Separate Spheres and Public Places: Reflections on the history of science popularization and science in popular culture," *History of Science* 32 (1994): 237-267.

Corbin, Alain, *Le miasme et la jonquille. L'odorat et l'imaginaire social, XVII-XIXe siècles* (Paris: Aubier, 1982).

Corvol, Andrée, transl. by Richard C. Holbrook, "The Forest," Pierre Nora and David P. Jordan, eds., *Rethinking France: Les lieux de mémoire*, vol. 2, *Space* (Chicago: University of Chicago Press, 2006).

Crawforth, Michael A., *Weighing Coins: English folding gold balances of the eighteenth and nineteenth centuries* (London: Cape Horn Trading, 1979).

Crosland, Maurice, "'Slippery Substances'. Some practical and conceptual problems in the understanding of gases in the pre-Lavoisier era," Frederic L. Holmes and Trevor H. Levere, eds., *Instruments and Experimentation in the History of Chemistry* (Cambridge: MIT Press, 2000), 79-89.

Crouzet, François, *The First Industrialists: The problem of origins* (Cambridge: Cambridge University Press, 1985).

Cullen, William R., *Is Arsenic an Aphrodisiac? The sociochemistry of an element* (Cambridge: RSC, 2008)

Curth, Louise Hill, ed., *From Physick to Pharmacology: Five hundred years of British drug retailing* (Aldershot: Ashgate, 2006).

Cushman, Gregory, *Guano and the Opening of the Pacific World: A global ecological history* (Cambridge: Cambridge University Press, 2013).

Dalmasso, Gilbert, "Présence de la 'chymie' dans la France du Nord de la deuxième moitié du XVIIIe siècle au premier tiers du XIXe siècle," PhD Thesis, Université de Lille III, 2005.

Darnton, Robert, *Mesmerism and the End of the Enlightenment in France* (Cambridge, MA: Harvard University Press, 1968).

Darnton, Robert, *The Business of Enlightenment* (Cambridge: Harvard University Press, 1979).

Darnton, Robert, *The Forbidden Bestsellers of Pre-Revolutionary France* (New York: W.W. Norton and Co., 1996).

Daston, Lorraine, ed., *Biographies of Scientific Objects* (Chicago: University Press, 2000).

Daston, Lorraine, ed., *L'économie morale des sciences modernes* (Paris : La Découverte, 2014).

Daston, Lorraine, ed., "The Moral Economy of Science," *Osiris* 10 (1995): 2-24.

Daston, Lorraine, ed., *Things That Talk. Object Lessons from Art and Science* (New York: Zone Books, 2008).

Daumas, Maurice, *Les instruments scientifiques aux XVII et XVIII siècles* (Paris: Presses Universitaires de France, 1953).

Dauphin, Alain and Dominique Pradeau, eds., *La pharmacie hospitalière dans les hôpitaux de Paris* (Reims: Imprimerie Sogedic-Landais, 1995).

Davies, Timothy, "British Private Trade Networks and Metropolitan Connections in the Eighteenth Century," Maxine Berg, ed., *Goods from the East, 1600-1800: Trading Eurasia* (Basingstoke: Palgrave Macmillan, 2015), 154-167.

Davison, Luke Antony Francis, "Raising up Humanity: A cultural history of rescucitation and the Royal Humane Society of London, 1774-1808," PhD thesis, University of York, 2001.

Davy, John, *Memoirs of the Life of Sir Humphry Davy, Bart*, 2 volumes (London: Longman, 1836).

Deforges, Yves, *Le graphisme technique. Son histoire et son enseignement* (Seyssel: Champ Vallon, 1981).

Delépine, Marcel, "Joseph Pelletier and Joseph Caventou," *Journal of Chemical Education* 28 (1951): 454-461.

Deligeorges, Stéphane, Alexandre Gady and Françoise Labalette, *Le jardin des plantes et le Muséum national d'histoire naturelle* (Paris: Monum, 2004).

Denys, Catherine, *Police et Sécurité au XVIIIe siècle dans les Villes de la frontière Franco-Belge* (Paris: L'Harmattan, 2002).

Devaux, Guy, «Marchands de mort-aux-rats,» *Revue d'histoire de la pharmacie* 92 (2004): 509-516.

Dhombres, Jean, and Nicole Dhombres, *Naissance d'un nouveau pouvoir: Sciences et savants en France, 1793-1824* (Paris: Payot, 1989).

Diederiks, H.A., "Hendrik de Heus. Een Amsterdamse ondernemer in het begin van de 19e eeuw," *Amstelodamum* 56 (1969): 58-65.

Dillemann, Georges, "Les établissements d'enseignement pharmaceutique de 1803 à 1994," *Annales pharmaceutiques françaises* 53 (1995) : 1-7.

Dillemann, Georges, ed., *La faculté de pharmacie de Paris, 1882-1982* (Paris: Éditions Comarco, 1982).

Douglas, Susan J, "Some Thoughts on the Question 'How Do New Things Happen?'" *Technology and Culture* 51 (2010): 293-304.

Drayton, Richard, *Nature's Government: Science, imperialism, and the "improvement" of the world* (New Haven: Yale University Press, 2000).

Driel, Joppe van, "Ashes to Ashes: The stewardship of waste and oeconomic cycles of agricultural and industrial improvement, 1750-1800," *History and Technology* 30 (2014): 177-206.

Driel, Joppe van, "The Filthy and the Fat: Oeconomy, chemistry and resource management in the age of revolution," PhD Thesis, University of Twente, 2016.

Driel, Joppe van and Lissa Roberts, "Circulating Salts: Chemical governance and the bifurcation of 'nature' and 'society'," *Eighteenth-Century Studies* 49 (2016): 233-263.

Durie, Alastair J., *The Scottish Linen Industry in the Eighteenth Century* (Edinburgh: John Donald, 1979).

Dutton, Harold Irvin, "The Patent System and Inventive Activity During the Industrial Revolution (1750-1852)," PhD Thesis, London University, 1981.

Dyer, G.P. and P.P. Gaspar, "Reform, the New Technology and Tower Hill, 1700-1966," C.E. Challis ed., *A New History of the Royal Mint* (Cambridge: Cambridge University Press, 1992), 398-606.

Ecole normale supérieure, *Le livre du centenaire. Edition du bicentenaire* (Paris: Presses de l'École normale supérieure, 1994).

Eddy, Matthew D., "The Sparkling Nectar of Spas; or, Mineral water as a medically commodifiable material in the province, 1770-1805," Ursula Klein and Emma Spary, eds., *Materials and Expertise in Early Modern Europe: Between market and laboratory* (Chicago: University of Chicago Press, 2010), 283-292.

Edgerton, David, *Shock of the Old: Technology and global history since 1900* (London: Profile Books, 2006).

Ellis, Kenneth, *The Post Office in the Eighteenth Century: A study in administrative history* (Oxford: Oxford University Press, 1958).

Elmpt, Theo van, *Netherlands East Indies. Paper currency, 1815-1827* (Uithoorn: Elran Express, 2009).

Elson, Robert E., "*The Floracrats* (book review)," *American Historical Review* 116 (2011): 1469.

Emerson, Roger and Paul Wood, "Science and Enlightenment in Glasgow, 1690-1802," Charles Withers and Paul Wood, eds., *Science and Medicine in the Scottish Enlightenment* (East Linton: Tuckwell Press, 2002), 79-142.

Erikson, Emily, *Between Monopoly and Free Trade: The English East India Company, 1600-1757* (Princeton: Princeton University Press, 2014).

Ewing, Heather, *The Lost World of James Smithson: Science, revolution, and the birth of the Smithsonian* (New York: Bloomsbury, 2007).

Faure, Henri, *Histoire de la céruse depuis l'antiquité jusqu'aux temps modernes* (Lille: Levebvre-Ducrocq, 1889).

Fichter, James R., *So Great a Proffit. How the East India trade transformed Anglo-American capitalism* (Cambridge: Harvard University Press, 2010).

Fisher, George Park, *Life of Benjamin Silliman M.D., LL.D.*, 2 vols. (New York, 1866).

Fissell, Mary E., *Patients, Power, and the Poor in Eighteenth Century Bristol* (Cambridge, Cambridge University Press, 1991).

Forbes, John S., *Hallmark. A history of the London Assay Office* (London: The Goldsmith's Company, 1999).

Forgan, Sophie, ed., *Science and the Sons of Genius: Studies on Humphry Davy* (London, Science Reviews, 1980).

Fors, Hjalmar, "The Knowledge and Skill of Foreigners: Projectors and experts at the early modern Swedish Board of Mines," Hartmut Schleiff and Peter Konečný, eds., *Staat, Bergbau und Bergakademie im 18. und frühen 19. Jahrhundert* (Stuttgart: VSWG, 2012), 53-62.

Fors, Hjalmar, *The Limits of Matter: Chemistry, mining and enlightenment* (Chicago: University of Chicago Press, 2015).

Foster, John Bellamy, "Marx's Theory of Metabolic Rift: Classical foundations for environmental sociology," *American Journal of Sociology* 105 (1999): 366-405.

Foucault, Michel, *Naissance de la biopolitique, Cours au Collège de France 1978-1979* (Paris: Gallimard Seuil Haute études, 2004).

Fox, Robert, *The Caloric Theory of Gases from Lavoisier to Regnault* (Oxford: Oxford University Press, 1971).

Frangsmyr, Tore, J.L. Heilbron, and Robin E. Rider, eds., *The Quantifying Spirit in the Eighteenth Century* (Berkeley: University of California Press, 1990).

Frank, Lawrence, *Victorian Detective Fiction and the Nature of Evidence: The scientific investigations of Poe, Dickens, and Doyle* (London: Palgrave, 2003).

Fressoz, Jean-Baptiste, *L'apocalypse joyeuse. Une histoure du risque technologique* (Paris: Le Seuil, 2012).

Friedman, Georges, "L'*Encyclopédie* et le travail humain," *Annales, histoire, sciences sociales* 8 (1953): 53-68.

Frindlund, Mats, "Affording Terrorism: Idealists and materialities in the emergence of modern terrorism," Max Taylor and P.M. Currie, eds., *Terrorism and Affordance* (London: Continuum, 2012), 73-92.

Fritschy, Wantje, *De patriotten en de financiën van de Bataafse Republiek. Hollands krediet en de smalle marges voor een nieuw beleid (1795-1801)* ('s-Gravenhage: Stichting Hollandse Historische Reeks, 1988).

Fujii, Kiyohisa, "The Berthollet-Proust Controversy and Dalton's Atomic Theory, 1800-1820," *British Journal for the History of Science* 19 (1986): 177-200.

Fullmer, J.Z., "Technology, Chemistry, and the Law in Early 19th-century England," *Technology and Culture* 21 (1980): 1-28.

Fullmer, June, "Humphry Davy: Fund raiser," Frank A.J.L. James, ed., *The Development of the Laboratory: Essays on the place of experiment in industrial civilisation* (London: Macmillan, 1989), 11-21.

Fullmer, June, *Sir Humphry Davy's Published Work* (Cambridge, MA: Harvard University Press, 1969).

Fullmer, June, *Young Humphry Davy: The making of an experimental chemist* (Philadelphia: American Philosophical Society, 2000).

Galbraith, John, "The 'Turbulent Frontier' as a Factor in British Expansion," *Comparative Studies in Society and History* 2 (1960): 150-168.

García Ballester, Luis and Juan L. Carrillo "The Repression of Medical Science in Absolutism Spain: The case of Juan Manuel de Aréjula, 1755-1830", *Clio Medica* 9 (1974): 207-211.

García Ballester, Luis and Juan L. Carrillo "Un ejemplo de represión de la ciencia en la España absolutista: La supresión del capitulo 15 de la 'Breve descripción de la fiebre amarilla' (1806) de J.M. Arejula," *Revista de Occidente* 134 (1974): 205-11.

García Belmar, Antonio, "The Didactic Use of Experiment: Louis-Jacques Thenard's lectures at the Collège de France," José Ramón Bertomeu Sánchez and Agustí Nieto Galán, eds., *Chemistry, Medecine and Crime. Mateu J.B. Orfila (1787-1853) and his times* (Sagamore Beach : Science History Publications, 2006), 25-53.

García Belmar, Antonio and José R. Bertomeu Sánchez, "Pedro Gutiérrez Bueno, los libros de texto de química, y los nuevos públicos de la química en el último tercio del siglo XVIII," *Dynamis* 2 (2001): 351-74.

García Belmar, Antonio, "España fumigada. Consensos y silencios en torno de las fumigaciones ácido-minerales en España, 1770-1804," (in progress).

García Belmar, Antonio, "Louis Jacques Thenard's chemistry courses at the Collège de France, 1804-1835," *Ambix* 57 (2010): 48-63.

Garnier, Arthur Edmund, *The Chronicles of the Garniers of Hampshire during Four Centuries, 1530-1900* (Norwich and London: Jarrold and Sons, 1900).

Gee, Brian, and William H. Brock, "The Case of John Joseph Griffin: From artisan-chemist and author-instructor to business leader," *Ambix* 38 (1991): 29-62.

Geiger, Reed G., *The Anzin Coal Company 1800-1833: Big business in the early stages of the French industrial revolution* (Newark: University of Delaware Press, 1974).

Gieryn, Thomas F., "Boundary-Work and the Demarcation of Science from Non-Science: Strains and interests in professional ideologies of scientists," *American Sociological Review* 48 (1983): 781-795.

Gillipsie, Charles Coulston, *Science and Polity in France. The end of the old regime* (Princeton: Princeton University Press, 1980).

Gillipsie, Charles Coulston, *Science and Polity in France: The revolutionary and Napoleonic years* (Princeton: Princeton University Press, 2004).

Gillipsie, Charles Coulston, "Science in the French Revolution," *Proceedings of the National Academy* 45 (1959): 677-84.

Gillipsie, Charles Coulston, *The Edge of Objectivity: An Essay in the History of Scientific Ideas* (Princeton: Princeton University Press, 1966).

Gillipsie, Charles Coulston, *The Montgolfier Brothers and the Invention of Aviation, 1783-1784* (Princeton: Princeton University Press, 1983).

Gillispie, Charles C. and Ken Alder, "Exchange: Engineering the Revolution," *Technology and Culture* 39 (1998): 733-54.

Golinski, Jan, *British Weather and the Climate of Enlightenment* (Chicago: University Of Chicago Press, 2011).

Golinski, Jan, "Chemistry," Roy Porter, ed., *The Cambridge History of Science: Volume 4, eighteenth-century science* (Cambridge: Cambridge University Press, 2003), 375-396.

Golinski, Jan, "Conversations on Chemistry: Talk about phlogiston in the coffee house society, 1780-1787," Trevor H. Levere and Gerard L'Estrange Turner, eds., *Discussing Chemistry and Steam: the minutes of a coffee house philosophical society, 1780-1787* (Oxford: Oxford University Press, 2002), 191-205.

Golinski, Jan, "'Fit Instruments': Thermometers in eighteenth-century chemistry," Frederic L. Holmes and Trevor H. Levere, eds., *Instruments and Experimentation in the History of Chemistry* (Cambridge: The MIT Press, 2000), 185-210.

Golinski, Jan, "Humphry Davy: The experimental self," *Eighteenth-Century Studies* 45 (2011): 15-28.

Golinski, Jan, "Precision Instruments and the Demonstrative Order of Proof in Lavoisier's Chemistry," *Osiris* 9 (1994): 30-47.

Golinski, Jan, *Science as Public Culture: Chemistry and Enlightenment in Britain, 1760-1820* (Cambridge: Cambridge University Press, 1992).

Gooday, Graeme, "Placing or Replacing the Laboratory in the History of Science?" *Isis* 99 (2008): 783-795.

Gooding, David, "'In Nature's School': Faraday as an experimentalist," David Gooding and Frank James, eds., *Faraday Rediscovered; Essays on the life and work of Michael Faraday, 1791-1867* (London: Macmillan, 1985), 105-136.

Goss, Andrew, *The Floracrats. State-sponsored science and the failure of enlightenment in Indonesia* (Madison: University of Wisconsin Press, 2011).

Graaf, Ton de, *Voor handel en Maatschappij. Geschiedenis van de Nederlandsche Handel-Maatschappij, 1824-1964* (Boom: Amsterdam, 2012).

Granjel, Luis, *Historia de la Real Academia Nacional de Medicina de Madrid* (Madrid: Real Academia Nacional de Medicina, 2006).

Grapi, Pere and Mercé Izquierdo, "*Berthollet's* Conception of Chemical Change in Context," Ambix 44 (1997): 113-130.

Griffin, Emma, *A Short History of the Industrial Revolution* (London: Palgrave, 2010).

Guerlac, Henry, *Lavoisier – The Crucial Year: The background and origin of his first experiments on combustion in 1772* (Ithaca: Cornell University Press, 1961).

Guerrini, Anita, "The Ghastly Kitchen," *History of Science* 54 (2016): 78-104.

Guignard, Laurence, *Juger la folie. La folie criminelle devant les Assises au XIXe siècle* (Paris: PUF, 2010).

Guignard, Léon, ed., *Centenaire de l'École supérieure de pharmacie de l'Université de Paris, 1803-1903* (Paris: A. Joanin, 1904).

Guillerme, André, "Enclosing Nature in the City: Supplying light and water to Paris, 1770-1840," *Construction History* 26 (2011): 79-93.

Guillerme, André, *La naissance de l'industrie à Paris entre sueurs et vapeurs (1780-1830)* (Paris: Champ Vallon, 2007).

Haan, Frederik de, "Personalia der periode van het Engelsch bestuur over Java 1811-1816," *Bijdragen tot de Taal-, Land- en Volkenkunde in Nederlandsch-Indië* 92 (1935): 477-681.

Hamlin, Christopher, "The City as a Chemical System? The chemist as urban environmental professional in France and Britain, 1780-1880," *Journal of Urban History* 33 (2007): 702-728.

Hamlin, Christopher, *A Science of Impurity: Water analysis in nineteenth-century Britain* (Berkeley, CA: University of California Press, 1990).

Hamlin, Christopher, "State Medicine in Britain," Dorothy Porter, ed., *The History of Public Health and the Modern State,* (Amsterdam: Rodopi, 1994), 132-164.

Harkness, Deborah E. "Managing an Experimental Household: The Dees of Mortlake and the practice of natural philosophy," *Isis* 88 (1997): 247-262.

Harris, Frances, "Living in the Neighbourhood of Science: Mary Evelyn, Margaret Cavendish and the Greshamites," Lynette Hunter and Sarah Hutton, eds., *Women, Science, and Medicine 1500-1700: Mothers and sisters of the Royal Society* (Stroud: Sutton Publishing, 1997), 198-217.

Harvey, Karen, *The Little Republic: Masculinity and domestic authority in eighteenth-century Britain* (Oxford: Oxford University Press, 2012).

Haudricourt, André-George, *La technologie, science humaine. Recherche d'histoire et d'ethnologie des techniques* (Paris: Maison des sciences de l'homme, 1987).

Hembry, Phyllis, *The English Spa 1560-1815: A social history* (London: Athlone, 1990).

Hentschel, Klaus, *Mapping the Spectrum: Techniques of visual representation in research and teaching* (Oxford: University Press, 2002).

Hentschel, Klaus, *Visual Cultures in Science and Technology* (Oxford: Oxford University Press, 2014).

Hilaire Pérez, Liliane, "Invention and the State in Eighteenth-Century France," *Technology and Culture* 32/3 (1991): 911-31.

Hilaire Pérez, Liliane, *L'invention technique au siècle des Lumières* (Paris: Albin Michel, 2000).

Hilaire Pérez, Liliane, *La pièce et le geste. Artisans, marchands et culture technique à Londres au XVIIIe siècle* (Paris: Albin Michel, 2013).

Hilaire Pérez, Liliane, "Technical Invention and Institutional Credit in France and Britain in the Eighteenth Century," *History and Technology* 16 (2000): 295-306.

Hilaire Pérez, Liliane and Marie Thébaud Sorger, "Les techniques dans l'espace public. Publicité des inventions et littérature d'usage au XVIIIe siècle (France, Angleterre)," *Revue de Synthèse* 127 (2006): 393-428.

Hobhouse, Hermione, ed., *Survey of London: Volumes 43 and 44, Poplar, Blackwall and Isle of Dogs* (London: London County Council, 1994).

Hobsbawm, E.J., *The Age of Revolution: Europe 1789-1848* (London: Weidenfeld and Nicholson, 1962).

Hoitsema, Copius and F. Feith, *De Utrechtsche munt uit haar verleden en heden* (Utrecht: Oosthoek, 1912).

Holloway, S.W.F., *Royal Pharmaceutical Society of Great Britain, 1841-1991: A political and social history* (London: The Pharmaceutical Press, 1991).

Holmes, Frederic L., *Eighteenth-century Chemistry as an Investigative Enterprise* (Berkeley, CA: Office for History of Science and Technology, University of California at Berkeley, 1989).

Holmes, Frederic L. and Trevor H. Levere eds., *Instruments and Experimentation in the History of Chemistry* (Cambridge, Mass.; London, England: MIT Press, 2000).

Homburg, Ernst, "The Rise of Analytical Chemistry and its Consequences for the Development of the German Chemical Profession," *Ambix* 46 (1999): 1-32.

Homburg, Ernst and de Vlieger, Johan H., "A Victory of Practice over Science: Failed innovations in the white lead industry (1780-1850)," *Archives internationales d'histoire des sciences* 46 (1996): 95-112.

Horn, Jeff, "Privileged Enclaves: Opportunities in eighteenth-century France," *Proceedings of the Western Society for French History* 32 (2004): 29-45.

Horn, Jeff, *The Path not Taken: French industrialization in the age of revolution, 1750-1830* (Cambridge, MA: MIT Press, 2006).

Horn, Jeff, Leonard N. Rosenband and Merritt Roe Smith, *Reconceptualizing the Industrial Revolution* (Cambridge, MA: MIT Press, 2010).

Hudson, Pat, *The Industrial Revolution* (London: Bloomsbury, 2014).

Hufbauer, Karl, *The Formation of the German Chemical Community 1720-1795* (Berkeley, CA: University of California Press, 1982).

Hughes, Thomas P., *American Genesis: A century of invention and technological enthusiasm, 1870-1970* (Chicago/London: The University of Chicago Press, 2004, first published in 1989).

Hughes, Thomas P., "The Evolution of Large Technological Systems," Wiebe E. Bijker, Thomas P. Hughes, and Trevor J. Pinch eds., *The Social Construction of Technological Systems: New directions in the sociology and history of technology* (Cambridge, Mass: MIT Press, 1987), 51-82.

Hunting, Penelope, *A History of the Society of Apothecaries* (London: Society of Apothecaries, 1998).

Hurlbert, William Henry, *France and the Republic* (Charleston: BiblioLife, 2007 [1890]).

Hutchby, Ian, "Affordances and the Analysis of Technologically Mediated Interaction," *Sociology* 37 (2003): 581-89.

Hutchby, Ian, "Technologies, Texts, and Affordances," *Sociology* 35 (2001): 441-56.

Ingold, Tom, "Materials Against Materiality," *Archeological Dialogues* 14 (2007): 1-16.

Isoré, Jacques, "De l'existence des brevets d'invention en droit français avant 1791," *Revue historique de droit français et étranger* 16 (1937): 94-130.

Jackson, Catherine M., "The 'Wonderful Properties of Glass': Liebig's kaliapparat and the practice of chemistry in glass," *Isis* 106 (2015): 43-69.

Jackson, Roland, "John Tyndall and the Early History of Diamagnetism," *Annals of Science* 72 (2015): 435-489.

Jacob, Margaret, *The First Knowledge Economy: Human capital and the European economy, 1750-1850* (Cambridge: Cambridge University Press, 2014).

Jacob, Margaret C. and Larry Stewart, *Practical Matter: Newton's science in the service of industry and empire* (Cambridge, MA: Harvard University Press, 2004).

James, Frank A.J.L., "'Agricultural Chymistry is at present in it's infancy': The Board of Agriculture, the Royal Institution and Humphry Davy," *Ambix* 62 (2015): 363-385.

James, Frank A.J.L., ed., *'The Common Purposes of Life': Science and society at the Royal Institution of Great Britain* (Aldershot: Ashgate, 2002).

James, Frank A.J.L., *'the first example ... of an extensive scheme of pure scientific medical investigation': Thomas Beddoes and the Medical Pneumatic Institution in Bristol, 1794 to 1799* (London: Royal Society of Chemistry Historical Group Occasional Publication, 2016).

James, Frank A.J.L., *Michael Faraday: A very short introduction* (Oxford: Oxford University Press, 2010).

James, Frank A.J.L. and Anthony Peers, "Constructing Space for Science at the Royal Institution of Great Britain," *Physics in Perspective* 9 (2007): 130-185.

Jankovic, Vladimir, *Confronting the Climate. British air and the making of environmental medicine* (New York: Palgrave Macmillan, 2010).

Jas, Nathalie, "Public Health and Pesticide Regulation in France Before and After Silent Spring," *History and Technology* 23 (2007): 369-388.

Jay, Mike, *The Atmosphere of Heaven: The unnatural experiments of Dr. Beddoes and his sons of genius* (New Haven: Yale University Press, 2009).

Jenner, Mark and Patrick Wallis, "The Medical Marketplace," Mark Jenner and Patrick Wallis, eds., *Medicine and the Market in England and its Colonies, c.1450-c.1850* (London: Palgrave Macmillan, 2007), 1-17.

Jensen, William B., "Machael Faraday and the Art and Science of Chemical Manipulation," *Bulletin for the History of Chemistry* 11 (1991): 65-75.

Johnston, Kenneth, *Unusual Suspects: Pitt's reign of alarm and the lost generation of the 1790s* (Oxford: Oxford University Press, 2013).

Johnston, W.T., *The Secret of Turkey Red* (Livingston: Officina W.T. Johnston, 1993).

Jones, Jean, "James Hutton and the Forth and Clyde Canal," *Annals of Science* 39 (1982): 255-263.

Jones, Jean, "James Hutton's Agricultural Research and his Life as a Farmer," *Annals of Science* 42 (1985): 573-601.

Jones, Peter M., *Industrial Enlightenment: Science, technology and culture in Birmingham and the West Midlands, 1760-1820* (Manchester: Manchester University Press, 2009).

Jones, Peter M., *The Peasantry in the French Revolution* (Cambridge: Cambridge University Press, 1988).

Jonker, J. and K.E. Sluyterman, *At Home on the World Markets: Dutch international trading companies from the sixteenth century until the present* (The Hague: SDU Uitgevers, 2000).

Julia, Dominique, "L'Ecole normale de l'an III : bilan d'une expérience révolutionnaire", *Revue du Nord* 78 (1996): 853-886.

Julien, Pierre, "Plaidoyer pour les notes et factures anciennes de pharmacie," *Revue d'histoire de la pharmacie* 37 (1990) : 81-92.

Jungnickel, Christa and Russell McCormmach, *Cavendish: The experimental life* (Lewisburg: Bucknell, 1998).

Keay, John, *The Honorable Company: A history of the English East India Company* (London: Harper Collins, 1991).

Kemp, Pieter Hendrik van der, "De Nederlandsche-Indische proefgulden van 1815," *Tijdschrift voor munt- en penningkunde* (1913): 21-60.

Kemp, Pieter Hendrik van der, "De zilveren Java-Ropijen van de jaren 1816-1817," *Bijdragen tot de Taal-, Land- en Volkenkunde van Nederlandsch-Indië* 67 (1913): 275-366.

Kemp, Pieter Hendrik van der, "Episodes uit de geschiedenis der aanmuntingen ten behoeve van Oost-Indië, 1802-1807," *Bijdragen tot de taal-, land- en volkenkunde van Nederlandsch-Indië* 70 (1915): 227-440.

Kemp, Pieter Hendrik van der, *De teruggave der Oost-Indische koloniën 1814-1816* ('s-Gravenhage: M. Nijhoff, 1910).

Kent, Andrew, ed., *An Eighteenth Century Lectureship in Chemistry: Essays and addresses relating to the chemistry department of Glasgow University* (Glasgow: Jackson Son & Co., 1950).

Kersaint, Georges, "L'usine de Fourcroy et Vauquelin," *Revue d'Histoire de la Pharmacie* 47 (1959): 25-30.

Kian, Kwee H., "The Expansion of Chinese Inter-insular and Hinterland Trade in Southeast Asia, c. 1400-1850," David Henley and Henk Schulte Nordholt, eds., *Environment, Trade and Society in Southeast Asia* (Brill: Leiden, 2015), 149-165.

Kim, Mi Gyung, *Affinity: That elusive dream* (Cambridge, MA: MIT Press, 2003).

Kinzler, Catherine, *Condorcet. L'instruction publique et la naissance du citoyen* (Paris: Folio/Essais, 1984).

Klein, Cornelia and Anthony Philpotts, *Earth Materials: Introduction to mineralogy and petrology* (Cambridge: Cambridge University Press, 2013).

Klein, Ursula, "Apothecary Shops, Laboratories and Chemical Manufacture in Eighteenth-century Germany," Lissa Roberts, Simon Schaffer and Peter Dear, eds., *The Mindful Hand: Inquiry and invention from the late Renaissance to early industrialisation* (Amsterdam: Royal Netherlands Academy of Arts and Sciences, 2007), 247-278.

Klein, Ursula, "Apothecary-Chemists in Eighteenth Century Germany," Lawrence Principe, ed., *New Narratives in Eighteenth-Century Chemistry* (Dordrecht: Springer, 2007), 97-137.

Klein, Ursula, ed., *Artisanal-Scientific Experts in Eighteenth-Century France and Germany*, special issue of *Annals of Science* 69 (2012): 303-433.

Klein, Ursula, "Blending Technical Innovation and Learned Natural Knowledge: The making of ethers," Ursula Klein and Emma Spary, eds., *Materials and Expertise in Early Modern Europe: Between market and laboratory* (Chicago: University of Chicago Press, 2010), 125-157.

Klein, Ursula, "Chemical Experts at the Royal Prussian Porcelain Manufactory," *Ambix* 60 (2013): 99-121.

Klein, Ursula, *Experiments, Models, Paper Tools: Cultures of organic chemistry in the nineteenth century* (Palo Alto: Stanford University Press, 2002).

Klein, Ursula, "The Laboratory Challenge: Some revisions of the standard view of early modern experimentation," *Isis* 99 (2008): 769-782.

Klein, Ursula, "The Prussian Mining Official Alexander von Humboldt," *Annals of Science* 69 (2012): 27-68.

Klein, Ursula, "A Revolution that Never Happened," *Studies in History and Philosophy of Science* 49 (2015): 80-90.

Klein, Ursula, "Technoscience *avant la lettre*," *Perspectives on Science* 13 (2005): 226-266.

Klein, Ursula and Carsten Reinhardt, eds., *Objects of Chemical Inquiry* (Sagamore Beach, MA: Science History Publications, 2014).

Klein, Ursula and Emma Spary, eds., *Materials and Expertise in Early Modern Europe: Between market and laboratory* (Chicago: Chicago University Press, 2010).

Klein, Ursula and Wolfgang Lefèvre, *Materials in Eighteenth-Century Science: A historical ontology* (Cambridge, MA: MIT Press, 2007).

Kneitz, Agnes, "'As if the River was not Meat and Drink to You': Social novels as a means of framing nineteenth-century environmental justice," *Interdisciplinary Studies in Literature and Environment* 1 (2015): 1-16.

Knight, David, *Humphry Davy: Science and power* (Oxford: Blackwell, 1992; Second Edition, Cambridge: Cambridge University Press, 1996).

Knight, Roger, "Rescued from the Myths of Time: Toward a reappraisal of European merchant houses in mid-nineteenth century Java, ca. 1830-1870," *Bijdragen tot de Taal,- Land- en Volkenkunde* 170 (2014): 313-341.

Knight, Roger and Malcolm Wilcox, *Sustaining the Fleet, 1793-1815: War, the British Navy and the contractor state* (Woodbridge: Boydell Press, 2010).

Knorr Cetina, Karin, "Objectual practice," T.R. Schatzki, K. Knorr Cetina, & E. von Savigny, eds., *The Practice Turn in Contemporary Theory* (New York: Routledge, 2001), 184-197.

Knox, Kevin C. "'The Deplorable Frenzy': The slow legitimisation of chemical practice at Cambridge University," Mary D. Archer, Christopher D. Haley, eds., *The 1702 Chair of Chemistry at Cambridge: Transformation and Change* (Cambridge: Cambridge University Press, 2005), 1-30.

Konečný, Peter, "The Hybrid Expert in the 'Bergstaat': Anton von Ruprecht as a professor of chemistry and mining and as a mining official, 1779-1814," *Annals of Science* 69 (2012): 335-347.

Kranakis, Eda, *Constructing a Bridge: An exploration of engineering culture, design and research in nineteenth-century France and America* (Cambridge MA: MIT Press, 1997).

Kranakis, Eda, "Hybrid Careers and the Interaction of Science and Technology,"

Peter Kroes and Martijn Baker, eds., *Technological Development and Science in the Industrial Age* (Dordrecht: Kluwer Academic Publishers, 1992), 177-204.

Laissus, Yves, *Le Muséum national d'histoire naturelle* (Paris: Gallimard, 1995).

Lantink, Frans Willem et al., *Voor stad en staat*, vol. 1: *Plattegrond* (Amsterdam: Joh. Enschedé, 2003).

Laparra López, Emilio, *La Alianza de Godoy con los revolucionarios* (Madrid: Consejo Superior de Investigaciones Científicas, 1992).

Laparra López, Emilio, "Ilustrados e Inquisición ante la Iglesia constitucional francesa," *Revista de Historia das Ideias* 10 (1988): 359-374.

Laparra López, Emilio, "La inestabilidad de la monarquía de Carlos IV," *Studia Histórica. Historia moderna* 12 (1994): 23-34.

Lavoisy, Olivier and Dominique Vinck, "Le dessin comme objet intermédiaire de l'industrie," Pierre Delcambre, ed., *Communications organisationnelles. Objets, pratiques et dispositifs* (Rennes: Presses Universitaires de Rennes, 2000), 47-63.

Law, John and Vicky Singleton, "Object Lessons," *Organization* 12 (2005): 331-355.

Le Roux, Thomas, "Accidents industriels et régulation des risques: l'explosion de la poudrerie de Grenelle en 1794," *Revue d'Histoire Moderne et Contemporaine* 58-3 (2011): 34-62.

Le Roux, Thomas, "Chemistry and Industrial and Environmental Governance in France, 1770-1830," *History of Science* 54 (2016): 195-222.

Le Roux, Thomas, "La chimie, support du développement de l'industrie perfectionnée sous la Révolution et l'Empire," Natacha Coquery, ed., *Les progrès de l'industrie perfectionnée* (Toulouse: Presses universitaires du Mirail, 2017), 26-35.

Le Roux, Thomas, "Déclinaisons du 'conflit'. Autour des atteintes environnementales de l'affinage des métaux précieux, Paris, années 1820," Thomas Le Roux and Michel Letté, eds., *Débordements industriels dans la cité et leurs conflits, XVIII-XXIe siècles* (Rennes: Presses Universitaires de Rennes, 2013), 179-198.

Le Roux, Thomas, "Guyton de Morveau et le grand basculement de l'expertise sanitaire et environnementale (1773-1809)," *Annales Historiques de la Révolution française* 383 (2016): 153-176.

Le Roux, Thomas, *Le laboratoire des pollutions industrielles, Paris, 1770-1830* (Paris: Albin Michel, 2011).

Le Roux, Thomas, "Risques et maladies du travail: Le Conseil de salubrité de Paris aux sources de l'ambiguïté hygiéniste au XIXe siècle," Anne-Sophie Bruno, Éric Geerkens, Nicolas Hatzfeld and Catherine Omnès, eds., *La santé au travail, entre savoirs et pouvoirs (XIXe-XXe siècles)* (Paris: Presses universitaires de Rennes, 2011), 45-63.

Lehman, Christine, "Innovation in Chemistry Courses in France in the Mid-Eighteenth Century : Experiments and affinities," *Ambix* 57 (2010) : 3-26.

Lehman, Christine, "Les multiples facettes des cours de chimie en France au milieu du XVIIIe siècle,» *Histoire de l'éducation* 130 (2011) : 31-56.

Leong, Elaine, "Collecting Knowledge for the Family: Recipes, gender and practical knowledge in the early modern English household," *Centaurus* 55 (2013): 81-103.

Leong, Elaine, "Making Medicines in the Early Modern Household," *Bulletin of the History of Medicine* 82 (2008): 145-168.

Leroi-Gourhan, André, *L'Homme et la Matière* (Paris: Albin Michel, 1943).

Lestel, Laurence, Anne-Cécile Lefort and André Guillerme, eds., *La céruse: usages et effets X^e-XXe siècles* (Paris: Centre d'histoire des techniques CNAM, 2003).

Levere, Trevor H., "Dr. Thomas Beddoes and the Establishment of His Pneumatic Institution: A tale of three presidents," *Notes and Records of the Royal Society of London* 32 (1977): 41-49.

Levere, Trevor H., "Dr. Thomas Beddoes at Oxford: Radical politics in 1788-1793 and the fate of the Regius Chair in Chemistry," *Ambix* 28 (1981): 61-69.

Levere, Trevor H., "Measuring Gases and Measuring Goodness," Frederic Lawrence Holmes and Trevor H. Levere, eds., *Instruments and Experimentation in the History of Chemistry* (Cambridge: Cambridge University Press, 2000), 105-136.

Levere, Trevor H., "The Role of Instruments in the Dissemination of the Chemical Revolution," *Endoxa* 19 (2005): 227-242.

Lindeboom, G.A., *Herman Boerhaave. The man and his work* (London: Methuen, 1968).

Lowood, Henry, *Patriotism, Profit, and the Promotion of Science in the German Enlightenment* (New York: Garland Press, 1991).

Lucassen, Jan, "Working at the Ichapur Gunpowder Factory in the 1790's, Parts I and II," *Indian Historical Review* 39 (2012): 19-56, 251-271.

Lucier, Paul, "Court and Controversy: Patenting science in the nineteenth century," *British Journal for the History of Science* 29 (1996): 139-154.

Lundgren, Anders, "The New Chemistry in Sweden: The debate that wasn't," *Osiris* 4 (1988): 146-168.

Lynch, Michael, ed., *Truth Machine: The contentious history of DNA fingerprinting* (Chicago: University of Chicago Press, 2008).

MacKenzie, Donald, Fabian Muniesa and Lucia Siu, eds., *Do Economists Make Markets? On the performativity of economics* (Princeton: Princeton University Press, 2007).

MacLean, J., "Koperindustrie in Nederland, 1750-1850," *Economisch en social-historisch jaarboek* 43 (1971): 39-63.

MacLeod, Christine, *Inventing the Industrial Revolution. The English patent system (1660-1800)* (Cambridge: Cambridge University Press, 1988).

Malcolm, Joyce Lee, *Guns and Violence: The English experience* (Cambridge, MA: Harvard University Press, 2009).

Mallett, Oliver, "Contesting the History of Enterprise and Entrepreneurship," *Work, Employment and Society* 29 (2015): 177-182.

Malm, Andreas and Alf Hornborg, "The Geology of Mankind? A critique of the Anthropocene narrative," *The Anthropocene Review* 1 (2014): 62-69.

Markus, Thomas, *Buildings & Power: Freedom and control in the origin of modern building types* (London: Routledge, 1993).

Martin Fugier, Anne, *La vie élégante ou la formation de Tout-Paris, 1815-1848* (Paris: Fayard, 1990).

Martinón Torres, Marcos, "Inside Solomon's House: An archaeological study of the old Ashmolean chymical laboratory in Oxford," *Ambix* 59 (2012): 22-48.

Massard Guilbaud, Geneviève, *Histoire de la pollution industrielle, France 1789-1914* (Paris: Editions de l'EHESS, 2010).

Matthias, Peter, *The First Industrial Nation: An economic history of Britain, 1700-1914*, second edition (London: Methuen, 1983).

McCosh, Frederic W.J., *Boussingault: Chemist and agriculturist* (Dordrecht: Kluwer, 1984).

McEvoy, John G., *The Historiography of the Chemical Revolution: Patterns of interpretation in the history of science* (London: Routledge, 2010).

McKeon, Michael, *The Secret History of Domesticity: Public, private, and the division of knowledge* (Baltimore: Johns-Hopkins University Press, 2009).

McKie, Douglas, "Priestley's Laboratory and Library and Other of his Effects," *Notes and Records of the Royal Society* 12 (August 1956): 114-136.

Meinel, Christophe, "Reine und angewandte Chemie," *Berichte zur Wissenschaftsgeschichte* 8 (1985): 25-45.

Melton, James van Horn, *The Rise of the Public in Enlightenment Europe* (Cambridge: Cambridge University Press, 2001).

Miller, A.H., revised by Ralph Harrington, "Reid, David Boswell (1805-1863)," *Oxford Dictionary of National Biography* (Oxford: Oxford University Press, 2004).

Miller, David, *James Watt, Chemist: Understanding the origins of the steam age* (London: Pickering and Chatto, 2009).

Millerand, Florence, "Les gestionnaires d'information 'invisibles' dans la production d'une base de données scientifiques," *Revue d'anthropologie des sciences* 6 (2012): 163-190.

Minard, Philippe, "Le métier sans institution: les lois d'Allarde-Le Chapelier de 1791 et leur impact au début du XIXe siècle," Steven Kaplan and Philippe Minard, eds., *La fin des corporations* (Paris: Belin, 2003), 81-95.

Misa, Thomas, *From Leonardo to the Internet: Technology and culture from the Renaissance to the present* (Baltimore: Johns Hopkins University Press, 2004).

Mitchell, Timothy, *Carbon Democracy: Political power in the age of oil* (London: Verso, 2011).

Mitchell, Timothy, "Rethinking Economy," *Geoforum* 29 (2008): 1116-1121.

Mitchell, Timothy, *Rule of Experts: Egypt, techno-politics, modernity* (Berkeley, 2002).

Mody, Cyrus and Michael Lynch, "Test Objects and other Epistemic Things: A history of a nanoscale object," *British Journal for the History of Science* 43 (2010): 423-458.

Mokyr, Joel, *The Enlightened Economy: An economic history of Britain, 1700-1850* (New Haven: Yale University Press, 2009).

Mokyr, Joel, ""Entrepreneurship in the Industrial Revolution," David Landes, Joel Mokyr and William Baumol, eds., *The Invention of Enterprise* (Princeton, NJ: Princeton University Press, 2012), 183-210.

Mokyr, Joel, *The Gifts of Athena: Historical origins of the knowledge economy* (Princeton: Princeton University Press, 2002).

Mol, Annemarie, *The Body Multiple: Ontology in medical practice* (Raleigh, NC: Duke University Press, 2002).

Moquette, J.P., "De munten van Nederlandsch-Indië," *Tijdschrift voor Indische Taal-, Land- en Volkenkunde* (1909): 33-96.

Morange, Claude, "Las estructuras de poder en el tránsito del Antiguo al Nuevo Régimen," Joseph Pérez and Armando Alberola, eds., *España y América entre la Ilustración y el liberalism* (Alicante- Madrid: Casa de Velázquez- Instituto de Cultura Juan Gil-Albert, 1993), 35-7.

Morrell, J.B., "The Chemist Breeders: the Research Schools of Liebig and Thomas Thomson," *Ambix* 9 (1972): 1-46.

Morrell, J.B., "Practical Chemistry in the University of Edinburgh, 1799-1843," *Ambix* 16 (1969): 66-80.

Morrell, J.B., "Thomas Thomson: Professor of Chemistry and University Reformer," *British Journal for the History of Science* 4 (1969): 245-286.

Morris, Peter J.T., "The Eighteenth Century: Chemistry allied to anatomy," Robert Williams, Joseph Paton, Allan Chapman, and John Shipley Rowlinson, eds., *Chemistry at Oxford: A history from 1600 to 2005* (London: Royal Society of Chemistry, 2008), 52-78.

Morris, Peter J.T., *The Matter Factory: A history of the chemistry laboratory* (London: Reaktion Books, 2015).

Morris, Peter J.T., Colin Russell and John Smith, *Archives of the British Chemical Industry: A handlist* (Oxford: British Society for the History of Science, 1988).

Morson, A.F.P., *Operative Chymist,* Clio Medica no. 45 (Amsterdam: Rodophi, 1997).

Morvan Becker, Frédéric, "L'école gratuite de dessin de Rouen, ou la formation des techniciens au XVIIIe siècle," PhD thesis, Université Paris VIII-Saint Denis, 2010.

Muir, John, *John Anderson, Pioneer of Technical Education and the College He Founded in Glasgow* (Glasgow: John Smith & Son, 1950).

Musson, Albert and Eric Robinson, *Science and Technology in the Industrial Revolution* (Manchester: Manchester University Press, 1969).

Narbeth, Colin, Robin Hendy and Christopher Stocker, *Historische bankbiljetten en aandelen* (Baarn: Moussault's Uitgeverij, 1979).

Netscher, Elisa and Jacobus A. van der Chijs, *De munten van Nederlands-Indië* (Batavia 1863).

Nieto Galàn, Agustí, *Colouring Textiles. A history of natural dyesuffs in industrial Europe* (Boston: Boston Studies in the Philosophy and History of Science, 2001).

O'Farrell, P.N., *Heriot Watt University: An illustrated history* (Harlow: Pearson Educational, 2004).

Oreskes, Naomi and Erik Conway, *Merchants of Doubt: How a handful of scientists obscured the truth on issues from tobacco smoke to global warming* (London: Bloomsbury Press, 2010).

Orlikowski, Wanda J. "Sociomaterial Practices: Exploring technology at work," *Organization Studies* 28 (2007): 1435-1448.

Ozouf, Mona, *L'homme régénéré. Essais sur la révolution française* (Paris: Gallimard, 1989).

Palladino, Paolo and Michael Worboys, "Science and Imperialism," *Isis* 84 (1993): 91-102.

Parascandola, John, *King of Poisons. A history of arsenic* (Washington: Potomac Books, 2012).

Paris, John, *The Life of Sir Humphry Davy*, 2 volumes (London: Colburn and Bentley, 1831).

Parker, Harold T., *An Administrative Bureau During the Old Regime: The Bureau of Commerce and its relations to French industry from May 1781 to November 1783* (London: Associated University Press, 1993).

Parssinen, Terry M., *Secret Passions, Secret Remedies: Narcotic drugs in British society, 1820-1930* (Manchester: MUP, 1983).

Parthasarathi, Prasannan, *Why Europe Grew Rich and Asia Did Not: Global economic divergence, 1600-1850* (Cambridge: Cambridge University Press, 2011).

Pascoe, W.H., *CCC: The History of the Cornish Copper Company* (Hayle: Haylebooks, 1981).

Pascual Artiaga, Mercedes, "La ciudad ante el contagio: medidas políticas y administrativas dictadas en la epidemia de fiebre amarilla de 1804 en Alicante," *Asclepio* 54 (2002): 125-153.

Pennell, Sara, "'Pots and Pans History': The material culture of the kitchen in early modern England," *Journal of Design History* 11 (1998): 201-216.

Pennell, Sara and Elaine Leong, "Recipe Collections and the Currency of Medical Knowledge in the Early Modern 'Medical Marketplace'," Mark Jenner and Patrick Wallis, eds., *Medicine and the Market in England and its Colonies, c.1450-c.1850* (Basingstoke: Palgrave Macmillan, 2007), 133-152.

Perkins, John, "Chemistry Courses, the Parisian chemical world and the chemical revolution, 1770-1790," *Ambix* 57 (2010): 27-47.

Perkins, John, "Creating Chemistry in Provincial France before the Revolution: The examples of Nancy and Metz," *Ambix* 51 (2004): 4375.

Perkins, John, ed., *Sites of Chemistry in the Eighteenth Century,* special issue of *Ambix* 60 (2013).

Perrin, Carleton, "Of Theory Shifts and Industrial Innovations: The relations of J.A.C. Chaptal and A.L. Lavoisier," *Annals of Science* 43 (1986): 511-542.

Peset, Mariano and José Luis Peset, *Muerte en España. Política y sociedad entre la peste y el cólera* (Madrid: Seminarios y Ediciones, 1977).

Pierrot, Nicolas, "Les images de l'industrie en France, peintures, dessins, estampes, 1760-1870," PhD thesis, Université Paris I Panthéon-Sorbonne, 3 vols., 2010.

Pirohakul, Teerapa and Patrick Wallis, "Medical Revolutions? The growth of medicine in England, 1660-1800," *Journal of Social History* 49 (2016): 510-531.

Pissarello, Giulia. "Industrialism as "Tragedy of Ugliness": D.H. Lawrence's ecological consciousness," *Griseldaonline* 10 (2011): 31-42.

Planchon, Gustave, *L'enseignement des sciences physico-chimiques au Jardin des apothicaires et à l'École de pharmacie de Paris* (Paris: Imprimerie Ernest Flammarion, 1897).

Plessis, Alain, ed., *Naissances des libertés économiques, 1791-fin XIXe siècle* (Paris: Institut d'histoire de l'industrie, 1993).

Pomeranz, Kenneth, *The Great Divergence: China, Europe and the making of the modern world economy* (Princeton: Princeton University Press, 2000).

Porter, Roy and Dorothy Porter, "The Rise of the English Drugs Industry: The role of Thomas Corbyn," *Medical History* 33 (1989): 277-95.

Porter, Theodore, *Trust in Numbers: The pursuit of objectivity in science and public life* (Princeton: Princeton University Press, 1995).

Potofsky, Allan, "Recycling the City: Paris, 1760s-1800," Ariane Fennetaux, Amélie Junqua and Sophie Vasset, eds., *The Afterlife of Used Things: Recycling in the long eighteenth century* (New York: Routledge, 2015), 71-88.

Powers, John C., "Leiden Chemistry in Edinburgh: Herman Boerhaave, James Crawford and Andrew Plummer," R.G.W. Anderson, ed., *Cradle of Chemistry: The early years of chemistry at the University of Edinburgh* (Edinburgh: John Donald, 2015), 25-58.

Principe, Lawrence M., "Apparatus and Reproducibility in Alchemy," Frederic L. Holmes and Trevor H. Levere, eds., *Instruments and Experimentation in the History of Chemistry* (Cambridge, MA: MIT Press, 2000), 55-74.

Pritchard, Sara B., "Mining Land and Labor," *Environmental History*, 10 (2005): 731-733.

Puig Pla, Carlos, "Els primers socis-artistes de la Reial Acadèmia de Ciències i Arts de Barcelona (1746-1842)," Agustí Nieto Galán and Antoni Roca Rosell, eds., *La Reial Acadèmia de Ciències i Arts de Barcelona als segles XVIII y XIX. Història, ciència i societat* (Barcelona: Institut d'Estudis Catalans, 2000), 287-310.

Puigvert, Joaquim, ed., *Bisbes, Illustració i jansenisme a la Catalunya del segle XVIII* (Vic: Biblioteca Universitaria, 2000).

Pyenson, Lewis, *Empire of Reason. Exact sciences in Indonesia 1840-1940* (Leiden: Brill, 1989).

Quarrell, W.H. and Margaret Mare, eds., *London in 1710 from the Travels of Zacharius Conrad von Uffenbach* (London: Faber and Faber, 1934).

Queruel, Alain, *Vauquelin et son temps (1763-1829)* (Paris: L'Harmattan, 1994).

Rainhorn, Judith, "La santé au risque du marché, Savoir médical, concurrence économique et régulation des risques entre blanc de zinc et blanc de plomb (France XIXe siècle)," Thomas Le Roux, ed., *Risques et accidents industriels, fin XVIIe-fin XIXe siècles* (Rennes: Presses universitaires de Rennes, 2016).

Ramsay, William, *The Life and Letters of Joseph Black* (London: Constable, 1918).

Rappert, Brian, "Technologies, Texts, and Possibilities: A reply to Hutchby," *Sociology* 37 (2003): 565-80.

Reichardt, Rolf and H.J. Lüsebrink, "Révolution à la fin du 18e siècle. Pour une relecture d'un concept-clé du siècle des Lumières," *Mots* 16 (1988): 35-68.

Rey, Alain, *Révolution: Histoire d'un mot* (Paris: Gallimard, 1989).

Rheinberger, Hans Jörg, "Experiment, Différence and Writing I," *Studies in History and Philosophy of Science* 23 (1992): 305-331.

Rheinberger, Hans Jörg, "A Reply to David Bloor: 'Toward a sociology of epistemic things'," *Perspectives on Science* 13 (2005): 406-410.

Rheinberger, Hans Jörg, *Toward a History of Epistemic Things* (Palo Alto: Stanford University Press, 1997).

Richmond, Lesley, Julie Stevenson and Alison Turton, eds., *The Pharmaceutical Industry: A guide to historical records* (Aldershot: Ashgate, 2003).

Riedman, Sarah Regal, *Antoine Lavoisier: Scientist and citizen* (London: Schuman, 1967).

Rip, Arie, "De Facto Governance of Nanotechnologies," Morag Goodwin, Bert-Jaap Koops, and Ronald Leenes, eds., *Dimensions of Technology Regulation* (Nijmegen: Wolf Legal Publishers, 2010), 285-308.

Roberts, Lissa L., "Accumulation and Management in Global Historical Perspective: An introduction," *History of Science* 52 (2014): 227-246.

Roberts, Lissa L., "'Le centre de toutes choses': Constructing and managing centralization on the Isle de France," *History of Science* 52 (2014): 319-342.

Roberts, Lissa L., "Condillac, Lavoisier, and the Instrumentalization of Science," *Eighteenth-Century: Essays and interpretation* 33 (1992): 252-271.

Roberts, Lissa L., "The Death of the Sensuous Chemist: The 'new' chemistry and the transformation of sensuous technology," *Studies in History and Philosophy of Science Part A* 4 (1995): 503-529.

Roberts, Lissa L., "Exploring Global History Through the Lens of History of Chemistry: Materials, identities and governance," *History of Science* 54 (2016): 335-361.

Roberts, Lissa L., "Geographies of Steam: Mapping the entrepreneurial activities of steam engineers in France during the second half of the eighteenth century," *History and Technology* 27 (2011): 417-439.

Roberts, Lissa L., "P.J. Kasteleyn and the "Oeconomics" of Dutch Chemistry," *Ambix* 53 (2006): 255-272.

Roberts, Lissa L., "Practicing Oeconomy During the Second Half of the Long Eighteenth Century: An introduction," *History and Technology* 30 (2014): 133-148.

Roberts, Lissa L., ed., *Practicing Oeconomy in the late Eighteenth Century*, Special issue of *History and Technology* 30 (2014): 133-279.

Roberts, Lissa L., "Producing (in) Europe and Asia, 1750-1850," *Isis* 106 (2015): 857-865.

Roberts, Lissa L., "A Word and the World: The significance of naming the calorimeter," *Isis* 82 (1991): 199-222.

Roberts, R.S., "The Early History of the Import of Drugs into Britain," F.N.L. Poynter, ed., *The Evolution of Pharmacy in Britain* (London: Pitman, 1965), 165-185.

Robinson, Eric, and Douglas McKie, eds., *Partners in Science: Letters of James Watt and Joseph Black* (London: Constable, 1970).

Roche, Daniel, "Encyclopedias and the Diffusion of Knowledge," Mark Goldie and Robert Wokler, eds., *The Cambridge History of Eighteenth-Century Political Thought* (Cambridge: Cambridge University Press, 2006), 172-194.

Rodríguez Ocaña, Esteban, "La cuestión del lazareto marítimo permanente en la España del siglo XVIII, de Cádiz a Mahon," *Asclepio* 40 (1988): 265-276.

Rotherham, Ian D. and David McCallam, "Peat Bogs, Marshes and Fens as Disputed Landscapes in Late Eighteenth-Century France and England," Louise Lyle and David McCallam, eds., *Histoires de la Terre: Earth sciences and French culture 1740-1940* (Amsterdam: Rodopi, 2008), 75-88.

Rouff, Marc, *Les mines de charbon en France au XVIIIe siècle* (Paris: Rieder et Cie, 1922).

Rouffaer, Gerret P., *De voornaamste industrieën der inlandsche bevolking van Java en Madoera* ('s-Gravenhage: Martinus Nijhoff, 1904).

Rowlinson, John, *Sir James Dewar, 1842-1923: A ruthless chemist* (Farnham: Ashgate, 2012).

Russell, Ben, *James Watt: Making the world anew* (London: Reaktion, 2014).

Russell, Colin A., Noel G. Coley and Gerrylynn K. Roberts, *Chemists by Profession. The origins and rise of the Royal Institute of Chemistry* (Milton Keynes: Open University Press, 1977).

Ruston, Sharon, "From "The Life of the Spinosist" to "Life": Humphry Davy, chemist and poet," Margaret Hagen and Margery Skagen, eds., *Literature and Chemistry: Elective affinities* (Aarhus: Aarhus University Press, 2013), 77-97.

Sadoun Goupil, Michelle, *Le chimiste C.L. Berthollet, 1748-1822, sa vie, son oeuvre* (Paris: Vrin, 1977).

Schabas, Margaret and Neil Di Marchi, "Introduction to Oeconomies in the Age of Newton," Margaret Schabas and Neil Di Marchi, eds., *Oeconomies in the Age of Newton. Annual Supplement to History of Political Economy* 35 (2003): 1-13.

Schabas, Margaret, *The Natural Origins of Economics* (Chicago: University of Chicago Press, 2005).

Schaffer, Simon, "Assay Instruments and the Geography of Precision on the Guinea Trade," Marie-Noëlle Bourget, Christian Licoppe and H. Otto Sibum, eds., *Instruments, Travel and Science. Itineraries of precision from the seventeenth to the twentieth century* (London: Routledge, 2002), 20-50.

Schaffer, Simon, "'The Charter'd Thames': Naval architecture and experimental spaces in Georgian Britain," Lissa Roberts, Simon Schaffer and Peter Dear, eds., *The Mindful Hand: Inquiry and invention from the late Renaissance to early industrialisation* (Amsterdam: Royal Netherlands Academy of Arts and Sciences, 2007), 279-305.

Schaffer, Simon, "Measuring Virtue: Eudiometry, enlightenment and pneumatic medicine," Andrew Cunningham and Roger French, eds., *The Medical Enlightenment of the Eighteenth Century* (Cambridge: Cambridge University Press, 1990), 281-318.

Schaffer, Simon, "Natural Philosophy and Public Spectacle in the Eighteenth Century," *History of Science* 21 (1983): 1-43.

Schatzberg, Eric, " From Art to Applied Science," *Isis* 103 (2012): 555-563.

Scheffers, Albert A. J., "Om de kwaliteit van het geld. Het toezicht op de muntproductie in de Republiek en de voorziening van kleingeld in Holland and West-Friesland in de achttiende eeuw," PhD Thesis, Leiden University, 2013.

Scheffers, Albert A. J., "Johan Sebastiaan van Naamen, Muntmeester van Batavia 1764-1768 en Utrecht 1782-1797 in Perspectief," *De Muntkoerier* 42:11 (2013): 18-19.

Schofield, Robert E., *The Enlightened Joseph Priestley: A study of his life and work from 1773 to 1804* (University Park: Pennsylvania State University Press, 2004).

Scholten, Cornelis, *De munten van de Nederlandsche gebiedsdelen overzee, 1601-1948* (Amsterdam: J. Schulman, 1951).

Schrader, Astrid, "Responding to Pfiesteria Piscicida (The Fish Killer): Phantomatic ontologies, indeterminacy, and the responsibility in toxic microbiology," *Social Studies of Science* 40 (2010): 275-306.

Schrøder, Michael, *The Argand Burner: Its origin and development in France and England, 1780-1800: an epoch in the history of science illustrated by the life and work of the physicist Ami Argand, 1750-1803* (Odense: Odense University Press, 1969).

Schumpeter, Joseph, *Capitalism, Socialism and Democracy* (New York: Harper, 2008).

Selin, Henrik, *Global Governance of Hazardous Chemicals: Challenges of multilevel management* (Cambridge: MIT Press, 2010).

Serrano, Elena, "Chemistry in the City: The scientific role of female societies in late eighteenth-century Madrid," *Ambix* 60 (2013): 139-159.

Serrano, Elena, "Making Oeconomic People: The Spanish Magazine of Agriculture and Arts for Paris Rectors (1797-1808)," *History and Technology* 30 (2014): 149-176.

Sewell, William, "Visions of Labour: Illustrations of the mechanical arts before, in and after Diderot's *Encyclopédie*," Steven Kaplan and Cynthia Koepp, eds., *Work in France. Representations, meaning, organization and practice* (Ithaca: Cornell University Press, 1986), 258-286.

Shapin, Steven, "The House of Experiment in 17th-Century England," *Isis* 79 (1988): 373-404.

Shapin, Steven, "The Invisible Technician," *American Scientist* 77 (1989): 554-63.

Shapin, Steven and Simon Schaffer, *Leviathan and the Air Pump: Hobbes, Boyle, and the experimental life* (Princeton: Princeton University Press, 1985).

Sibum, H. Otto, "Les gestes de la mesure. Joule, les pratiques de la brasserie et la science," *Annales. Histoire, sciences sociales* 4-5 (1998): 745-774.

Simmons, Anna, "Medicines, Monopolies and Mortars: The chemical laboratory and the pharmaceutical trade at the Society of Apothecaries in the eighteenth century," *Ambix* 53 (2006): 221-36.

Simmons, Anna, "Stills, Status, Stocks and Science: The laboratories at Apothecaries' Hall in the nineteenth century," *Ambix* 61 (2014): 141-161.

Simon, Jonathan, *Chemistry, Pharmacy and Revolution in France, 1777-1809* (Aldershot: Ashgate, 2013).

Simon, Jonathan, "La chimie et la pharmacie en 1800," *Revue d'Histoire de la Pharmacie* 49 (2001): 175-182.

Slater, A.W., "Howards, Chemical Manufacturers, 1797-1837: A study in business history," M.Sc. Econ. Thesis, London University, 1956.

Slater, A.W., ed., "Autobiographical Memoir of Joseph Jewell, 1763-1846," *Camden Miscellany* 22 (1964): 113-78.

Slijkhuis, Henri, *De kopermolen in Zuuk* (Vorchten: De Bekenstichting, 2015).

Slinn, Judy, "Research and Development in the UK Pharmaceutical Industry from the Nineteenth Century to the 1960s," Mikuláš Teich and Roy Porter, eds., *Drugs and Narcotics in History* (Cambridge: Cambridge University Press, 1995), 168-186.

Smeaton, William A., "Guyton de Morveau, Louis Bernard," Charles C. Gillispie, ed., *Dictionary of Scientific Biography* (New York: Cribner, 1976), 600-604.

Smeaton, William A., "Platinum and Ground Glass: Some innovations in chemical apparatus by Guyton de Morveau and others," Frederic L Holmes and Trevor H. Levere, eds., *Instruments and Experimentation in the History of Chemistry* (Cambridge: MIT Press, 2000), 211-238.

Smith, Annette M., "State Aid to Industry: An eighteenth-century example," T.M. Devine, ed., *Lairds and Improvement in the Scotland of the Enlightenment* (Glasgow: 9th Scottish Historical Conference, 1978), 21-30.

Smith, John Graham, *The Origins and Early Development of Heavy Chemical Industry in France* (Oxford: Clarendon Press, 1979).

Smith, Michael Stephen, *The Emergence of Modern Business Enterprise in France* (Cambridge, MA: Harvard University Press, 2005).

Sobieniak, Chantal, *Rebondissements dans l'affaire Lafarge* (Paris: Lucien Souny, 2010).

Spang, Rebecca, *Stuff and Money in the Time of the French Revolution* (Cambridge, MA: Harvard University Press, 2015).

Speight, James G., *The Chemistry and Technology of Coal* (Boca Raton: Taylor and Francis, 2013).

Stander, S., "Transatlantic Trade in Pharmaceuticals during the Industrial Revolution," *Bulletin of the History of Medicine* 43 (1969): 326-343.

Stander, Simon S., "A History of the Pharmaceutical Industry with Particular Reference to Allen and Hanbury, 1775-1843," M.Sc. Econ. Thesis, London University, 1965.

Stansfield, Dorothy, *Thomas Beddoes M.D., 1760-1808: Chemist, physician, democrat* (Dordrecht: Reidel, 1984).

Star, Susan Leigh and James Grisemer, "Institutional Ecology, 'Translations' and Boundary Objects: Amateurs and professionals in Berkeley's Museum of Vertebrate Zoology, 1907-39," *Social Studies of Science* 19 (1998): 387-420.

Steele, I.K., *Atlantic Merchant Apothecary: Letters of Joseph Cruttenden, 1710-1717* (Toronto: University of Toronto Press, 1977).

Steffen, Will, Jacques Grinevald, Paul Crutzen and John McNeill, "The Anthropocene: Conceptual and historical perspectives," *Philosophical Transactions of the Royal Society A* 369 (2011): 842-867.

Stein, Claudia, "The Birth of Biopower in Eighteenth-Century Germany," *Medical History* 55 (2011): 331-37.

Stevens, Thomas, *Van der Capellen's koloniale ambitie op Java. Economisch beleid in een stagnerende conjunctuur, 1816-1826* (Amsterdam, 1982).

Stewart, Larry, "'Ordinary' People and Philosophers in the Laboratories and Workshops of the Early Industrial Revolution," Margareth C. Jacob and Catherine Secretan, eds., *In Praise of Ordinary People: Early modern Britain and the Dutch Republic* (New York: Palgrave Macmillan, 2013), 95-122.

Stiles, Margaret, "The Quakers in Pharmacy," F.N.L. Poynter, ed., *The Evolution of Pharmacy in Britain* (London: Pitman, 1965), 113-30.

Stockland, Etienne, "'La guerre aux insectes': Pest control and agricultural reform in the French enlightenment," *Annals of Science* 70 (2013): 435-460.

Sutton, Geoffroy, *Science for a Polite Society: Gender, culture, and the demonstration of enlightenment* (Colorado: Westview Press, 1995).

Swann, John P., "The Pharmaceutical Industries," Peter J. Bowler and John V. Pickstone, eds., *The Cambridge History of Science Volume 6: Modern Life and Earth Sciences* (Cambridge: Cambridge University Press, 2009), 126-140.

Tana, Li, "Cochinchinese Coin Casting and Circulating in Eighteenth-Century Southeast Asia," Eric Tagliacozzo and Wen-chin Chang, eds., *Chinese Circulations. Capital, commodities, and networks in Southeast Asia* (Durham: Duke University Press, 2011), 130-148.

Taylor, George V., "Types of Capitalism in Eighteenth-Century France," *The English Historical Review* 79 (1964): 478-97.

Taylor, J.G., *The Social World of Batavia. Europeans and Eurasians in colonial Indonesia* (Madison: University of Wisconsin Press, 2009).

Thébaud Sorger, Marie, *L'aérostation au temps des lumières* (Rennes: Presses Universitaires de Rennes, 2009).

Thébaud Sorger, Marie, "Amitiés, entraides et circulations techniques: les affinités électives de l'entrepreneur Argand," Michel Cotte, ed., *Circulations techniques, en amont de l'innovation: Hommes, objets et idées en mouvement* (Montbeliard: Presses Universitaires de Franche-Comté – UT Belfort-Montbéliard, 2004), 111-128.

Thébaud Sorger, Marie, "Innovation and Risk Management in Late Eighteenth-Century France: The administration of inventions in French cities at the end of the *ancien régime*," Christelle Rabier, ed., *Fields of Expertise. A comparative history of expert procedures in Paris and London. 1600 to present* (Cambridge: Cambridge Scholar Press, 2007), 261-289.

Todericu, Doru, "Les mines du pays de Liège dans les papiers du savant Français Jean Hellot (1685-1766)," *Technologia* 6 (1983): 61-68.

Tomic, Sacha, "Le cadre matériel des cours de chimie dans l'enseignement supérieur à Paris au XIXe siècle," *Histoire de l'éducation* 130 (2011): 57-83.

Tomic, Sacha, *Comment la chimie a transformé le monde. Une histoire en sept tableaux* (Paris: Le Square, 2013).

Tomic, Sacha, *Aux origines de la chimie organique. Méthodes et pratiques des pharmaciens et des chimistes (1785-1835)* (Rennes: Presses Universitaires de Rennes, 2010).

Tomory, Leslie, *Progressive Enlightenment: The origins of the gaslight industry, 1780-1820* (London; Cambridge, MA: MIT Press, 2012).

Traill, Thomas, "Memoir of Dr. Thomas Charles Hope, late Professor of Chemistry in the University of Edinburgh," *Transactions of the Royal Society of Edinburgh* 16 (1848): 419-434.

Travis, N.J. and E.J. Cocks, *The Tincal Trail: A history of borax* (London: Harrap, 1984).

Tresch, John, *The Romantic Machine. Utopian science and technology after Napoleon* (Chicago: University Press, 2012).

Tribe, Keith, "Oeconomic History: An essay review," *Studies in the History and Philosophy of Science* 36 (2005): 586-597.

Trompette, Pascale and Mélanie Lemonnier, "Funeral Embalming: The transformation of a medical innovation," *Science Studies* 22 (2009): 9-30.

Tweedale, Geoffrey, *At the Sign of the Plough: Allen and Hanburys and the British pharmaceutical industry, 1715-1990* (London: John Murray, 1990).

Uittenbogaard, Roland. *Evolution of Central Banking? De Nederlandsche Bank 1814-1852* (Dordrecht: Springer, 2015).

Underwood, E.A., Cecil Wall and H.C. Cameron, eds., *A History of the Worshipful Society of Apothecaries of London, vol. 1, 1617-1815* (London: Oxford University Press, 1963).

Viallaneix, Paul and R. Elmoznino, eds., *L'apprentissage du savoir savant* (Paris: PUF, 1995).

Vickery, Amanda, *Behind Closed Doors: At home in Georgian England* (New Haven: Yale University Press, 2009).

Vries, B.W. de, *De Nederlandse papiernijverheid in de negentiende eeuw* ('s-Gravenhage: Martinus Nijhoff, 1957).

Vries, Jan de, *Consumer Behavior and the Household Economy, 1650 to the Present* (Cambridge: Cambridge University Press, 2008).

Waal, A.M. van de, "De oudste bankbiljetten. Eerste relatie van de Nederlandsche Bank met Joh. Enschedé en Zonen," *Ontwikkelings- en ontspanningsvereniging "De Nederlandsche Bank"* 8 (1953): 4-14.

Wachsmuth, David, "Three Ecologies: Urban metabolism and the society-nature divide," *The Sociological Quarterly* 53 (2012): 506-523.

Wakefield, Andre, "Police Chemistry," *Science in Context* 13 (2000): 231-267.

Walker, Brett, *Toxic Archipelago: A history of industrial disease in Japan* (Seattle: University of Washington Press, 2010).

Wallis, Patrick, "Exotic Drugs and English Medicine: England's drug trade, c.1550 – c.1800," *Social History of Medicine* 25 (2012): 20-46.

Wallis, Patrick, "Medicines for London: The trade, regulation and lifecycle of London apothecaries c.1610-1670," PhD Thesis, Oxford University, 2002.

Ward, Candance, *Desire and Disorder: Fevers, fictions, and feelings in English Georgian Culture* (Lewisburg: Bucknell University Press, 2007).

Watson, G.M., "Some Eighteenth Century Trading Accounts," F.N.L. Poynter, ed., *The Evolution of Pharmacy in Britain* (London: Pitman, 1965), 45-78.

Watson, Katherine, *Poisoned Lives: English poisoners and their victims* (London: Hambledon, 2004).

Weber, A. "Bitter Fruits of Accumulation: the Case of Caspar Georg Carl Reinwardt (1773-1854)," *History of Science* 52 (2014): 297-318.

Werrett, Simon, "From the Grand Whim to the Gasworks: Philosophical fireworks in Georgian England," Lissa Roberts, Simon Schaffer and Peter Dear, eds., *The Mindful Hand: Inquiry and invention from the late Renaissance to early industrialisation* (Amsterdam: Royal Netherlands Academy of Arts and Sciences, 2007), 325-348.

Werrett, Simon, "Matter and Facts: Material culture and the history of science," Robert Chapman and Alison Wylie, eds., *Material Evidence: Learning from archeological practice* (London: Routledge, 2015), 339-351.

Werrett, Simon, "Recycling in Early Modern Science," *British Journal for the History of Science* 46 (2013): 627-646.

Werrett, Simon, *Thrifty Science* (Chicago: University of Chicago Press, forthcoming).

Whorton, James C., *The Arsenic Century: How Victorian Britain was poisoned at home, work, and play* (Oxford: Oxford University Press, 2010).

Wigelsworth, Jeffrey R., *Selling Science in the Age of Newton: Advertising and the commoditization of knowledge* (Farnham: Asghate, 2010).

Wylie, Alison and Robert Chapman, eds. *Material Evidence: Learning from archaeological evidence* (London: Routledge, 2014).

Wille, Robert-Jan, "The Coproduction of Station Morphology and Agricultural Management in the Tropics: Transformations in botany at the botanical garden at Buitenzorg, Java 1880-1904," Denise Phillips and Sharon Kingsland, eds., *New Perspectives on the History of Life Sciences and Agriculture* (Dordrecht: Springer, 2015), 253-275.

Williams, Michael, *Deforesting the Earth: From prehistory to global crisis* (Chicago: University of Chicago Press, 2003).

Willies, Lynn, "Derbyshire Lead Mining in the Eighteenth and Nineteenth Centuries," *Mining History* 14 (1999): 31-33.

Wilson, Renate, "Trading in Drugs through Philadelphia in the Eighteenth Century: A transatlantic enterprise," *Social History of Medicine* 26 (2013): 352-363.

Withers, Charles, *Urban Highlanders: Highland-lowland migration and urban Gaelic culture, 1700-1900* (East Linton: Tuckwell Press, 1998).

Wolfe, John, *Brandy, Balloons, and Lamps: Ami Argand, 1750-1803* (Carbondale: Southern Illinois University Press, 1999).

Wolters, Willem G., "The 'Doit Infestation in Java': Exchange rates between silver and copper coins in Netherlands India in the period, 1816-1854," Hans Ulrich Vogel, ed., *Money in Asia (1200-1900): Small currencies in social and political contexts* (Leiden: Brill, 2015), 108-139.

Woolgar, Steve and Daniel Neyland, *Mundane Governance: Ontology and accountability* (Oxford: Oxford University Press, 2015).

Wrigley, E.A., *Continuity, Chance and Change: The character of the industrial revolution in England* (Cambridge: Cambridge University Press, 1988).

Wrigley, E.A., *Energy and the English Industrial Revolution* (Cambridge: Cambridge University Press, 2010).

Zappey, W.M., *De economische en politieke werkzaamheid van Johannes Goldberg* (Brussel: N. Samson, 1967).

www.ingramcontent.com/pod-product-compliance
Lightning Source LLC
Chambersburg PA
CBHW071327230926
51849CB00002B/332
9789004325494